MW01054512

Disclosing the World

Disclosing the World

On the Phenomenology of Language

Andrew Inkpin

The MIT Press
Cambridge, Massachusetts
London, England

This book was set in Stone Sans and Stone Serif by Toppan Best-set Premedia Limited. Printed and bound in the United States of America.

Library of Congress Cataloging-in-Publication Data

Names: Inkpin, Andrew, author.
Title: Disclosing the world : on the phenomenology of language / Andrew Inkpin.
Description: Cambridge, MA : MIT Press, [2015] | Includes bibliographical references and index.
Identifiers: LCCN 2015038266 | ISBN 9780262033916 (hardcover : alk. paper)
Subjects: LCSH: Language and languages—Philosophy. | Phenomenology. | Linguistic analysis (Linguistics)
Classification: LCC P106 .I47 2016 | DDC 401—dc23 LC record available at http://lccn.loc.gov/2015038266

10 9 8 7 6 5 4 3 2 1

Für Silvia

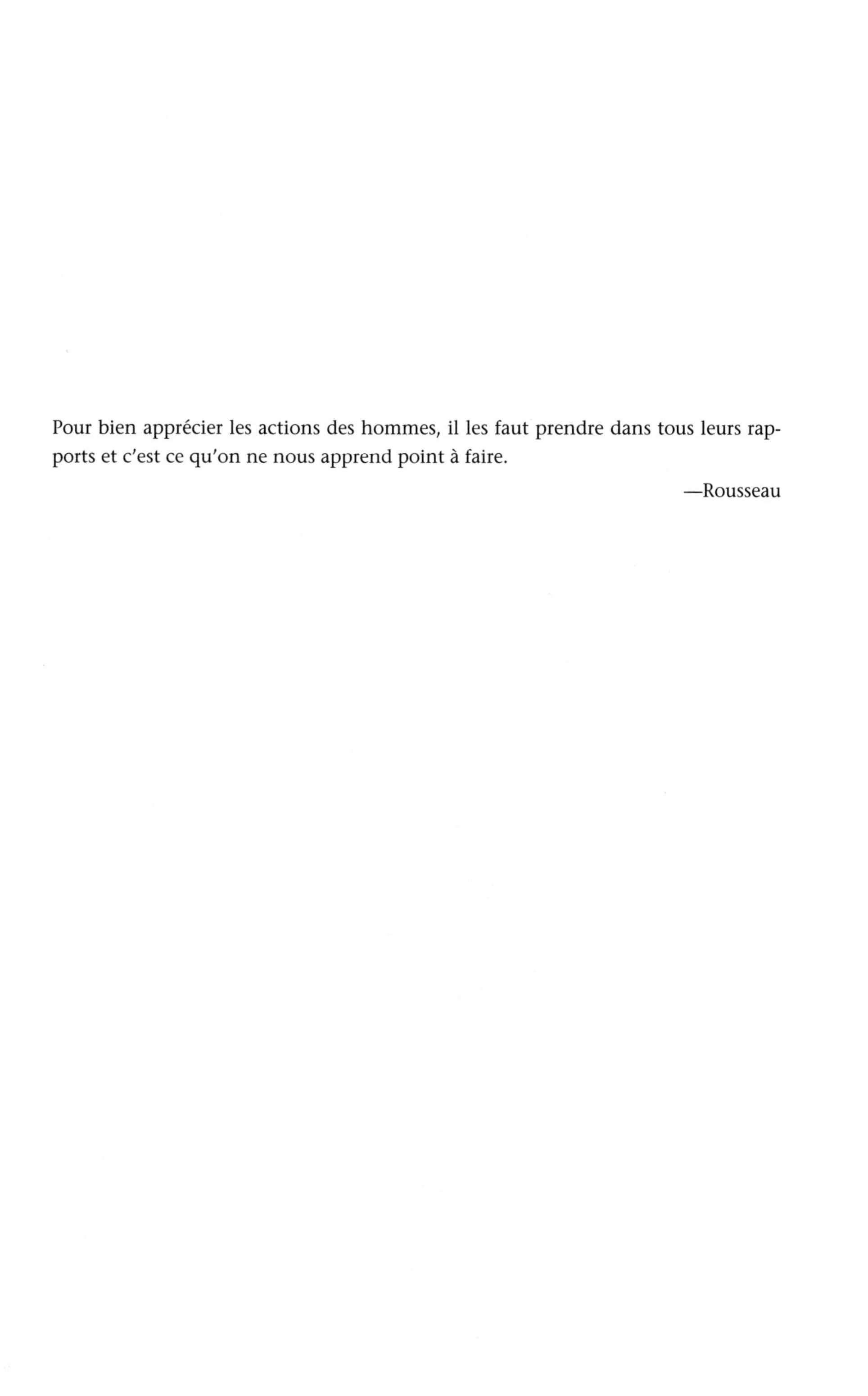

Pour bien apprécier les actions des hommes, il les faut prendre dans tous leurs rapports et c'est ce qu'on ne nous apprend point à faire.

—Rousseau

Contents

Acknowledgments

Much of the material in this book has been a part of my life for a long time, far longer no doubt than it should have been before its publication. Most of it either formed part of or is based on my PhD thesis. Prior to that, for reasons outlined in the introduction, many of the ideas developed here haunted my daily life long before I ever thought of them as part of a philosophical project. Subsequently, the thesis material has been rediscovered, reorganized, substantially revised, extended, and hopefully improved.

As with any other work of its kind, there is a story of indebtedness to be told about this book. As these things go, it is a relatively short story, a story almost too brief to have had a happy ending. Fewer debts, so to speak, but deeper indebtedness. To begin at the end, I would like to thank Philip Laughlin at the MIT Press for his enthusiasm for the project, and Christopher Eyer, Judith Feldmann, and Elizabeth Judd for their efficient and professional help through the production process. My thanks also to the anonymous reviewers of the draft manuscript for their constructive and detailed comments and criticisms, consideration of which helped me to address at least some of its inadequacies. Some of the work in preparing the manuscript they read was facilitated by teaching relief through a Faculty Teaching Fellowship at the University of Melbourne. I am grateful to my colleagues in the philosophy discipline at Melbourne for preserving (*malgré tout*) an amicable and supportive working environment. I would particularly like to thank Karen Jones and Greg Restall, who have often gone far beyond the call of duty, in terms of both professional guidance and personal support, as well as François and Laura Schroeter for their unobtrusive willingness to help out in matters that matter.

Looking back further I wish to thank Stephen Mulhall and Peter Gallagher, my PhD examiners, for their subsequent support throughout the attritional and uncertain process of academic job hunting. Without funding from the UK Arts and Humanities Research Council my graduate studies would not have been possible and this book would not exist. This is clearly the right place, albeit belatedly, for me to gratefully acknowledge that support. As a graduate student at University College London I had the privilege and pleasure of being supervised by Sarah Richmond and Sebastian Gardner, both of whom introduced me to the—then novel—experience that academic philosophers can also be good people. Sarah's expert guidance and tact kept me on track (just about!), despite the fact that when I arrived at UCL the rules of philosophy language-games had suddenly changed and I had become—*pour autrui*—a "continental" philosopher. Sebastian was an inspiration as a supervisor and without his generous, unstinting, and no doubt undeserved support I would not be where I am now, and might never have ventured into the Southern Hemisphere. I am deeply grateful to both Sarah and Sebastian. My greatest all-round debt, however, is to Silvia Seja, who has accompanied this book from start to finish, and has suffered with it *almost* as much as I have. It is dedicated to her, of course, in the hope that in the future we will be climbing more literal and fewer metaphorical mountains.

Abbreviations

Martin Heidegger

Logik — *Logik: Die Frage nach der Wahrheit* (Heidegger 1976)
[*Logic: The Question of Truth*]

PAA — *Phänomenologie der Anschauung und des Ausdrucks* (Heidegger 1993b)
[*The Phenomenology of Intuition and Expression*]

PIA — *Phänomenologische Interpretationen zu Aristoteles* (Heidegger 1985)
[*Phenomenological Interpretations of Aristotle*]

SZ — *Sein und Zeit* (Heidegger 1993c)
[*Being and Time*]

Maurice Merleau-Ponty

PdM — *La prose du monde* (Merleau-Ponty 1969)
[*The Prose of the World*]

PdP — *La phénoménologie de la perception* (Merleau-Ponty 1945)
[*The Phenomenology of Perception*]

S — *Signes* (Merleau-Ponty 1959)
[*Signs*]

SdC — *La structure du comportement* (Merleau-Ponty 2002)
[*The Structure of Behavior*]

Ludwig Wittgenstein

BGM — *Bemerkungen über die Grundlagen der Mathematik* (Wittgenstein 1989a)
[*Remarks on the Foundations of Mathematics*]

BPP I *Bemerkungen über die Philosophie der Psychologie I* (Wittgenstein 1989b)
[*Remarks on the Philosophy of Psychology I*]
LSPP *Letzte Schriften über die Philosophie der Psychologie* (Wittgenstein 1989c)
[*Last Writings on the Philosophy of Psychology*]
PG *Philosophische Grammatik* (Wittgenstein 1989d)
[*Philosophical Grammar*]
PU *Philosophische Untersuchungen* (Wittgenstein 1989e)
[*Philosophical Investigations*]
ÜG *Über Gewißheit* (Wittgenstein 1989g)
[*On Certainty*]

Introduction: A Phenomenological Approach to Language

The view of language presented in this book was motivated by certain kinds of experience. All of us, of course, have experience of language. But familiarity with our linguistic tools and the circumstances of their use means that words usually come to us as effortlessly as the actions they accompany or embody, so that by default we lack an articulate grasp of the various ways language mediates understanding of the world and what it is about language that enables it to do this.

In my own personal case, such default experience of language was interrupted and put in a different perspective by biographical circumstances. Shortly after the fall of the Berlin Wall I arrived in the former East Germany, where I was to spend several months teaching English, never suspecting that I would end up living, working, and studying there for nearly ten years. It was a fascinating time and place to be for many reasons, not least of which was the challenge of getting by initially with mediocre school German in a country where few spoke English (Russian perhaps). Thus over many years and in countless situations an awareness of my own linguistic competence—usually a lack of it—became a pervasive and salient part of everyday experience. One aspect of this were the difficulties and frustration that accompany an inadequate grasp of a foreign language. The inflexibility and brittleness characterizing second-language competence impact on both your ability to act in common practical situations and the ability to formulate your thoughts. Rather than being strictly correlated, however, the link between pragmatic and linguistic competence is an ambivalent one. On the one hand, very often syntactic and semantic errors do not impede you in doing what you want to do, and an interlocutor—if need be, a sympathetic one—will understand "what you are saying" even if your own words fail to say it. On the other hand, there are occasions when even slight changes in circumstances or unexpected variations on standardly "scripted" procedures bring an abrupt breakdown in pragmatic fluency by requiring linguistic knowledge that is off your beaten track. Such problems persist even with a high degree of fluency. Long after you can make yourself understood without difficulty, limitations in grammatical flexibility and a restricted vocabulary can mean that your powers of expression lack subtlety and precision, limiting your ability to articulate thoughts and leading you to sense linguistic constraints in the exercise of your intelligence.

Default experience of language contrasts not only with cases in which we are palpably deficient as second-language speakers, but also with situations in which we are forced to reflect on and extend understanding of our first language. This typically becomes particularly clear in what might be called "interfacing situations" between two languages. For example, in teaching your first language to foreigners, you rapidly become aware of the difficulties of explaining the way we use words, in terms of both grammar and meaning. Or again, when translating from a foreign language into a first language, you often encounter lacunae and vulnerabilities in your ability to find the right words—those carrying relevant connotations and conveying appropriate contrasts. These kinds of experience have a sobering effect. You become aware that you have only a limited grasp of your first language too, no matter how effortless "mastery" of it might otherwise seem. However, sensitivity to such limitations need not depend on the contrast with a second language. There are other examples of careful, skilled, or creative use of language in which reflective awareness of the expressions to be used plays a central role. Poetry or legal writing would be obvious examples in which the choice of particular words is of the essence, but similar demands can arise with advertising slogans, humor, and in personal or emotionally delicate communication. In such cases we are careful about our choice of expression because we expect the potential implications of our words to matter—implications that might otherwise never cross our minds.

Though clearly an extreme case, the difficulties encountered in translation and in reflective use of language came for me to be epitomized in Martin Heidegger's philosophical prose, which is characterized by the inseparability of the articulation of his thoughts from the interpretation of individual words. Unsurprisingly, this is difficult to convey in translation, because the difficulties concerned emerge precisely through Heidegger's exploitation of peculiarities of German. But to give some indication of what is involved consider, as one example of many, Heidegger's characterization of modern technology as "das *Ge-stell*" (Heidegger 1954, 23). Being based on the idiomatic German word *Gestell*, meaning framework, rack, or shelf, this term is immediately suggestive of both constraint by a rigid framework and the commodification of an environing world that is to be stored up on shelves. Beyond this, however, Heidegger weaves a rich web of associations relying on the underlying verb *stellen* (to put or place). First, he contrasts the way earlier generations had "bestellt"—meaning to tend or care for—the land with the way modern technological practices make demands on (*herausfordern*) nature. For these too the land is something to be "bestellt," but now in the word's sense of ordering or demanding something (Heidegger 1954, 18). Above all, the term *Ge-stell* serves to signal an intimate link between the nature of modern technology and the modern worldview, a link Heidegger takes to be already expressed in a series of established idiomatic locutions involving *stellen*. Thus rather than tending and caring, modern technology is a way of producing (*her-stellen*) that exhibits a confrontational, predatory mindset: entities are to be tracked down or hunted (*nach-stellen*), fixed by observation (*fest-stellen*), seized or secured for our ends (*sicher-stellen*).[1] This in turn is linked with modern thinking's ontologization of the subject-object distinction, following which the way objects of thought are presented or modeled (*dar-stellen*) takes on the specific character of putting objects before (*vor-stellen*, "representing" to) consciousness.[2]

Throughout his philosophical writing Heidegger typically, as in this example, both exploits the established meaning of words and simultaneously gives them a highly idiosyncratic meaning within the framework of his theory, often relying on—so to speak, "etymological"—clues in the form of words as an indication of their meaning. The intended effect is that his key terminology should acquire a peculiar expressive intensity, with Heidegger's suggested interpretations of terms apparently resonating with their familiar established sense. Using language in this way gives the connections Heidegger discerns a perceptible quality that is usually (unavoidably) lost in translation and that underlies the poetic feel of his German prose. This approach has both its admirers, who are sympathetic to Heidegger's claims to be bringing out the deep "original" (or "primordial") sense of words, and its detractors, who dismiss it as mere punning or "wordplay." Without wanting to adjudicate here the general expediency or otherwise of this approach to philosophy, it seems to me that the way Heidegger exploits expressive potentials in the (German) language does reveal something significant about the way language works more generally—although not, *pace* Heidegger, anything that has to do with the recovery of "original" meanings.

As already pointed out, this brief catalog of nondefault linguistic circumstances is drawn from personal experience. Although in itself that fact is of no philosophical interest, the experiences just sketched are. First, because they are *kinds* of experience that are generally available, so that anyone finding themselves in such circumstances is likely to have corresponding experiences. Second, although these experiences contrast with default, prereflective experience of language, they involve no resources beyond those available in all language use. That is, although they arise in unusual situations, or usual situations experienced in unusual ways, in each case it is an experience of language as such, language in its natural habitat so to speak. Accordingly, the factors that determine failures of linguistic competence and those on which we draw in extending our understanding of language, or in making highly reflective or expressive use of language, should all somehow be either in play or latent in all language use. Such nondefault experiences can therefore shed light on what is involved in language use more generally, and in this sense might reasonably be expected to be of interest to and/or illuminated by philosophy of language.

It seems to me that many common approaches in contemporary philosophy of language do not live up to this expectation. To take perhaps the most obvious example, one would look in vain to Quine's views about translation for insight into the difficulties posed by understanding or rendering in another language the nuances of Heidegger's language use. Quine's behavioristic view of radical translation simply does not articulate the kind of factors that would enable one to understand Heidegger's linguistic practices—nor indeed to explicate the requirements of pragmatic adequacy or expressive subtlety alluded to in the examples above.[3] To be sure, Quine is not alone. Indeed most contemporary philosophical approaches to language are guided by antecedent assumptions about what language should be like—the degree of systematicity or the form it should have, the epistemological function it is desired or perceived to play, how it connects with philosophy of mind or metaphysics, and so on—in a way leaving them peculiarly insensitive to the factors that speakers might perceive to be at work in different kinds of linguistic phenomena. So, while I am not suggesting that such approaches lack value, it seems there is a need at least to ground such theories by setting out how they relate to the various kinds of experience speakers can have of language.

A significant step in the right direction is provided by Charles Taylor's (1985, 218) distinction between "designative" and "expressive" conceptions of language. Whereas designative conceptions concentrate on referential or extensional aspects of language, "expressive" conceptions focus on language's "power to make things manifest" (Taylor 1985, 238). While stressing that, since they address quite different questions, there is no conflict between these two broad approaches to language, Taylor (1985, 252) nonetheless suggests as a deficiency of the designative view that it does not "take account of the matrix of activity within which the connections between words and their referents arise and are sustained." By contrast the nondesignative approach—which he also refers to as the "HHH" (Herder, Humboldt, Hamann) conception—holds out the prospect of a richer philosophical understanding of language by treating language as a "speaking activity" that Taylor characterizes in terms of three key features. Thus language is taken to be a medium through which (a) things are formulated or articulated, and so brought to explicit awareness, (b) subject matter is "put in the public space," and in which (c) characteristically human concerns are constituted (Taylor 1985, 256–263). Taylor goes on to associate the nondesignative approach with a series of further theses—for example, that language is holistic, is always the common property of a "speech community," or is somehow mysterious due to a link with "subject-related properties"—which strike me as less compelling. Nonetheless, I think that the particular virtue of the nondesignative approach, in the minimal form just sketched, is to identify a general perspective and a corresponding set of questions for philosophical study of language. Language, in this perspective, is a process of articulation in the public space that plays a constitutive role in human actions and thought. The right questions to ask in understanding its speaking activity are: What does linguistic articulation consist in? What is it for language to express or realize meaning? And wherein does the power of language to make things manifest lie?

An alternative way of capturing these concerns, one that will resonate helpfully throughout this book, is to adopt Heidegger's term *disclosure* and to say that Taylor's focus is on the disclosive function of language, what it is that language does in revealing or disclosing the world. Part of Heidegger's motivation for using this term is that he reserves the term *understanding* for a more specific use. However, through its close link with the idea of revelation, the term *disclosure* helpfully signals both that it is the world itself we are aware of—not merely our own mental representations—and that the focus is on our basic, "domain-opening" (Seel 2002, 50) awareness of things rather than an intellectualized end product.

The overall aim of the present book is thus to address the broad questions just identified about the disclosive function of language, and to do so in a way that accords properly with the various kinds of experience speakers can come to have of language. This latter requirement establishes a sense of priorities that I think is best summed up by describing the approach to be taken as a "phenomenological" one. Before outlining the strategy

and position to be developed in the following chapters, it will therefore be appropriate to say something about what it is for a conception of language to be "phenomenological."

1 Getting Phenomenology Right

The term *phenomenology* is commonly used in a variety of ways. It is perhaps most frequently used in relation to the intellectual tradition initiated by Edmund Husserl and associated with some basic methodological commitment, signaled by talk of a return to "things themselves," "intuition" as a source of validity, or reliance on the "phenomenological reduction." Matters are complicated by the fact that even among this tradition's authors there is no clear consensus as to what that commitment is, making it something of a commonplace that there are as many kinds of "phenomenology" as there are "phenomenologists."[4] Beyond this, the term *phenomenology* also has an established use in English-speaking philosophy, epitomized in the expression "getting the phenomenology right," on which it simply conveys the need for accounts of something to be "true to appearances"—that is, to tally with our experience of that something.

Although this book will draw extensively on authors in the phenomenological tradition, the latter use seems to me to capture most transparently the emphasis I have been suggesting is needed in philosophy of language. However, at the same time it identifies a minimal commitment that is shared by most, perhaps all, of the positions usually described as phenomenology.

My aim in the following will therefore be to work with a correspondingly *minimalist* conception of phenomenological method, defined by the basic commitment—subsequently referred to as *phenomenological accountability*—to describe accurately how things appear or manifest themselves. In general this commitment implies that a phenomenological conception of some X should accurately reflect the way(s) that X is experienced. In the particular case of language, this translates into the requirement that a (minimally) phenomenological conception of language should describe accurately the way language is experienced by speakers.

What are we supposed to be describing? What are the linguistic "phenomena" that we experience? It is important first to distinguish the case of language from a common use of the term *phenomenology* in (analytic) philosophy of mind. For the concern here is not with phenomenal aspects of consciousness (such as pain, yellowness)—its felt quality, qualia, "what it is like to be . . . ," and so on. Nor is it with any kind of subject-internal goings on, to which an agent may have "introspective" or "privileged" access in the "first-personal" realm. It is sometimes highlighted (correctly in my view) that such inwardly directed uses of the term *phenomenology* are more limited than that intended in the phenomenological tradition, for which phenomena include the self-showing of features of an agent's interpersonal or impersonal surroundings.[5] But no matter what one thinks about such subject-internal goings-on, the case of language is different, because linguistic phenomena—as most famously argued by Wittgenstein—take place in a public or shared space. Even if words had a special feel to me (something like a characteristic smell or color), or they felt different to everyone else, this would be irrelevant to their function in language. Accordingly, it is to experience of phenomena in this generally available, interpersonal space that a phenomenological conception of language answers.

A related point is that although experience of language concerns how language "appears" to speakers, this does not imply there is a domain of appearances that stands in a principled contrast to reality. This is again a standard view in the phenomenological tradition, one emphasized particularly by Heidegger's definition of a phenomenon as "that which shows itself in itself" (*SZ* 28), while Sartre (1943, 11) already considered it a commonplace of phenomenology that the being-appearance dualism had been overcome. This point is again particularly clear with regard to language, because in general the experience agents have of language will be interwoven with experience of the surrounding world, such that experience of language is simply a particular aspect of our experience of reality.

Linguistic phenomena thus take place in the same space as our shared everyday experience of the real world. Despite this, specifying what our experience of language is like might still appear problematic. Our default prereflective experience, as already noted, is that we usually fail to notice it: rather than striking us in any specific way, our experience "of language" is usually submerged in our broader experience of the world. Yet, at the same time, the idea of phenomenological accountability appears to imply that our default prereflective experience (of language) has a determinate character of its own that we can be made aware of.[6] There is no real conflict here. Although we might lack or have lacked an awareness of certain features of language, it is quite possible we would acknowledge the presence of these features when they are pointed out to us. In the same way, we often fail to notice other aspects of the world that are available to us in the field of our experience. Think of the way we can discover new features of an artistic work as we learn more about it, or learn to appreciate the subtleties of a sport, natural habitat, type of cuisine (etc.) we previously knew little or nothing about. In all these cases we can become sensitive to the presence of features that we had previously failed to notice, but which, we then concede, had been there—were experienceable—all along.

Another important consideration is that linguistic experience comes in different varieties. Each of us has a limited experience of language, and not everyone has the same default prereflective experience—for example, because we live in different countries and live in different ways. At an individual level this suggests a need for modesty about the extent of our own "mastery" of language. Particularly as philosophers we should be wary of mistaking our own province for the whole linguistic world, and remain open to the possibility that considering kinds of experience we lack by default might enrich our understanding by drawing attention to otherwise unnoticed features of language. A phenomenological conception of language faces a parallel challenge. Rather than picking out one kind of linguistic situation or theoretical construct and creating "language" in its image, this should reflect the variety and complexity of speakers' experience. At the same time, it faces a unifying constraint, as in each case it is language that is experienced. The aim of phenomenological description of language will therefore be to identify factors that make language what it is, the interplay between which issues in both the diversity and the unity of linguistic phenomena.

Our experience of language can thus be inattentive and limited in just the same way as our experience of the world more generally, and in both cases phenomenological description can help by highlighting what we have previously been missing. But what is it to "describe" phenomena? The basic idea is to focus on features that are straightforwardly available or accessible, those that can be "just seen" or recognized straightaway—at least once they are pointed out—in experience by appropriately engaged language users. The intended contrast is with features that require deliberation, reasoning, or other thought processes that involve inferences. It is this link with *noninferential verification* that defines phenomenology as a descriptive undertaking.[7]

The link with description, so defined, determines both the significance and the limitations of a minimalist phenomenological approach. On the limitations side, the reliance on noninferential verification implies a contrast with certain kinds of theory. This contrast is clearest with scientific theories, which are typically based on reflection, inferences, systematic theory construction, complex processing of empirical data, and so on—in short, all manner of evidence that cannot be "just seen." This limitation can be helpfully characterized by saying that phenomenological description contrasts with explanation. In saying this, the underlying thought is that description can tell us what or how things are, whereas explanation has the task of telling us *why* things are the way they are. Now, this contrast may be challenged, as there is significant disagreement over whether or not noninferential verification—and hence phenomenological description—can be exploited in some way to answer explanatory questions and whether such questions are better left to science. This is an issue I return to in chapter 10. Nevertheless, it remains the case that a *minimalist* phenomenology will be of very limited explanatory use at best. For, while it seems reasonable to expect that some features elicited (noninferentially) from our experience of language will be of relevance in developing explanatory accounts of how language works (see below), it is clearly implausible to think that a complete understanding of its causal underpinnings (say) could be gained in this way.

The limitation just described can also be characterized metaphorically in terms of a distinction between surface and depth. In these terms, noninferential verification secures access to the experiential "surface" of phenomena, allowing us to map the experienceable features of—in the present case—language. Conversely, attempts to understand theoretically the underlying processes will lead below this surface, yielding a "deeper" understanding of the phenomenon in question. One advantage of this metaphor is that it expresses not only varying degrees of understanding of phenomena, but also the thought that there will in general be a need to move beyond minimalist phenomenology in some way. In the wrong hands this metaphor might therefore be taken to suggest that phenomenological features are merely superficial and hence unimportant. That is obviously not how it is intended here, because the conviction underlying a phenomenological approach is precisely that the experiential surface *is* important. Rather, insofar as any body with depth is delimited or defined by its surface, the metaphor is better seen as hinting at the specific importance of the "surface" of experience—to which I now turn.

The significance of a minimalist phenomenology is to provide a basic means for defining what we are trying to understand, in the present case by building up a suitably detailed picture of linguistic phenomena. The claim that phenomenology defines the explanandum or subject matter for empirical inquiry is a commonplace of traditional phenomenology, and is standardly taken to justify its primacy over empirical science. Thus Husserl thought that empirical science needs phenomenology to overcome the "naïveté" of its appeals to experience: although the "experimental method is indispensable," he says, "it presupposes what no experiment can accomplish, the analysis of consciousness itself" (Husserl 1965, 25). Corresponding claims are found in the work of Brentano, Heidegger, and Merleau-Ponty, to name just three.[8] Underlying these claims is the idea that empirical explanation *presupposes* a clear and complete conception of whatever it is trying to explain, and that phenomenological description can provide such a conception. This idea is open to challenge. To take the obvious example, given the causal autonomy and various scales (e.g., microscopic or cosmic) of natural entities, it seems plausible that our experiential perspective will not always allow these to be defined adequately. Accordingly, one might think that the task of progressively defining its explanandum is an integral part of scientific research, rather than something the latter presupposes. Hence it might be doubted that in general phenomenology can play this role of defining explananda.

No matter what one thinks about this general question, in the case of language the need for phenomenological description and its importance are more securely anchored. One reason for this is that language is an artifact, existing only in virtue of speakers' having some understanding and hence experience of it. Another is that language is a cognitive tool, the very means by which we express and exchange many of our thoughts. For both these reasons it is difficult to make sense of the idea that linguistic phenomena, unlike nature, could in principle be detached from the perspective of experience. Instead they imply that language has an in-built experiential aspect, so that a phenomenological conception of language must be expected to form at least part of a full philosophical understanding of language. Once this is recognized, however, it is clear that any explanatory theory one has of language will ultimately need to cohere with this phenomenological conception—identifying which causal or functional features sustain or underlie the experience speakers have of language. As a consequence of this need for coherence, a minimalist phenomenological conception of language will serve to articulate conditions of adequacy for theories that probe "below" the experiential surface of linguistic phenomena. In this sense, even if one doubts it can do this more generally, in the case of language phenomenology does play the role traditionally ascribed to it of defining the explanandum.

All the same, it is to some extent implicit in a phenomenological approach of the kind outlined here that philosophical theses are the end result rather than the starting point of inquiry. Although a given description of phenomena may be loaded with metaphysical or epistemological implications, the whole point of a phenomenological approach is to extract such theses from a descriptive understanding of phenomena. In view of this, the philosophical interest of a phenomenological approach to language should be expected to emerge gradually over the course of the present work, allowing its implications and significance for philosophy of language to be addressed directly and at greater length in the final part of the book.

2 Getting Phenomenology Historically Right

At this point it might be thought that the minimalist conception of phenomenology just outlined is difficult to reconcile with the term's historical use in the phenomenological tradition. In particular it might be objected that the kinds of experience emphasized so far are too mundane, too embedded in the real world, or that they render phenomenology indistinguishable from empirical inquiry, so that the distinctive nature and force of traditional phenomenological claims have been given up. One might even suspect that "minimalist" phenomenology amounts to an attempt to assimilate the phenomenological tradition to the term's definition in English-speaking philosophy. As should become clear over the following chapters, that is far from my intention. Although it cannot be shown in detail here, I therefore want to indicate very briefly how a minimalist conception of phenomenology can be plausibly reconciled with the term's use in the phenomenological tradition.

The impression that minimalist phenomenology is not phenomenology in a traditional sense can be traced back to two interconnected features of Husserl's view. The first is the notion of experience relevant to phenomenology. Although in some sense committed to describing the structures of conscious experience, Husserl distances himself from straightforward appeals to experience of the kind I have so far described. He does this because both everyday and scientific experience are characterized by what he terms the "natural attitude," the background assumption that the world as a whole exists (Husserl 1992c, 61 [§30]). The problem, as Husserl saw it, with this assumption is its implication that inquiry remains empirically conditioned—that is, in some sense answerable to facts, or the contingencies of the actual world. Husserl believed, however, that the foundational task of philosophy cannot be bound in this way and instead requires a modified form of experience—"transcendental experience"—in what he calls the "transcendental field of experience" (Husserl 1992d, 28ff. [§12]). This modified form of experience is to be distinguished first by its "purity," the fact that it is not accountable to the actual state of the world, or that "no experience, qua experience . . . can take on *the function of justification*" (Husserl 1992c, 20–21 [§7]). It also involves an extension of the field of phenomenological experience, to encompass the realm of possibilities or merely imagined ("as if") intentional acts, and of the notion of description, so as to apply over this extended field of pure phenomena.[9]

The second feature of Husserl's position relevant here is the supposed ontological basis of intentionality. In his *Logical Investigations* Husserl had aspired to talk about the essential structures of intentionality without making any ontological assumptions (Husserl 1992a, 240). Later, however, he came to consider phenomenology as a form of transcendental idealism and so to postulate an ontological realm of transcendental subjectivity. The transcendental subject is assumed to be the bearer of all the intentional acts a given empirical subject actually has, along with all the intentional acts it might conceivably have (its "horizons" of possible experience). Put simply, it is the form of subject that has transcendental experience.

Both these features of Husserl's position involve an extension from the realm of the actual and empirical to that of the possible. Hence the need for and the importance of the "phenomenological reduction": its "bracketing" or "neutralization" of validity (*Geltung*) assumptions—assumptions about what (intentional) objects exist and which states of affairs pertain—is a necessary step in extending the supposed realm of inquiry in the way Husserl intends.[10] It is also these two features—along with Husserl's emphasis on the reduction—that underlie the thought that phenomenology is not a straightforward matter of describing experience and that it is not concerned with agents like us.

The situation is different with Heidegger and Merleau-Ponty, each of whom rejects the features of Husserl's position just described. First, both authors eschew the phenomenological reduction and instead emphasize the accountability of phenomenology to the experience humans usually (in the natural attitude) have.[11] Thus *Being and Time* is built around—what claims to be—a description of everyday ("inauthentic") disclosure of the world and acknowledges that empirical (or "vulgar") intuition is an "unavoidable presupposition" in securing the "mode of access that genuinely belongs" to ontological inquiry (*SZ* 37, cf. 31). Similarly, Merleau-Ponty insists on the general need to found phenomenological analysis in the "lived world": "in a word phenomenology" is "the decision to demand from experience itself its proper sense" (*PdP* iii, 338).

Second, both Heidegger and Merleau-Ponty reject Husserl's view of the ontological basis of intentionality. Indeed, one of Heidegger's principal objections to Husserl is that he needed, but lacked, some conception of the mode of being (*Seinsart*) of what he was calling "transcendental subjectivity."[12] The crux of Heidegger's objection, which he saw as a consequence of Husserl's requirement of "presuppositionlessness," is that radical inquiry into the nature of intentionality—that is, inquiry that claims to be discovering its own conditions of possibility—cannot fail to address the nature of its ontological setting (cf. Heidegger 1979, 123). Heidegger was moved by this *ontological realizability condition* to reject the idea of a transcendental subject and instead to think of intentionality as being constituted by concrete mortal agents (Dasein). This move is again paralleled in Merleau-Ponty's rejection of the "image of a constituted world" implicit in the idea of an "absolute constituting consciousness" in favor of a focus on embodied agency.[13] "No philosophy," he explains, "can ignore the problem of finitude on pain of not knowing itself as philosophy" (*PdP* 48). In each case transcendental subjectivity is rejected as an ultimately unintelligible assumption, and replaced by a notion of concrete and finite subjectivity whose experience is recognizable—without "purification" via the reduction—as corresponding to our own in the "natural attitude."

As a historical thesis it might be argued that these differences are part of Heidegger's, and perhaps Merleau-Ponty's, attempt to be more consistently phenomenological than Husserl himself was. But for the present purposes, as the following chapters will draw on their views extensively, I want simply to highlight that it is plausible to think of Heidegger and Merleau-Ponty as sharing commitment to phenomenological accountability in the dual sense that they (a) take phenomenology to be accountable to an "unpurified" notion of experience in the "natural attitude," and (b) posit forms of agency in which we can readily recognize ourselves. In this respect their positions are consistent with minimalist phenomenology, showing that the latter can be reconciled with a historically common use of the term *phenomenology*.

Underlying the above differences is a fundamental disagreement about the relationship between the actual and the possible that bears on the kind of claims made by a phenomenological position. On the one hand, Husserl holds that actuality is grounded in possibilities, that "'in itself' the science of pure possibilities is prior to that of actual realities and makes this [science of actual realities] possible as a science."[14] This view underlies his belief that extending the notions of experience and subjectivity beyond the merely actual to include possibilities allows phenomenology to make transcendental claims—that is, claims about what "makes possible" the real or actual world. On the other hand, both Heidegger and Merleau-Ponty reverse Husserl's order of priority and instead conceive possibility as grounded in the actual. Thus for Heidegger humans (Dasein) not only cannot be thought of as actualizing timeless or unconditioned possibilities, but themselves form the ontological precondition for the being of possibilities (as Dasein's projects or plans).[15] And because Dasein is always conditioned by history and empirical facts, this implies—against Husserl—that possibilities, experience, and phenomena are *always* empirically conditioned. In much the same way, Merleau-Ponty's approach "preserves" from empiricism "the character of facticity": "That is also why phenomenology is a phenomenology, that is to say, studies the *appearance* of being to consciousness, instead of supposing its possibility to be given in advance."[16] In Husserl's view, assuming a priority of the actual over the potential in this way would render phenomenology incapable of making transcendental claims. Despite this, both Heidegger and Merleau-Ponty continue to present their own claims as transcendental ones, presumably intending their projects as a modified form of transcendental philosophy.

For my present purposes it is not necessary to adjudicate who is right on this matter for two reasons. First, what is important here is that Heidegger's and Merleau-Ponty's positions share a commitment to phenomenological accountability of the kind that defines minimalist phenomenology. That commitment does not in itself exclude the possibility of a phenomenological approach yielding more powerful explanatory or transcendental claims. I say a little more about this in the book's final chapter. However, for the moment it suffices to note that any attempt to advance transcendental phenomenological claims (about what makes . . . possible) is more ambitious than, but *consistent with*, the kind of descriptive claims (about what . . . is) that I have suggested define a minimalist phenomenology. To put it another way, in itself a minimalist phenomenology lacks the transcendental touch, and any attempt to establish a phenomenological position as a transcendental one will require some kind of further assumptions or methodological provisions. Second, it is important to keep in mind (from the last section) that in the case of language at least a minimalist approach can play the role traditionally assigned to phenomenology of defining the explanandum for empirical inquiry, quite independently of the question of whether or not phenomenological claims are taken to be "transcendental."

3 The Path Ahead

In view of the motivations set out above, the overall aim in the following chapters is to develop a phenomenological conception of the context and processes of linguistic disclosure and to identify some of the ways the resultant position is of philosophical interest. The development of this view will draw extensively on three authors—Martin Heidegger, Maurice

Merleau-Ponty, and Ludwig Wittgenstein—with the aim of fusing their respective conceptions of language into a unified view. This strategy, which might seem odd or at least unduly indirect, is guided by three convictions. First and foremost it seems to me that each of these authors' conceptions of language captures something of major importance about the phenomenology of language, so that a phenomenological approach has much to learn from each of them. Second, although according recognizably with certain aspects of our experience of language, each is also characterized by corresponding inadequacy or one-sidedness that yields a picture of the whole that is either vague or distorted. Finally, third, it seems to me that the relative strengths and weaknesses of these three authors' positions are such that they have the potential to complement one another. Accordingly, my aim here will be to critically combine these three visions of language into a single overall view that is more balanced than each of its respective component positions.

This exegetically committed strategy also requires a second kind of balance, between fidelity to and instrumentalization of the respective authors' views. Generally speaking my aim has been to capture accurately key aspects of each author's conception of language, but to be guided by the task at hand rather than by the specific philosophical (or metaphilosophical) aims of each author. The phenomenological position I want to develop will therefore emerge by working through three views of language toward which—as should become clear—I am broadly, but not totally, sympathetic. This approach clearly has the potential to disappoint in different ways. To some it might seem that I am cutting corners and failing to expound the respective authors' views in sufficient detail, or failing to adhere to their views with sufficient fidelity. To others it might seem that I am spending too much time on exegesis rather than focusing on philosophical issues. My aim, at least, has been to strike the right balance—that is, to do justice to the merits of each position while maintaining sufficient direction to avoid mere commentary. To put it another way, my aim has been to identify in some detail the interesting features of each author's views, without losing sight of why those features are indeed of philosophical interest.

The first part of the book centers on Heidegger's conception of language in *Being and Time* (1927) with the aim of drawing out the basic framework for a phenomenological conception of language. As the label "framework" suggests, the task here will be to identify a set of views about language that seem insightful and correct as far as they go, but which stand in need of further development. This "Heideggerian framework," as I call it, comprises two levels, one very general, the other more specific. The first chapter works at the very general level, aiming to build up a clear and tenable overall picture of what Heidegger at one point calls the "ontological 'place' of language," that is, the role of language in the broader phenomenon of disclosure. Having introduced its main features, this chapter highlights two apparently conflicting tendencies in Heidegger's analysis that threaten its coherence and render unclear the role of language in developing an articulate understanding of the world. I argue that careful attention to Heidegger's view of the connection between language and content allows these tendencies to be reconciled, so that the general contours of Heidegger's analysis can be preserved. An important upshot of this discussion is that Heidegger's emphasis on the foundational role of purposive understanding (roughly: practice) also applies to language, such that the content of propositions—taken to consist in predicative judgments—is founded in a more basic "prepredicative" grasp of the world. Heidegger's conception of language thus has the highly distinctive feature of allowing that language use can be based on prepredicative understanding alone, so that the disclosive function of language cannot be understood reductively in terms of operations at the predicative (propositional) level.

As Heidegger's approach is motivated by the need for phenomenological inquiry to cohere with lived experience, it can reasonably be expected to exhibit features that generally characterize a phenomenological conception of language. The aim of chapter 2 is to identify a pattern of commitments that play this role. One of the main features to emerge is that Heidegger is led to a view of language that eschews any contrast between an "inside" and an "outside" and instead sees it as embedded in or distributed over the wider world. A second is Heidegger's antireductionist outlook, in particular his rejection of the idea that language can be understood reductively as a "formal" system of signs. In view of the close parallels with his nondualist view of Dasein as "being-in-the-world," I sum up these commitments by saying that Heidegger provides a general picture of language as "language-in-the-world."

In critically assessing the philosophical significance of a phenomenological position, however, particularly in relation to other philosophical approaches to language, a more specific focus is required. This is provided by Heidegger's distinctive emphasis on the "derivative" nature of predication and the possibility of prepredicative language use, views that are at odds with mainstream philosophy of language, which usually takes propositional content to be explanatorily primary. The second chapter therefore goes on to look more closely at what the claim that there is a basic prepredicative level of meaning amounts to. I argue that such claims should be interpreted as an appeal to factors that are functionally and structurally presupposed in propositional content, but which remain irreducible to the latter.

Chapter 3 develops the Heideggerian framework at a more specific level, focusing on the complex function Heidegger sees linguistic signs as having in disclosing the world—how linguistic signs work, so to speak. This is approached by considering the ambivalent attitude *Being and Time* takes toward routine language use in everyday life ("idle talk"), which despite sufficing for day-to-day practical engagement is presented as a somehow deficient mode of disclosure. To understand its deficiency I draw on Heidegger's earlier view of philosophical concepts as "formal indications," because this sheds light on what he sees as the proper revelatory function of signs. His later ambivalence about language can then be explained, I argue, by the development of this earlier view into a conception of signs as instruments in a dual sense. Linguistic signs, by the time of *SZ*, are instruments in both the generic sense of being tools for pointing out features of the world in an articulate manner, and in the differentiated sense of being tools used to perform specific tasks in practice. Corresponding to these different modes of instrumentality I propose a distinction between *presentational sense* and *pragmatic sense*, reflecting two different ways the use of language is involved in articulating our understanding of the world. Accordingly, on this Heideggerian model the disclosive feats of linguistic signs are to be understood in terms of these two kinds of sense that they connect.

The following two parts of the book operate in parallel to develop, in different directions, the views adumbrated by the Heideggerian framework at both the general and specific level. Part II focuses on Merleau-Ponty's conception of language up to and around 1950. Thus chapter 4 first discusses some of the main features of his approach to language as a form of embodied expressive behavior and shows how this fits into the Heideggerian framework's overall picture of language. Having rejected any appeal to a completely constituted and fully determinate pattern of meaning, Merleau-Ponty identifies the task of specifying how language both connects us intimately with the world and plays a constitutive role in the formulation of thought. I show how his attempts to do this are based on thinking of language as expressing lived sense, where "lived" means both lived through by a biological body and forming part of speakers' experience. The attempt to explain the role of language in the formulation of thought also leads Merleau-Ponty to emphasize the phenomenon of creative expression and to develop the notion of "indirect" sense. I argue here that although he

exaggerates the importance of creative expression, Merleau-Ponty's discussions do shed light on some kinds of language use, before introducing his notion of indirect sense by considering how it contrasts with putatively "direct" sense.

Chapter 5 turns to the more specific level of the disclosive function of linguistic signs and explores three ways in which the notion of indirect sense explicates that of presentational sense. First it focuses on Merleau-Ponty's integration of Saussure's conception of signs into his earlier position, arguing that this provides a detailed picture of the *structure* of presentational sense. It then looks at Merleau-Ponty's use of modern painting as an analogy for the disclosive feats of language, showing how this provides an innovative model for the presentational *function* of linguistic signs. Part of the aim of these discussions is to bring out how the science of linguistics and the art of painting complement each other, particularly in the way they conceive linguistic/painted forms functioning as prepredicative factors underlying what Merleau-Ponty calls "conceptual" meaning. This characteristic focus of Merleau-Ponty's approach establishes a clear parallel with that of Heidegger. Hence chapter 5 also, third, undertakes the task of clarifying how and why we should think of indirect sense as having a kind of inchoate rationality that is consistent with the characterization of prepredicative factors in chapter 2. The chapter finishes by setting out how the notion of presentational sense, qua indirect sense, provides a phenomenologically plausible way both of dealing with certain difficulties inherent in Heidegger's view of language and of describing several kinds of linguistic experience, such as careful or innovative use and the relevance of etymology to linguistic meaning.

A feature common to both Heidegger's and Merleau-Ponty's conceptions of language is that while both point to the foundational importance of pragmatic use of language, neither provides a detailed account of the relationship between human practices and linguistic articulation. Part III makes up for this omission by drawing on the late Wittgenstein's conception of language, the defining feature of which is taken to be the emergence of the language-game analogy. Chapter 6 begins by clarifying my approach to reading Wittgenstein, arguing that he *was* committed to a conception of language, which, as the language-game analogy matures, takes on a form that can be plausibly integrated within a phenomenological approach. The "praxeological" conception of language he arrived at is based on recognition of the intrinsic link between language use and patterns of practical activity. However, in contrast to his earlier model of language as a calculus, it is also characterized by what I describe as a more "relaxed" view of language as a rule-governed activity. One aspect of this relaxation is a reconfigured notion of rules. Thus Wittgenstein starts to think of linguistic rules as having a shape or constitution like that typical of statistical distributions, where individual instances are scattered, more or less sharply, around a central or average value. Further, once the calculus model's ideal of full determinacy has been rejected, the standards for determinacy of linguistic rules become a function of the practical requirements inherent in the respective language-game. A second aspect of Wittgenstein's more relaxed attitude to rules is to acknowledge that the role of rules in linguistic practices is constrained in several ways, with the implication that more diverse forms of linguistic practices and competence can be accommodated than are allowed by the model of a fully rule-governed algorithm. Through these developments, I argue, Wittgenstein provides suitably versatile means to describe the many forms of linguistic practice and the many specific practical functions linguistic signs have, and so to explicate the notion of pragmatic sense in a phenomenologically appropriate manner.

A potential problem in assimilating Wittgenstein to my approach here lies in applying the idea of prepredicative sense suggested by the Heideggerian framework to his position. To show how this can be done, chapter 7 looks at Wittgenstein's conception of rule-following, in particular the way that linking this with customs or institutions encompasses activities lacking the rational transparency of paradigmatically intellectual activities such as the use of mathematical calculi. Wittgenstein's view of the limits of justification is then taken to provide a model for understanding how language is grounded in a stratum of language-games that are prior to justification. Together these considerations give rise to the idea of a linguistic form of knowing-how according to which language is grasped prepredicatively.

A further task for chapter 7 is to summarize how Wittgenstein's views complement those of Merleau-Ponty and Heidegger to yield a unified phenomenological conception of language within the Heideggerian framework. This unified conception provides an answer to the kind of questions Taylor identified by offering both a general picture of linguistic activity as language-in-the-world and a specific view of the disclosive function of linguistic signs, as instruments characterized by both presentational and pragmatic sense. Having endeavored to maintain a grounding in linguistic phenomena, it also sheds light on the catalog of experiences outlined above by identifying the kind of features that can be appealed to both in ascertaining a speaker's pragmatic fluency and in justifying the expressive or articulatory appropriateness of choosing particular words.

One aim of phenomenology as a method is to begin with experience-based description that avoids bias due to theoretical prejudices. Nevertheless, as it takes shape a phenomenological view of the kind developed here can hardly fail to have broader philosophical implications. Without claiming to exhaust such possibilities, part IV of the book therefore attempts to identify some of the broader philosophical and discursive significance of the conception of language built up here by addressing three issues. Chapter 8 aims to clarify the sense in which language, on the phenomenological conception developed here, can be said to disclose the world. The central question in this respect is whether language genuinely reveals what is around us, just as it is, or whether what language reveals to us as the "world" is somehow refracted by human projects and purposes. A more conventional way of expressing this concern would be to ask whether or not my phenomenological conception of language has realist or idealist implications. Given the obvious parallel between this conception of language and Heidegger's being-in-the-world, I take his discussion of realism and idealism in *Being and Time* as a guide. The two main lessons of this discussion, I claim, are that Heidegger sees traditional questions about realism and idealism as based on mistaken ontological presuppositions, and that on his view there is no principled gap between the way the world appears "for us" in the perspective of our projects and the way the world is "in itself." Applying the second lesson to the case of language, I show how the constitutive and mediative roles of language can be understood as genuinely disclosing the world without introducing a potentially refractive or distortive breach in our contact with our surroundings. Returning to the first lesson, I then distinguish the phenomenological conception of language developed here from some familiar forms of both realism and nonrealism, and argue that by rejecting an inside-outside opposition it has the virtue of moving beyond such conventional alternatives.

Chapter 9 focuses on the relationship between my phenomenological approach and more mainstream positions in philosophy of language under the auspices of what I call the "semantics approach." The latter is characterized by general—broadly Fregean—assumptions about the primacy of the propositional and about the subpropositional workings of language that are more or less standard in contemporary ("analytic") philosophy of language. To assess the relationship between these two approaches I return to Heidegger's claim that propositional content is founded in prepredicative factors. Having clarified the differences between these two approaches, and various possibilities for interpreting Heidegger's foundation claim, I set out how the semantics approach might try to assimilate a phenomenological conception of language as a partial theory of the background abilities that speakers require to exhibit semantic competence. A preferred strategy for the semantics approach is to construe such abilities as a "weak" foundation—that is, as causal or enabling factors that do not challenge the accuracy or completeness of the semantics approach. Against this, however, I argue for a somewhat stronger claim—a "moderate functional foundation" claim—on behalf of the phenomenological conception developed here. By indicating how presentational and pragmatic sense resist assimilation to the functional picture implicit in the semantics approach, I claim that these capture aspects of linguistic functioning not adequately covered by the semantics approach and so make a distinctive contribution to philosophical understanding of language. To highlight this distinctiveness, I then situate my position in the context of a debate between Hubert Dreyfus and John McDowell about the nature of prereflective activity, arguing that it allows for modes of intelligent behavior that avoid the two extremes of nonconceptual coping and pervasive conceptualism that they respectively advocate. Finally, I use an analogy with twentieth-century physics to sum up how the overall relationship between the phenomenological and semantics approaches should be seen.

My metaphorical characterization of minimalist phenomenology as a descriptive approach concerned with the experiential "surface" of language hints at the need to go beyond this and attain "deeper" understanding. The book's final chapter therefore considers what kind of explanatory approach would be best suited to complement the conception of language provided by a minimalist phenomenology. It begins by canvassing two classic strategies for developing a more ambitious and richer conception of phenomenological method, namely Husserl's transcendentalism and Heidegger's hermeneutic phenomenology. Both are rejected on the grounds that they result in tension with the requirement of phenomenological accountability that is basic to minimalist phenomenology. I go on to argue that the explanatory perspective best able to respect this requirement is instead provided by recent work in cognitive science—"4e" cognitive science—focusing on the embodied, embedded, enacted, and extended nature of cognition. Two characteristic themes of 4e cognitive science—the use of features of the environment as cognitive tools, or so-called scaffolds, and the role of representation—are then considered to show that this branch of inquiry shares with a phenomenological approach both the general commitments summed up by the "language-in-the-world" label and a specific focus on prepredicative factors. Aligning it with 4e cognitive science raises the question of whether any philosophical

task remains for a minimalist phenomenology of language. In answer to this I argue that far from eliminating the need for phenomenology, or in some way allowing it to be "naturalized," the need to articulate and converge with a phenomenological conception of language is a requirement inherent in the role that 4e cognitive science attributes to scaffolds. Consequently, rather than simply moving beyond phenomenology, the explanatory claims of 4e cognitive science should both recognize and be seen as complementing the descriptive claims of a minimalist phenomenology of language.

I A Heideggerian Framework

1 The "Place" of Language

Over the course of its long development language became increasingly central to Martin Heidegger's philosophical work, in terms of both his own distinctively crafted use of language and the intimate connection he came to see between language and thinking. In his earlier work *Being and Time* (*SZ*) Heidegger had treated language as peripheral to his overall philosophical project and discussed it in only a dispersed and fragmentary manner. Indeed he there concedes that language's mode of being has been left largely open, claiming only to identify its "ontological 'place,'" and would later hint that *SZ*'s treatment of language comprises its "basic deficiency" (*SZ* 166; Heidegger 1959, 93). Despite this, it is to *SZ* that I will look here to extract a basic framework for a phenomenological conception of language. Apart from the fact that it is here that Heidegger's commitment to phenomenology is most salient, the main reason for this is *SZ*'s emphasis on the foundational role of understanding equipment in everyday practices. For this yields an account of the pragmatic aspect of language, which is surely central to lived human experience of language, but which disappears completely from view in Heidegger's later writings.

My talk of a Heideggerian "framework" is intended to reflect a broadly affirmative approach to Heidegger's views, while signaling that these stand in need of further development—a task undertaken in subsequent chapters. At a very general level—the focus of this chapter and the next—this framework comprises an overall picture of the role Heidegger sees language having in the broader phenomenon of disclosure and a pattern of commitments that characterize his conception of language as a phenomenological one. At a more specific level—the focus of chapter 3—this framework also encompasses Heidegger's view of the complex function linguistic signs have in disclosing the world.

I describe this framework as "Heideggerian" rather than Heidegger's own to allow for the possibility of divergences. For example, I assume here that Heidegger's conception of language can be understood and assessed independently of his own overall project (the "question of being"). Further, because *SZ*'s treatment of language is peripheral and somewhat schematic, there is inevitably a need to fill in some gaps and for some interpretation. The label "Heideggerian" therefore serves as a convenient reminder that, while aiming for exegetic fidelity to Heidegger's texts, the primary concern here is with their value for a phenomenological conception of language.

Heidegger's crafted and idiosyncratic use of language often generates its own difficulties. At one extreme, some of Heidegger's more sympathetic readers adhere to his terminology as closely as possible, which—especially in translation—does little to communicate, and risks obscuring, the philosophical value of his views.[1] At the other extreme, many English-speaking readers immediately paraphrase Heidegger's views using "standard" vocabularies from contemporary philosophy of mind or language, and so risk effacing their distinctive configuration.[2] My aim here will be to steer a middle path by preserving the contours of Heidegger's analyses without relying exclusively on his terminology or emulating his style. While using familiar translations of Heidegger's terminology where there is no reason not to, I also depart from established translations where this seems appropriate. Further, where there is significant risk of confusion with less specific or more common concepts, I mark Heidegger's technical usage of certain terms through capitalization (e.g., Understanding, Attunement, Articulacy).

The aim of this chapter is to clarify Heidegger's overall picture of the role language plays—the "place" of language (*SZ* 166)—in disclosure. The first of its five sections provides requisite background by introducing Heidegger's phenomenological conception of the world. The second outlines his account of how human understanding of the world takes on determinate form, culminating in language, and identifies two problems in understanding where language fits into this account. The third section connects these problems with the interpretation of Heidegger's notion of Articulacy (*Rede*) and with the underlying philosophical question of whether all intelligent behaviors are inherently linguistic in form. The fourth section sheds light on Heidegger's unorthodox view of the relation between language and content by examining his discussion of predicative judgments ("Statements"). The final section exploits this unorthodox view so as to interpret Heidegger's notion of Articulacy and solve the two problems left open in the second section.

1 The World of Significance

The context within which *SZ* locates language is set out in its "preparatory" analytic of Dasein. The term *Dasein* is obviously central to Heidegger's thinking and its precise interpretation is a correspondingly delicate matter, but for the present purposes it can be taken simply to refer to humans as bearers of understanding of the world. The proximal aim of the "preparatory" analysis is to set out the general structure of the understanding—in Heidegger's terms, the disclosure of the world—that Dasein enacts and embodies. The topic of language makes a fairly late entry into his discussion, featuring as the terminus ad quem of the processes in which what Heidegger calls "Significance" takes on determinate form. To understand both the strengths and limitations of Heidegger's conception of language, it will be necessary to retrace the various steps of his discussion.

The starting point for Heidegger's analysis is the concept of "Significance" (*Bedeutsamkeit*), which is introduced in his discussion of the "worldliness of the world" (*SZ* 63). This discussion centers on a contrast between "Things" and "Equipment," as two different ways of individuating entities, and aims to show that the being of Equipment is somehow prior to that of Things. Heidegger understands a "Thing" (*Ding, res*) to be an entity individuated in terms of (determinate) properties and terms the being, or mode of existence, of Things *Vorhandenheit*, usually translated as presence-at-hand (or "occurrentness"; *SZ* 42 and passim). On his view, present-at-hand "Things" are entities that are characterized in terms of an essence (i.e., a properties-based determination of what the entity is) and existence (i.e., whether or not it is part of the actual world).[3] Although this characterization might seem unremarkable, it is important to be aware that for Heidegger presence-at-hand, or equivalently "reality" (*Realität*, "res-ality"; cf. *SZ* 67–68), is an important *technical term*. In appropriating these familiar terms—*thing* and *reality*—one of Heidegger's aims is to suggest that they have a more specific sense than we usually realize. More importantly, however, he believes that this properties-based essence + existence model cannot be intelligibly applied to all kinds of entity, most notably to humans (*SZ* 42, 151), so that a broader palette of ontological concepts is needed to characterize adequately other ways for entities to be.

One of the requisite further ontological concepts is that which Heidegger contrasts with Things, namely that of Equipment. As he defines it, "Equipment" (*Zeug*, πράγματα) refers to entities as encountered in (practically) dealing with one's environment and as individuated with regard to "what they are for"—that is, their purpose.[4] Heidegger labels the Equipmental way of being *Zuhandenheit*, "readiness-to-hand" (or "availability"), and treats the hammer as paradigmatic of this way of being. One might be inclined to question this distinction: Why is what an entity "is for" not a "property"? The answer to this—as will be seen more fully in section 4—is that Heidegger *defines* "properties" as context-independent determinations whereas what an entity "is for" entails reference to contexts of use, so that the two modes of individuation are distinct (mutually irreducible).

Employing the Equipment/Things distinction, Heidegger somewhat polemically rejects the idea that "the world" is to be understood as the totality of Things, hence as a naturalistic (physicalistic) whole, and instead urges that the primary being of entities—their *An-sich-sein* or "in-itselfness" (*SZ* 71, 75, 87)—lies in their Equipmental being, or readiness-to-hand. This is reflected in the claim that "the world" is an instrumental nexus, not the sum total of present-at-hand Things, but a whole of instrumental relations (*Verweisungen*[5]) between entities. It is this "structure of the world," as the purposive or instrumental whole in terms of which humans make sense of their environment and themselves, that Heidegger calls "Significance" (*SZ* 86–87).

According to this definition of *world*—as an instrumental rather than a naturally causal nexus—"worldliness," as Heidegger puts it, "is ... itself an existentiale"—that is, a structural feature of Dasein's being. Hence, in talking of the "world" as "that 'within which' a factical Dasein ... 'lives'" (*SZ* 64–65), what Heidegger means is the lived world, the "world of our involvements" (Taylor 1995, 107), or the fabric of this world *as experienced*. In this sense, Heidegger's conception of the world is clearly a phenomenological one, a conception of the world attuned to how it appears to agents.

Given this phenomenological conception of the world, together with Heidegger's focus on everyday tool use, it might be wondered how the distinction between presence-at-hand and readiness-to-hand is to be interpreted. Heidegger is presumably making a genetic claim about the order in which entities are encountered in lived experience of the world.[6] He also seems to be making an epistemic claim about the way entities are understood in lived experience, either purposively as Equipment or objectively as Things. Moreover, Heidegger himself presents the distinction as an ontological one, as a distinction between two "kinds of being" (*Seinsarten*; e.g., *SZ* 42, 69, 104). Yet even while insisting that readiness-to-hand "may not ... be understood as a mere character of grasping [*als bloßer Auffassungscharakter*]," he also somewhat enigmatically suggests that claims to ontological priority are bound to respect the "ontological sense of knowing," and with this the "sequence of discovering and appropriating dealings with the 'world'" (*SZ* 71). Perhaps his aim in saying this is simply to highlight the need for a phenomenological ontology to cohere with lived experience, or to warn against the mistaken ontology he thinks results from the traditional focus on knowing. Be that as it may, it is important to note that there is no obvious inconsistency between the three kinds of claim just attributed to Heidegger. The ontological distinction reflects the thought that specific uses are essential to the being of some entities (Equipment), such that (context-independent) properties alone do not suffice to define them. The best example of this are useful artifacts, such as hammers or tools more generally, which are entities produced by humans for a particular purpose and are centrally determined by what they are for. These contrast most obviously with naturally occurring entities (volcanoes, trees, animals), which are not human made and not necessarily useful, and which are more plausibly characterized as present-at-hand. Heidegger then consistently draws on this ontological distinction in describing prereflective experience, particularly to develop the genetic/epistemic thought that default awareness of our environment involves a distinct kind of practical intelligence or know-how rather than propositional knowing-that.[7]

Nevertheless, in terms of its broader philosophical implications several concerns might remain about Heidegger's phenomenological conception of the world. In particular, it might still be wondered whether it can deal satisfactorily with natural entities, which are supposed to be as they are "in themselves," independently of and prior to human understanding. How are such entities supposed to depend on readiness-to-hand? Indeed Heidegger's view might seem more generally to resemble an idealist notion of the world in focusing merely on how the world appears "for us" rather than capturing the world as it is "in itself."[8] Accordingly, it might be thought that Heidegger sometimes relies on equivocation or ambiguity in putting his conception of worldhood to work. In his brief and dismissive treatment of the traditional problem of external-world skepticism, for example, one might suspect that asserting the existence of the world in Heidegger's sense does not directly bear on the supposed problem.

I want to postpone discussion of such concerns until chapter 8, where they will be discussed at greater length in clarifying the more general question of what it means to say that the "world" is "disclosed"—a question likely to be faced by any phenomenological conception of language. There I will argue that concerns such as those just listed are dependent on an opposition between realism and idealism resulting from basic ontological assumptions that Heidegger rejects. For the moment, however, it will suffice to note that these concerns are not of immediate relevance to Heidegger's conception of language. For linguistic entities (e.g., words, sentences) are clearly materially present entities that (nevertheless) have an irreducible ontological dependence on the way they are produced, used, and experienced by humans. This fact, without yet implying anything about whether and how the present-at-hand/ready-to-hand distinction applies to language, means that any problems Heidegger's position might have in dealing with entities beyond the realm of human experience will not directly impact on his conception of language itself.

2 The Articulation of Significance

The most general characterization in *SZ* of the role of language in disclosure is the analysis of the "existential constitution of 'the there'" in §§28–34. Since Dasein *is* its respective "there" ("*ist* selbst je sein 'Da'") and "*Dasein is its disclosedness*" (*SZ* 132–133), these sections amount to an analysis of the structure of disclosure. Overall they undertake to tell us how the grasp humans have of the world around them takes on a determinate form—that is, how particular entities are picked out or individuated against the background of Significance.

On Heidegger's analysis, disclosure has two principal aspects, which he describes as being equally basic ("original") or "equiprimordial" (*SZ* 133). The first, *Befindlichkeit,* is intended to convey that Dasein matters to itself, *that* Dasein *finds itself* in the world and that it constitutively, and prereflectively, feels or "finds itself" *in some way.*[9] This dual sense of "finding itself" would perhaps be best captured in English by the term *affectivity,* which suggests both the passivity of factical thrownness (being affected by the world) and a link with feelings or moods (affects). For convenience, however, I will use the more familiar translation "Attunement" in the following. Corresponding to its dual sense, as felt facticity, Heidegger sees Attunement as having a specific function in that various of its modes—especially "anxiety"—reveal Dasein to itself, as mattering to itself, by presenting it as the "entity which is answerable to its being" (*SZ* 134).

The second principal aspect of disclosure is that it always involves some kind of comprehension or cognition. For this Heidegger adopts the common term "Understanding" (*Verstehen*), which he gives a specific and somewhat unorthodox sense distinguished by three features from both the (Kantian) discursive "faculty of concepts" and Heidegger's contemporaries' debates about the distinction between understanding and explanation, supposedly reflecting the respective methodologies of the humanities and sciences.[10] The first is that for Heidegger, Understanding is purposive: it is concerned with that "for the sake of which" Dasein exists (its *Worumwillen*; *SZ* 143). By this he means that Understanding pertains not to entities as such, but to the "projected possibility" or "project" as which Dasein lives (*SZ* 144, 145). The purposiveness of Understanding thus relates to human endeavors and aims, rather than to cosmic teleological constants, and it is in relation to such human aims that entities encountered in the environment are potentially useful or expedient. In Heidegger's words: "The primary 'for what' is a for-the-sake-of-which. The 'sake for which' always concerns the being of *Dasein*" (*SZ* 84). The second distinctive feature of Heidegger's Understanding is that it is holistic, encompassing the "entire basic constitution of being-in-the-world" (*SZ* 144; cf. 152). The third is that Understanding is indeterminate or inarticulate: unlike a "plan," Heidegger tells us, Understanding entails no "thematic" grasp of the possibilities it addresses (*SZ* 145). This characterization again relies on a technical term, *thematization*, which Heidegger defines as an "articulation of the understanding of being," a "delimitation of the area of subject matter" that prefigures the concepts appropriate to the relevant entities and befits this subject matter for "objectivizing" scientific inquiry (*SZ* 363). Thus rather than being directed at specific entities, or indeed at entities at all, Heidegger's "Understanding" refers to an overall sense of purpose—a purposive, holistic, and indeterminate (hence preobjective) grasp of the world—that forms the foundation of all modes of disclosure.

Heidegger contrasts this overall sense of purpose, the "pre-structure of Understanding," with "the development of Understanding," as a level or stage within disclosure at which "the structure of *something as something*" arises.[11] He calls this level or stage of disclosure *Auslegung*, a term usually rendered as "interpretation." This translation invites various associations as well as potential confusion with Heidegger's own term *Interpretation*, which he applies only to reflective theoretical activity (e.g., *SZ* 130, 357). So as to mark less ambiguously the specific sense Heidegger links with *Auslegung*, I will translate this term in the following as "Setting-out"—an expression that also has the advantage of getting closer to the (trans)literal sense of "laying out" in *Auslegung*. Setting-out is the process in which features within Understanding become individuated "as" a such-and-such, and it is with this move that differences between entities are to become expressly or "explicitly understood"—that is, capable of being marked with an expression (linguistic or otherwise).[12] Two points are worth noting about the relationship between Understanding and Setting-out. First, there is an important difference in that whereas Understanding pertains to—is of—projected possibilities or aims, Setting-out concerns the entities addressed. Through this distinction Heidegger is claiming that entities are only individuated (in disclosure) in relation to a presupposed purposive horizon, no matter how vague the latter may initially be. A corollary of this, second, is that the way entities are picked out in Setting-out and the concepts subsequently developed are constrained by the purposive projection of Understanding (*SZ* 150).

While always "grounded in Understanding" (*SZ* 148), it is distinctive of Heidegger's position that Setting-out is to occur in different modes, corresponding to his distinction between the ready-to-hand (*Zuhandenes*) and the present-at-hand (*Vorhandenes*). To begin with, Heidegger foregrounds "circumspective" Setting-out, which individuates entities as ready-to-hand or "Equipment" (*Zeug*)—that is, entities as encountered in (practically) dealing with one's environment and as individuated with regard to their purpose or "what they are for" (their "*Um-zu*" or "*Wozu*"). As Heidegger puts it, to the question of what a certain entity is, circumspective Setting-out responds "it is for ...."[13] This is how, on his view, we are aware of entities when using them: a hammer is picked out *as* an entity *for* hammering in the context of a corresponding practical activity. Subsequently in §33 Heidegger further identifies a "derivative" mode of Setting-out, "predicating articulation," in which entities are individuated as present-at-hand or "Things" (*Ding, res*; *SZ* 42) —that is, entities individuated in terms of (determinate) properties (*SZ* 155, 158). This distinction between two modes of Setting-out turns out to be of great importance in understanding Heidegger's conception of language and is discussed in greater detail below in section 4.

So far—in §§28–33—Heidegger seems to be developing a stratified picture of what I will call *progressive determination*, according to which Dasein's disclosure moves progressively from lower to higher levels of determinacy, with the former corresponding to foundational or more "primordial" structures.[14] However, this analysis is given an additional, and potentially problematic, twist by the further claim in §34 that Attunement and Understanding are "equiprimordially determined" by *Rede*, a further ontological structure that forms the "existential-ontological foundation of language" (*SZ* 133, 160). Although briefly anticipated in §28 (*SZ* 133), this feature had not been mentioned throughout §§29–33, so it comes as something of a surprise when Heidegger now claims to have "constantly made use" of this notion while "suppressing" it in his "thematic analysis" (*SZ* 161). More importantly, as we will now see, this new feature appears to conflict with the story of progressive determination told in §29–33 and to suggest instead that disclosure is always structured by some kind of *underlying determination* linked with the notion of Articulacy.

Heidegger defines *Rede* as "the articulation [*Artikulation*] of intelligibility" (*SZ* 161). For reasons that will emerge in the sequel, I want to avoid the common translations of the term *Rede* as "Discourse" or "Talk" and will instead, following this definition, use the term "Articulacy."[15] The reference in this definition to "intelligibility," which is glossed as the "Significant whole" (*Bedeutungsganze*), establishes a link with the notion of "Significance" discussed above, so that Articulacy can be understood simply as the articulation of Significance.[16] Heidegger further characterizes Articulacy in terms of four "constitutive elements": (a) that which is being talked about or referred to (*das Worüber*); (b) what is said (*Geredetes, das Gesagte als solches*); (c) communication (*Mitteilung*), by which Heidegger means the sharing of disclosure (i.e., of Attunement and Understanding), in the public space; and (d) intimation (*Bekundung*), that which Dasein reveals of itself in performing Articulate acts (*SZ* 161–162).

It is only against this background that Heidegger turns explicitly to language (*Sprache*), which he characterizes as the "spokenness" or "spoken-out-ness of Articulacy" and as the "totality of words" (*SZ* 161, 167). While Heidegger's terminology here—*(Hin)Ausgesprochenheit*—literally refers to the state of having been pronounced aloud, his further comments (*SZ* 168–169) make clear that this is to encompass the written use of words. *Language* thus has a somewhat narrow sense for Heidegger as the sum of actually pronounced or written linguistic utterances, the public use of signs, or the perceptible presence of Articulacy in the world. At the same time Heidegger assures us that it is the ontological structure of Articulacy that determines the meanings or "significations" (*Bedeutungen*) of words: "divided up in Articulating articulation [*redenden Artikulation*]," "intelligibility's whole of signification *advenes to the word* [*kommt zu Wort*]. Words grow onto the significations [*den Bedeutungen wachsen Worte zu*]" (*SZ* 161). These comments, together with Heidegger's narrow definition of language, can give the impression that he sees the disclosive role of language as peripheral, as merely labeling the language-like meanings ("significations") that are themselves determined by the ontologically prior structure of Articulacy.

This results in two problems in understanding where language fits into Heidegger's overall picture. The first is the apparent tension introduced by §34, which concerns the level or the way in which determinate meanings are generated and threatens the overall stability of Heidegger's analysis. On the one hand is the story of progressive determination told in §§28–33, according to which disclosure takes on a determinate form by moving through different levels of "originality," from inarticulate Understanding, through Setting-out as ready-to-hand Equipment, and finally to Setting-out as present-at-hand Things. On the other hand is the story of underlying determination suggested by §34: for in claiming that Articulacy is "equiprimordial" with Attunement and Understanding Heidegger seems also to suggest that linguistic meanings are articulated independently of, and indeed prior to, any form of Setting-out. Thus "intelligibility," he assures us in this context, is "already divided-up [*gegliedert*] prior to appropriating Setting-out," such that Articulacy "already underlies Setting-out and [the predicative] Statement" (*SZ* 161). The resultant problem is that the claims of §34 appear to collapse the foundational hierarchy set out in §§28–33. For Heidegger's claim that Articulacy is "equiprimordial" with Understanding and Attunement seems to imply that a language-like structure and semantics already pervades Understanding and so underlies all Setting-out. More specifically, his characterization of Articulacy's "constitutive elements" (cf. (a) and (b) above) suggests that this involves identifying entities as such and such and saying something about them, rendering obscure both Heidegger's general claims about Setting-out and his view that purposive or Equipmental Setting-out is basic. Heidegger's account of progressive determination thus appears to be undermined by the claims he makes about underlying determination by Articulacy.

The second problem concerning the role Heidegger supposes language to have is that his discussion might seem undecided as to whether language—in the narrow sense of public sign use—plays a constitutive role in disclosure. For according to the progressive determination story Understanding and Setting-out (and perhaps Articulacy) look like nonlinguistic feats, such that language has only a peripheral role. Hence, in Guignon's (1983, 118) view, there is "clearly the intimation that there could be a fully articulated sense of the world ... prior to or independent of the mastery of language." Yet, quite apart from the intrinsic difficulties of such a view, this seems to be inconsistent with the role Heidegger attributes to language in setting up the everyday ("inauthentic") understanding of historically situated Dasein. Hence the apparent indecision, which Guignon (1983, 117–118) sums up by suggesting that *SZ* remains "torn between two incompatible views of the nature of language," which he terms the "instrumentalist" and "constitutive" views.[17]

These problems can be seen as resulting from two desiderata Heidegger is trying to respect. The first is to think of language as acquiring significance in much the same way as other entities (such as hammers) might, as continuous with other meaningful activities. This aim is reflected in the progressive determination story, which consequently runs the risk of assimilating language to an essentially nonlinguistic account of meaning. The second desideratum is to acknowledge that linguistic articulation has a specific form—being about something, and saying something of it (cf. (a) and (b) above)—that defines it in contrast to other forms of intelligent behavior and representation. This acknowledgment is reflected in Heidegger's characterizations of Articulacy and in the underlying determination story, which then seems to imply problematically that all articulate understanding shares the specifically linguistic shape. The question is whether Heidegger succeeds in balancing these two desiderata, whether he can simultaneously account for the continuity of language with nonlinguistic phenomena and for the specificity of language. The answer to this question will depend on how Heidegger's notion of Articulacy is interpreted, specifically on whether this is taken to be essentially linguistic in form and whether it entails that a language-like articulation of Significance is inherent in all disclosure. To address the two problems just identified, and so to get clearer about the role of language in Heidegger's overall view of disclosure, the next step will therefore be to consider more closely what exactly Articulacy is supposed to be and in what sense it is the "existential-ontological foundation of language."

3 Linguistic versus Pragmatic Articulacy

In the literature on Heidegger the conflicting tendencies just described have led to two opposing views of what it is to be articulate being attributed to Heidegger. Adopting Taylor Carman's apt distinction, the first is the "linguistic model," which takes Articulacy "as something very closely bound up, if not indeed identical, with language"; the second is the "pragmatic model," which concentrates on Heidegger's definition of Articulacy as the "articulation of intelligibility," seeing this as applying to any differential behavior and so as not essentially linguistic.[18]

This exegetic dispute should be seen in connection with an underlying philosophical issue: whether humans have a capacity for articulate or intelligent behavior that cannot be assimilated to the model of language, and more generally how linguistic articulacy relates to other forms of intelligent behavior. The linguistic model of Articulacy is supported by the plausible idea that any articulate action can in principle be described in language, which encourages the thought that all such actions are inherently language-like.[19] Here the thought is that language can of course be used to describe any action, including apparently nonlinguistic actions such as Heidegger's example of hammer use. Further, unless we are to think of such descriptions as constantly misrepresenting those actions, it seems that the actions themselves must embody the same content (albeit tacitly or implicitly) as linguistic descriptions of them. Thus we are encouraged to assume an underlying semantic structure—common to linguistic and nonlinguistic acts—that would presumably form the proper object of a theory of meaning, and to think language has a special status in making explicit this underlying semantic structure.

Conversely, the pragmatic model of Articulacy is supported by the likewise plausible idea that many forms of intelligent behavior are possible without involving linguistic competence in any way. One kind of example are nonlinguistic behaviors that exhibit an articulate grasp of the world, such as making and understanding gestures, but also any number of bodily actions (swimming, looking for shelter from the rain, lighting a fire, etc.) and goal-directed animal behavior (e.g., hunting, mating rituals, nest building).[20] These behaviors clearly involve recognition of structure in the surrounding world—that is, the ability to pick out specific features of the world and to behave differentially toward them. Further, many of them can be, and indeed are, performed perfectly adequately by nonlinguistic animals. Another kind of example are activities such as music, dance, or the use of visual media (e.g., painting or film), which, while no doubt to some extent describable in language, involve articulate actions in media that in themselves are by no means obviously linguistic in form. For the present purposes it will be convenient to group together the two kinds of example just distinguished and to refer to them collectively as "intelligent nonlinguistic behaviors."

Thus the philosophical issue underlying the interpretation of Heidegger's notion of Articulacy is what role linguistic behaviors play in understanding intelligent behavior more generally. More precisely, the issue is whether language can be seen as paradigmatic of all intelligent behavior, or whether it is necessary to distinguish some more basic (more "primordial") capacity for articulate behavior, of which language use—no matter how important—is but one form of manifestation. In the terms introduced above, this translates into the question of whether all intelligent behaviors essentially have a linguistically Articulate form (linguistic model) or whether linguistic articulacy is one aspect of a more general notion of Articulacy (pragmatic model).

Returning to the exegetic issue, it might initially seem surprising that there should be such disagreement over the interpretation of Heidegger's notion of Articulacy. For if one focuses narrowly on what Heidegger writes *whenever* characterizing or defining Articulacy, it seems obvious to read him as advocating the linguistic model. A specific link with language is suggested, for example, not only by Heidegger's commonly used term *Rede*, which means "talk," but also by the fact that he discusses Articulacy and Language in the same section (§34). Further, in his extended discussion there of Articulacy Heidegger's wording and examples make clear his focus on what others might call speech acts and propositional attitudes, with no reference being made to intelligent nonlinguistic behaviors (cf. *SZ* 161–162). The same focus is clearly in evidence in §7, which states that concrete acts of Articulacy have the "character of speaking" (*Charakter des Sprechens*), of being pronounced by the voice in words (*stimmliche Verlautbarung in Worten*) (*SZ* 32). This is also echoed in Heidegger's introduction of Articulacy in §34 as the "existential-ontological foundation of language" (*SZ* 160). All Heidegger says is "of language." He does not explicitly claim that *Rede* is similarly the "existential-ontological foundation" of any other form of intelligent behavior, such as the kind of "coping" involved in using hammers, playing basketball, driving, or skiing. (As he could easily have done, if this were his view.) Last but not least, its linguistic cast is clearly signaled in Heidegger's characterization of Articulacy in terms of "what is said" (*Geredetes*) or "that which is said as such" (*das Gesagte als solches*). In this light Articulacy seems to be simply a generic notion of *linguistic ability* and to comprise, according to Heidegger's four-point characterization, the ability of an agent to pick out something referred to, to say something of that referent, to say it in the public domain, and simultaneously to express something about itself.[21]

Those who see Heidegger as committed to the pragmatic model often tend to disregard or downplay the features linking Articulacy with language. Thus Dreyfus's (1991) chapter on "Telling"—as he translates *Rede*—makes no reference to the linguistic features (a) and (b) in terms of which Heidegger characterizes it.[22] Blattner (1999, 72; cf. 67–75) acknowledges that Heidegger's characterizations of Articulacy are "shot through with linguistic suggestions," yet ignores this textual evidence in arguing that linguistic articulation is merely a "derivative" form of a more basic (and Dreyfusian) capacity to discern differential structure in the world. Similarly, while taking note of features (a)–(d) and observing that Heidegger "invites us to understand discourse [*Rede*] as something like a prelinguistic set of conditions precisely anticipating the semantic and pragmatic structures of language," Carman (2003, 227) dismisses this invitation as "obviously misleading" given Heidegger's view of Assertions (*Aussagen*) as a derivative mode of Setting-out.[23] By contrast, Wrathall (2011, 105ff.) subtly registers the connection with language that Heidegger indicates by translating *Rede* as "conversation." However, he takes this to mean "being conversant with" things in a way that, again, extends to nonlinguistic practices, and sees the function of *Rede* as "a gathering of meaningful elements into a unified structure, a meaningful, but prelinguistic articulation of the world."[24]

This tendency may not in itself be problematic, despite Heidegger's acute sensitivity to linguistic nuances. There are several reasons that might lead one to disregard the apparently linguistic connotations of the terminology he typically uses in characterizing *Rede*. The main reason is no doubt the attempt to situate those passages in a broader exegetic perspective. For Heidegger's discussion of the worldliness of the world, particularly his treatment of hammer use as a model of meaningful action, focuses on intelligent nonlinguistic behavior and appears to be urging a broad conception of meaning that does not center on language. And if this is to be consistent with the claim of §34 that Articulacy is equiprimordial with Understanding and Attunement, it might seem that Articulacy cannot be linguistic in kind, just as the pragmatic model claims. A second reason for resisting the linguistic model is that it might seem to license a reductive, language-centered attitude of a kind Heidegger clearly opposes. For if all meaningful actions were implicitly linguistic in their semantics, there would presumably be little point either in focusing on nonlinguistic behaviors or in opposing the study of "meaning categories" (*Bedeutungskategorien*; *SZ* 165) centering on propositional meaning. Finally, given the apparent tension in Heidegger's position (set out in section 2 above), it might be felt that we are free to disregard the linguistic characterizations of *Rede* and exploit those aspects that seem most philosophically interesting.

We therefore appear to have a choice between two approaches, each of which has some plausibility, both exegetic and philosophical, but neither of which is completely satisfactory. On the one hand, a narrow exegetic focus on Heidegger's definitions suggests attributing to him the linguistic model of Articulacy. Yet this model appears to be linked with the underlying determination story—as set out in section 2—and so to threaten the overall coherence of Heidegger's position by undermining his account of progressive determination. On the other hand, taking a broader exegetic perspective, the pragmatic model coheres with the progressive determination story and with Heidegger's emphasis on intelligent nonlinguistic behavior, but is difficult to reconcile with his explicitly linguistic characterizations of Articulacy. As a result it is not clear whether Heidegger is best considered as an advocate of the linguistic model or the pragmatic model, and hence how his views contribute to philosophical debates between the two.

The remainder of this chapter will offer a limited defense of the "linguistic model." In doing this I will assume that Heidegger's own characterizations of Articulacy constitute prima facie evidence that he thought of Articulacy as closely linked with language, so that if possible this thought should be preserved. A major obstacle to preserving it is the apparent risk of incoherence this thought brings. My aim will therefore be to remove this obstacle by showing how Heidegger's position can be interpreted without committing him to the underlying determination story, thus allowing a coherent reading of §§28–34. Affirming the link between Articulacy and language prompts the question of where intelligent nonlinguistic behaviors fit into Heidegger's overall picture, and how to explain his emphasis on such behaviors. Although nonlinguistic behaviors are not my main concern here, I outline briefly in section 5 how I think these questions should be responded to. Before considering these issues, however, some preparatory work is needed.

4 The Heterogeneity of Sentences

To progress further in understanding how language is conceived in *SZ* it is necessary to consider the discussion in §33 of what Heidegger calls a "derivative mode" of Setting-out. This discussion is of great importance for two reasons. First, it sheds light on a central and distinctive aspect of Heidegger's conception of language by clarifying his view of the relation between language and content. Second, it provides the background for an interpretation of the notion of Articulacy that allows §§28–33 to be read consistently.

It will help to begin by considering how the linguistic and pragmatic models conceive the relation between language and content. While the linguistic model clearly relies on a notion of propositional content, a defense of the pragmatic model might appeal to a similar line of thought by claiming that the first two features defining Heidegger's Articulacy—namely, (a) that something is talked about, and (b) that something is said—can be construed in terms of a generic notion of content, so that the apparent linking of Articulacy with language is misleading. On this view, "what is said" would not entail a relation to language, since the order of explanation is rather that language can be used to say things precisely in virtue of its representing content. Further, though it is perhaps less obvious how, nonlinguistic states or actions might in principle also embody such content ("tacitly"). And in that case there would be nothing specifically linguistic about Articulacy and we would be left with the more general pragmatic model.

An obvious response to this defense would be to say that, because it is assumed to be expressible in language, the kind of "generic" content appealed to here is simply propositional content. It is perhaps worth noting that if this is all the difference between the two models amounts to, then they appear to disagree only over terminology, specifically over what to call linguistically expressible content. Much more important, however, is the fact that both the above defense and this response miss a central feature of Heidegger's position, which is precisely that it excludes the appeal to any kind of uniform content, whether propositional in form or generic. It is this distinctive and interesting feature of Heidegger's position that the discussion of a "derivative mode" of Setting-out in *SZ* §33 allows us to see.

The intended focus of this discussion can easily be obscured by a combination of translation, the passage of time, and Heidegger's intentionally unorthodox position. The common English translation of Heidegger's term *Aussage* as "assertion" somewhat misleadingly suggests that he has in mind a kind of speech act or illocutory force. I will instead render it here as "Statement" to reflect the fact that—as is usual in German philosophical parlance—Heidegger uses the terms *Aussage, Urteil,* and *Satz* interchangeably to refer to judgments expressed in propositional form.[25]

Heidegger characterizes Statements in terms of three features (*SZ* 154–155): Statements *point out* (*aufzeigen*) features of the world, allowing entities to be "seen" just as they are; their "mode of pointing out," as *predication*, makes what is (already) manifest "explicitly manifest in its determination"; finally, Statements are *communicative* acts of "allowing to be seen" (*Mitsehenlassen*) that involve "spoken-out-ness" (i.e., the public use of language).[26]

As the title of §33 indicates, Heidegger considers Statements to be a "derivative" mode of Setting-out, involving what he calls a "modification" in the "*as*-structure"—that is, in the way we are expressly or explicitly aware of what is understood—from the "hermeneutic *as*" to the "apophantic *as*."[27] This modification, in Heidegger's words, "first opens access to something like properties," to "the what" or "predicating articulation" of the entity concerned, a transition in which the entity is "cut off" from the "instrumental relations [*Verweisungsbezüge*] of Significance" (*SZ* 158, 155, 158). What he has in mind is thus a transition from an awareness of entities as ready-to-hand to an awareness of entities as present-at-hand. In characterizing the latter as "derivative" Heidegger's thought is clearly not only that purposive awareness of the ready-to-hand is importantly different from properties-based awareness of the present-at-hand, but that the latter presupposes the former (and not vice versa). This thought might seem surprising, because an awareness of entities as ready-to-hand—that is, of what they are for (cf. *SZ* 149)—is clearly relative to particular ends and particular contexts of use. By contrast, as instrumental relations are "cut off" entities come to be understood without relation to particular ends, in terms of determinations that are context-independent and hence proper to them (literally: proper-ties, or *Eigen-schaften*).

A standard response to this "modification" would presumably be to see it as a gain, as constituting progress toward objective knowledge. In what sense then is predicating articulation "derivative"? What sense of loss does Heidegger's talk of "derivativeness," "leveling off," or "suppression" (*SZ* 158) allude to? His concern is clearly that the switch to predicative awareness obscures the matrix of purposive activity that it presupposes as its "ontological origin" (*SZ* 158). In general terms this would mean it fails to provide a complete picture of how humans disclose the world, but Heidegger also sees it as involving a more specific sense of loss. An important clue to the latter is provided by his somewhat obscure claim that that "with which" we are dealing in a ready-to-hand manner (*das zuhandene Womit des Zutunhabens*) becomes that "about which" a Statement points something out (*das Worüber der aufzeigenden Aussage*; *SZ* 158). While this might initially sound like a mere change in labels, there is an underlying thought that emerges more clearly in other writings immediately prior to *SZ*. Thus in his 1925–1926 lectures this transition from the "with which" to the "about which" is explained as follows: "the as-what is obtained not from what a performance is for [*Wozu einer Verrichtung*], but from that about which the Statement is made" (*Logik* 155). Further, a Statement in Heidegger's sense is not "as such primarily related, in terms of its content, to the dealing with [*das Zutunhaben*]" (*Logik* 157). Similarly in *Prolegomena toward a History of the Concept of Time*, the early draft of *SZ*, Heidegger explains that the usual task of making manifest through talk (*redende Offenbarmachung*) is "appresentation of the environment" that forms the object of concern—rather than being "tailored to knowledge, research, theoretical propositions," etc.—and suggests that Articulacy (*Rede*) has the function of drawing attention to the instrumental relations inherent in Significance.[28] The thought common to these claims is that there is a difference between an awareness of entities *as connected with* the context(s) in which they are used or encountered ("with which") and an awareness that focuses on specific entities *as distinguished from* contexts of use ("about which").

In light of these comments, the point of Heidegger's claims about Statements can be understood more clearly. First, the shift from the hermeneutic *as* to the apophantic *as* stands for a change in referent: whereas the former relates to a complete context of action, the practical environment, the latter picks out a particular entity in abstraction from its context(s) of use. Second, the shift brings a change in the kind of attribute concerned: whereas the hermeneutic *as* is concerned with purposive determinations, such as appropriateness or expediency (*Geeignetheit, Dienlichkeit*; *SZ* 83), the apophantic *as* is concerned with properties. Here it is important to note that Heidegger has defined these two kinds of determination in a way that makes them independent and mutually irreducible: expediency is essentially relative to contexts of action, whereas properties are taken precisely to be invariant over contexts—that is, context-independent determinations. Thus on Heidegger's view the use of Statements involves a *semantic discontinuity* between two different kinds of "content" (marked by a change in both what is referred to and the perspective in which it is ascribed a sense).

The fact that he both explicitly discusses them and identifies them as linguistic communication could give the impression that Heidegger thinks of Statements—judgments expressed in propositional form—as paradigmatic of linguistic meaning. Yet, as signaled by the label "derivative," nothing could be further from the case. The motive for Heidegger's explicit discussion is rather his belief that the focus on propositional judgments (Statements) in traditional logic was due to a mistaken overemphasis on theoretical uses of language. Accordingly in §33 he is *opposing* the idea that ("theoretical") Statements provide a model for properly understanding how language functions in general.[29] This can be seen in the following revealing passage:

> Between Setting-out still completely enveloped in concernful understanding and the extreme countercase of a theoretical Statement about something present-at-hand [*Vorhandenes*] there is a multitude of intermediate steps: Statements about occurrences in the environment, accounts of the ready-to-hand [*Zuhandenes*], "reports on situations," recording and registering "facts of the matter," describing a state of affairs, recounting what has happened. These "sentences" cannot be reduced to

theoretical Statements without essentially distorting their sense. They, like theoretical Statements themselves, have their "origin" in circumspective Setting-out. (*SZ* 158; cf. *Logik* 156n.)

This passage has several important implications. First it makes clear that Heidegger's claims about the foundational role of purposive (circumspective) awareness include language. In particular it implies that the difference between hermeneutic and apophantic *as*-ness is such that different kinds of sentence[30] cannot be assimilated to predicative Statements without "essentially distorting their sense." The point is thus not to distinguish "assertion" as a particular kind of speech act, as opposed to other kinds of illocution or force.[31] It is rather that the distinction between the hermeneutic *as* and the apophantic *as* marks a semantic discontinuity—that described above—which would persist even if all the sentences referred to were both declarative in form and assertoric in force. The claim is not simply (and trivially) that the content of sentences can vary, but that the *kind* of content sentences express is heterogeneous: that language use can be linked with purposive awareness at one extreme, properties-based awareness at the other, or any admixture of the two.

Perhaps the most striking aspect of this claim is Heidegger's talk of a "multitude of intermediate steps" (*mannigfache Zwischenstufen*). Although he offers a list of examples, Heidegger does not really explain his thought fully. So how might it be understood? One obvious possibility is that a propositionally expressed judgment (Statement) might combine various terms, some of which relate to their referents as ready-to-hand while others relate to them as present-at-hand. However, while this might provide part of the explanation, it does not seem sufficient to justify talk of a "multitude" of intermediate possibilities. A second possibility is that context independence comes in varying degrees. This would be a consequence of Heidegger's view that the (purposive) use of language embedded in contexts of action is basic, and that context-independent determinations are gained—the hard way, so to speak—by "cutting off" instrumental relations. For there is no reason to think of this severing of context relativity as an all-or-nothing matter. Indeed, it seems more plausible to think that context independence is a variable attribute, corresponding to generalization over larger or smaller numbers of contexts of use. Thus we might begin by distinguishing at one extreme ad hoc words used in only a single context, specialist words used in only a single kind of practice, multipurpose words used in a wider range of situation types, and universal words that have the same meaning over all contexts (standing for Heideggerian "properties") at the other extreme. But once the pattern in these distinctions is recognized, it is clear that it admits a continuum of possibilities, depending on how large the class of contexts is over which the meaning of a given term remains invariant.[32] I suggest that Heidegger's position should be interpreted in this way as it would underwrite his explicit claim that a "multitude of intermediate steps" lie between the two extremes of a purely Equipmental and a purely theoretical grasp of entities.

A second implication of the above passage is to clarify that the distinction between the hermeneutic *as* and apophantic *as* does not coincide with that between the linguistic and nonlinguistic. Rather, on Heidegger's view, the delimitation between purposive and predicative awareness falls within language use. Although the two distinctions do not coincide, Heidegger's treatment of Statements as both a mode of Setting-out and linguistic introduces an asymmetry in that whereas predicative Statements are necessarily linguistic, language use does not necessarily involve predicative awareness. A striking consequence of this asymmetry is Heidegger's view that even a sentence that looks exactly like a predicative statement ("The hammer is heavy") can be an "expression" of "concernful deliberation" without being a predicative judgment.[33] Hence the possibility that language can be bound up in the constitution of everyday understanding without entailing (full) awareness of its predicative import.[34]

Finally, the above passage highlights that what Heidegger calls predicative "Statements" have a point distinguishing them from other uses of language. To borrow Carman's (2003, 211) fitting term, their use comprises "overtly demonstrative practices": "a kind of exhibiting or showing," the public letting-be-seen of determinations, that must be understood as "a mode of practical comportment, a doing." Yet Heidegger's aim here is precisely to emphasize that not all language use is like this, that much use of language is not such an act of "overt" exhibiting or showing, but is simply immersed or embedded in other actions. And while it is true that for Heidegger signs serve generally to draw attention to the surrounding world (cf. *SZ* 79–80), he clearly intends to distinguish between "overtly demonstrative practices" and practices in which what might be called "embedded showing" occurs. For example, while drawing attention to the environing world, car indicators (as signs) function for Heidegger at the level of practical responses and Equipmental understanding. That is, rather than functioning as part of a practice that has the *point* of "exhibiting or showing" as "Statements" do, their showing is embedded in and subordinate to the ends of other actions.[35] Conversely, the use of sentences qua Statements to manifest purpose-independent awareness of entities—to make what is already manifest "explicitly manifest in its determination" (*SZ* 155)—is a paradigmatically theoretical mode of language use, with a specific, overtly demonstrative aim. In this way, the semantic discontinuity between the hermeneutic *as* and the apophantic *as* also corresponds to a distinction between two different kinds of demonstrative role—embedded showing and overt demonstration—that language can play in practice.

Against the background of this discussion of Statements it is easier to see what both the claims contrasted at the beginning of this section miss about Heidegger's position. For the specific point of §33 is that the notion of propositional content is founded in purposive Setting-out, and can be made sense of only in relation to a feat of semantic organization exemplified in theoretical uses of language. Accordingly, Heidegger thinks this notion should not be overgeneralized and applied to all kinds of intentionality, or even to all kinds of language use. Yet Heidegger's view that sentences are contentful in heterogeneous ways also has the more general implication that language—and hence intentionality more generally—cannot be understood in terms of any kind of uniform content.

5 Linguistic Articulacy

At the end of the second section of this chapter we were left with two problems concerning the role language plays in Heidegger's analysis. I went on to suggest that to solve these problems a better understanding of Heidegger's notion of Articulacy would be required, and that for this an understanding of Heidegger's view of the relation between language use and content would be needed. This section pulls together the various loose ends by exploiting the discussion of the last section to interpret the notion of Articulacy and to solve the two problems previously left open.

I begin by returning to the question of how to interpret the notion of Articulacy in *SZ* and my defense of the claim that for Heidegger Articulacy is closely linked with language. A major difficulty with this claim is that it may seem to be inseparable from the underlying determination story and the risk of incoherence described in section 2 above. The problem there—the first of the two left open—was that the foundational hierarchy set out in §§28–33 appears to collapse in light of the claims about Articulacy in §34. For it appeared that Articulacy implants a language-like structure and semantics at the same basic level as Understanding and Attunement, thus undermining Heidegger's account of Setting-out.

With this problem in mind, it is particularly interesting that nothing in Heidegger's characterization of Articulacy (*SZ* 161; cf. section 2 above) aligns either of the features—what is talked about (a) and what is said (b)—to either circumspective or predicative Setting-out, or to a grasp of entities as ready-to-hand Equipment or present-at-hand Things. I want to suggest that the reason for this is that (a) and (b) are intended to apply to *both* purposive and properties-based understanding, and that recognizing this is the key to a consistent reading of §§28–34. The thought is thus that the notion of Articulacy is itself subject to the distinction between readiness-to-hand and presence-at-hand, so that it could be said to have different "modes."[36] Rather than being bound to a uniform notion of (propositional) content, (a) and (b) should then be taken as common characteristics of the different kinds of content expressed by sentences. That is, while in both cases words are used and entities picked out, the way they are picked out and in which something "is said" of those entities varies according to the mode of Articulacy, or the type of intelligent behavior in which the utterance is involved.[37]

Although Heidegger does not explicitly distinguish two modes of Articulacy in the way just suggested, this distinction is implicit in his discussion of Statements, most notably in the passage of *SZ* 158 discussed in section 4.[38] Indeed the two modes are exemplified in the contrast between overt demonstration and embedded showing, which differ both in their mode of showing and in the kind of articulate grasp of the world they exhibit.

Interpreting the notion of Articulacy as having different modes corresponding to the distinction between Equipment and Things, or readiness-to-hand and presence-at-hand, allows the apparent tension in §§28–34 to be eliminated. This can be seen by bringing together three thoughts. The first—from the preceding section—is the recognition that Equipmental Setting-out is to be construed not as a level of "prelinguistic" or "nonlinguistic" articulation, but as a basic purposive articulation of Understanding potentially linked with linguistic expression. The second thought is that since Articulacy is itself subject to the purposive-predicative distinction, there is no implication that predicative structure is projected back into or prior to purposive awareness via the notion of Articulacy. The third thought is that Heidegger's view of the hermeneutic-apophantic transition implies, indeed he explicitly allows, that the use of linguistic forms indicating predicative structure does not entail predicative awareness. Whereas the first thought clarifies that Heidegger's account of progressive determination does not exclude language, the latter two make clear that the notion of Articulacy does not shortcircuit the stratified picture suggested by that account.

Against this background, the connection between Articulacy and Language can be preserved without introducing incoherence to §§28–34. Heidegger's linguistic characterizations of Articulacy can then be seen as consistent with and complementing the progressive determination story—presumably just as he thought. At the same time, it is possible to see why Heidegger's analysis might *appear* incoherent. The source of the difficulty was the underlying determination story that §34 appears to suggest, specifically its implication that language-like content pervades all disclosure. However, §34 has this implication only if it is assumed that language use corresponds to a uniform type of "linguistic" content, an assumption that—as shown in section 4—Heidegger opposes. Finally, distinguishing different modes of Articulacy affects the interpretation of Heidegger's claim in §34 that Articulacy is a basic feature of all disclosure. Rather than entailing that language-like content pervades all disclosure, this should be taken to mean that disclosure will always involve Articulacy in one of its (content-like) modes.

If it is assumed that the discussion of §§28–34 can be rendered consistent in this way, then it becomes more difficult to discount the exegetic evidence that Heidegger sees a close connection between Articulacy and language. However, the question (mentioned at the end of section 3) then arises of how to accommodate intelligent nonlinguistic behaviors in his overall account. For it might seem that acknowledging such a connection is inconsistent, or at least in tension, with Heidegger's emphasis on hammer use and its supposed basic role in Dasein's disclosure. For example, the claim that Articulacy is "equiprimordial" with Understanding and Attunement might still seem to suggest that all intelligent behavior is in some way linguistic. Similarly, it might seem unclear, on my reading, where (if at all) nonlinguistic behaviors fit into the progressive determination story. Another potential concern is that my (limited) defense of the "linguistic model" has relied merely on prima facie evidence of Heidegger's characterizations of Articulacy together with an argument against a potential inconsistency, neither of which constitutes an argument against the tenability of the pragmatic model.

Such difficulties will appear particularly acute if one's interest is in applying Heidegger's analysis to nonlinguistic behaviors. Although that is not my focus here, an inability to accommodate plausibly intelligent nonlinguistic behaviors would clearly constitute a serious objection to any reading of Heidegger. I will therefore briefly outline how this can be done, provided the close connection between Articulacy and Language is understood in the right way. The connection, I suggest, is such that the full range of functions Heidegger attributes to Articulacy can be realized in language use but not in intelligent nonlinguistic behaviors, so that the former but not the latter provides an adequate model for defining Articulacy. For example, Heidegger's discussion of Statements (section 3 above) makes clear that he thinks predicative Setting-out can be realized only by using language, whereas purposive Setting-out can be realized both in language use and in intelligent nonlinguistic behaviors. He also links Articulacy with the function of pointing out features of the world (chapter 3 below), a function that can be plausibly attributed to all language use—as either embedded showing or overt demonstration—but is atypical of nonlinguistic behaviors. Thus language serves as a model for the various functions of Articulacy, defining a field of possibilities for Articulate activity, some—but not all—of which can also be realized in intelligent nonlinguistic behaviors.

In this light, the relation of intelligent nonlinguistic behaviors to the notion of Articulacy modeled on language use is like that of inauthentic to authentic Dasein: the former is in both cases a limited realization—albeit an indispensable, foundational one—of the available possibilities.[39] There is also no problem in explaining Heidegger's emphasis on intelligent nonlinguistic behaviors. He does this not because he wants to suggest it is more properly human to wield hammers than to wield propositions, but because focusing on purposive understanding serves as a corrective to traditional misrepresentations of the human-world relation. Finally, seeing the connection between the definition of Articulacy and language use in the way just outlined also indicates a sense in which focusing on intelligent nonlinguistic behaviors as a paradigm is inadequate, thus providing a reason to prefer the "linguistic model" (as construed here) to the pragmatic model.

The second problem left open at the end of section 2 was that Heidegger's discussion might seem undecided as to whether Language (public sign use) plays a constitutive role in disclosure. More specifically, the concern was that Heidegger's analysis ascribes only a peripheral role to language and so intimates that there might be "a fully articulated sense of the world" independently of language. Having looked at Heidegger's discussion of the transition from purposive to predicative awareness—that is, from the hermeneutic *as* to the apophantic *as*—it can now be seen that this concern is misplaced and that Heidegger is far from making any such intimation.

Recall that Heidegger thinks this transition takes place with a specific mode of language use, while denying that predication provides a general model for the functioning of language. Accordingly, he holds that purposive or circumspective articulation of the world ("Significance") can manifest itself both in language use and in nonlinguistic behaviors. In this respect he sees words as one kind of instrument among others and it is difficult to avoid the conclusion that they will play a constitutive role in at least some forms of practice—just as hammer use is "constitutive" of various kinds of manual task. Furthermore, in claiming that Statements involve the public use of "spoken-out" signs, Heidegger excludes the manifestation of predicative articulation in nonlinguistic behaviors and so implies that it depends on language use. Indeed, independently of Heidegger's stated commitments, his view of the relationship between readiness-to-hand and presence-at-hand seems to require this. For getting beyond immediate concerns with the expediency of individual entities to particular ends and attaining ends-independent predicative generalizations presumably requires symbolic representation of entities, so as to identify them in a stable way over variation in the functional context.[40] It also seems at least plausible to suggest that organizing the attribution of properties requires the complex but finite kind of syntactic framework that linguistic representation embodies.[41]

Even without considering the role of *Gerede* ("idle talk")—a task undertaken in chapter 3—it is therefore difficult to see Heidegger as denying that language has a constitutive role in disclosing the world. For he not only thinks that language is already used in Equipmental Setting-out, but also identifies a disclosive feat—predicative Setting-out—that is specific to language and surely forms part of any "fully articulated" view of the world. It is in this way that Heidegger can claim to be balancing the two desiderata, mentioned at the end of the second section of this chapter, of thinking of language as continuous with other meaningful phenomena while acknowledging its specificity. Rather than suggesting that the disclosive function of language can always be assimilated to that of nonlinguistic behaviors or structures, he simply sees some overlap insofar as both can be involved in Equipmental Setting-out. Hence it seems to me that Heidegger's apparent indecision over whether language plays a constitutive role results from a misunderstanding of his position, in particular of the various possibilities he sees for the expression of meaning through language use. Nevertheless, it would be fair to say that Heidegger encourages this misunderstanding insofar as *SZ* says very little about *how* language does constituting work. Accordingly, one of the tasks for the coming chapters will be to bring out more clearly the specific ways in which language plays a constitutive role in disclosing the world.[42]

One problem nonetheless remains concerning the notion of Articulacy. As mentioned above (at the end of section 2), Heidegger briefly suggests that Articulacy involves some kind of Articulate structure (*Gliederung*; *SZ* 161) that is prior even to Setting-out. How is this claim to be understood? Although at one point mentioning the need to ask about "the basic forms" of an "articulation of what can be understood altogether," Heidegger does not address this question directly, suggesting only that an answer presupposes his analytic of Dasein (*SZ* 165–166). However, from what has been said here two constraints can be discerned. First, given Heidegger's claims that Articulate structure precedes Setting-out, and that Setting-out effects the individuation of entities, Articulate structure should precede individuation. Or equivalently, given that Articulacy has both purposive and predicative modes, whatever intrinsic structure it has must be compatible with different ways of individuating entities. This suggests that the Articulate structure Heidegger invokes should be thought of as differential structure, a manifold exhibiting discernible differences without being organized into discrete units. Second, given the specific link between Articulacy and language, the structure involved should be discernible as linguistic. Although Heidegger does not explicate what I am calling Articulate structure any further, Saussure's conception of linguistic structure—discussed later (chapter 5, section 1)—can be seen to meet these constraints, and can to this extent be considered to play the role of Heidegger's Articulate structure.

To conclude this chapter I want to highlight two features of the view I am attributing to Heidegger that will be important in developing a phenomenological conception of language over the following chapters. The first is that Heidegger's conception of language is embedded in a broader discussion of disclosure in which purposive understanding has a foundational function. On this picture the ultimate context for language is a holistic and pragmatic notion of the world as experienced, the entire nexus of instrumental relations that Heidegger calls "Significance." We have an overall grasp of this Significance in the form of a purposive, holistic, inarticulate background awareness (Understanding) that can be developed (in Setting-out) into different forms of articulate understanding. For Heidegger, as has been seen, the most basic individuation of entities is the product of purposive understanding, which determines them (as ready-to-hand Equipment) in relation to specific ends. By contrast, predicative understanding, which individuates entities (as present-at-hand Things) according to purpose-indifferent properties, is "derivative" in the sense that it presupposes purpose-relative Equipmental understanding as the basis for discerning higher-order structures.

The second feature to be highlighted is that the founding of predicative in purposive awareness also applies to language. This was clearly central to Heidegger's view of Statements, as discussed in the penultimate section of this chapter. However, it is also reflected in Heidegger's treatment of both words and sentences/Statements as ready-to-hand (*SZ* 161, 224) and in his polemics against treating language as something present-at-hand (*SZ* 161). As will become clearer in the following chapters, the thought that language has Equipmental uses that cannot be functionally assimilated to predicative understanding is a highly distinctive and consequential feature of Heidegger's conception of language.

2 Phenomenological Commitments

Phenomenological positions are often thought of not only in terms of a commitment to a particular methodology but also as being distinguished by a typical pattern of commitments at the level of content, an overall shape that characterizes their view as a phenomenological one. This is a natural expectation since, if the commitment to some philosophical methodology works as a genuine constraint, it ought to be reflected in the resultant conception of its subject matter. In the introduction I suggested that Heidegger's approach in *SZ* is motivated by the goal of accurately describing lived experience, or phenomenological accountability, in which case it might reasonably be expected to exemplify the typical contours of a phenomenological position. Before looking more specifically at Heidegger's conception of the function of linguistic signs and linguistic competence in the next chapter, I therefore want to highlight briefly several general features of his view of language that characterize it as a phenomenological view. The point in doing this will be in part to identify what is distinctive about a phenomenological conception of language, what such a conception looks like. However, it will also provide a basis for critically assessing the significance of such a conception by identifying factors that govern whether it is compatible or in conflict with other views of language. I begin, in the first section of the chapter, by drawing attention to several features that characterize Heidegger's embedded and nonreductive conception of both language and intentionality more generally. The second section looks more closely at the idea of prepredicative foundation, which provides a specific focus in assessing the critical relevance of a Heideggerian conception of language.

1 Language as Language-in-the-World

Heidegger's conception of language belongs to an overall picture of intentionality that centers on and is shaped by his idea of Dasein as being-in-the-world. As a result, what makes his conception of language phenomenological can be discerned not only directly from his discussions of language, but also indirectly through what he says about Dasein. With this in mind, the first general feature I want to highlight is that on Heidegger's view language cannot be thought of as something with an inside-outside topology, as something "within which" intentional relations are contained. In one respect this might seem obvious, since some parts of Heidegger's account of the structures of disclosure made language look like an altogether peripheral aspect of intentionality. However, the point applies equally if (as chapter 1 argued) language is constitutively involved in the various modes of Heidegger's Setting-out, and is best appreciated in light of his shift away from Husserlian phenomenology. A central aspect of this shift, as pointed out in the introduction (section 2), was Heidegger's insistence on the need for phenomenological inquiry to examine the ontology required for its own realization. It was this requirement that led him to challenge the intelligibility of postulating transcendental subjectivity as a closed ("immanent") realm of constitution, to insist on the primacy of lived experience, and ultimately to reject basic subject-object dualism in favor of Dasein's essential "being-in-the-world."

Heidegger's rejection of an inside-outside topology in the case of Dasein can be seen clearly in two respects. In discussing knowledge of our material surroundings he emphasizes that rather than needing to break out of some "inner sphere," Dasein is in "its primary way of being always already 'outside'" in the world (*SZ* 62; cf. 162). Further, because he also characterizes Dasein as a historically and socially situated kind of agency, there is clearly no intimation that all of intentionality is either "inside" or the correlate of any one such agent. As a result, intentionality as a whole is pictured as having a *distributed* topology: rather than being contained "within" or corresponding to a single constituting entity, intentional relations are spread out over time and between agents. Thus on Heidegger's picture there is no sense of inside or interiority—be it physical, psychological ("mental"), individual, or social (e.g., cultural)—that essentially defines human being as Dasein.

Heidegger's objections to the idea of a constituting medium must, however, be thought of as generic. Any objections bearing on the conceivability of transcendental subjectivity as a closed, fully present realm of constituted eidetic relations would apply equally to attempts to think of language in the same way. Hence the only plausible assumption is that rather than substituting language-world dualism for mind-world dualism, as some adherents of the linguistic turn appear to do, Heidegger is opposed to any such basic dualism.

A second general feature of Heidegger's approach to language—one clearly motivated by the idea of "getting the phenomenology right"—is its *antireductionism*. This is just one example of his more general opposition to studying any given phenomenon by picking out one particular aspect or element and using it to explain ("reductively") all the others. Accordingly, Heidegger often claims to be guided by the idea of preserving the integrity of phenomena, the idea that in describing some phenomenon one must respect its character as a unified whole, and often criticizes others by invoking the imagery of "unitary" phenomena being "shattered," "demolished," or simply "passed over."[1]

Antireductionism is a recurrent theme in Heidegger's discussions of language. In addition to the claim that analysis in terms of "word Things" "shatters" language, it emerges with particular clarity in his rejection of attempts to conceive of language in terms of "symbolic forms," "categories of meaning," or the logic of propositions (*SZ* 161, 163, 165). Indeed this antireductionist attitude was encountered in chapter 1 as one of the central motives in Heidegger's view of Statements. In making these claims Heidegger's point is simply that picking out *one* of these aspects as the "essence of language" cannot yield a philosophically adequate understanding of language. The implication is that, because language is essentially embedded in a broader phenomenal context, to understand what or how language is instead requires consideration of the various processes and phenomenal relations it is involved in.[2]

One particularly important manifestation of this antireductionist attitude to language is Heidegger's *antiformalism*, a feature salient in *SZ*'s discussion of signs. He states that both the instrumental relations comprising the world and the relations between different kinds of signs can be "easily formalized," but that the problem with such formalizations is that they "basically say nothing" because they "level off the phenomena to such an extent that the proper phenomenal content gets lost" (*SZ* 78, 88). It is perhaps worth noting that, although he would object to it in the same way, Heidegger's talk of the "formal" and "formalization" is not specifically directed to the now common use of this term as standing for formal-logical quantification. Rather, it relies on a general distinction between symbolic representations (forms) and their fulfillment in phenomena, roughly analogous to the Kantian contrast between empty concepts and the material of intuitions.[3] (Hence for Heidegger modern formal logic would be just one variant of "formalism.") The point Heidegger is making is that any theory of signs focusing exclusively on formal and relational properties will yield no insight into the phenomenal basis from which its abstractions proceed, and in this sense will tell us nothing about what it is actually to be or to function as a sign. Further, in concentrating on one kind of similarity, such theories constitutively disregard other differences between phenomena. So Heidegger is not denying that such a formalized conception of semiotic phenomena (for example) can be developed. Indeed he claims this is "easy." The point of his antiformalism is rather to insist that an abstract formal conception of whatever kind of relationship—be it semiotic, syntactic, or semantic—can provide only an uninformative schematism. To be revealingly informative such a conception must be complemented by an account of what it is for those kinds of relationship to be realized in actual phenomena.

This leads to a further feature I want to highlight here. Heidegger thinks that one problematic aspect of traditional ontological concepts—especially the essence + existence model of entities—is that they neglect the temporal dimension and represent their objects in a timeless manner. However, for Heidegger the mode of being language has, just as that of subjectivity, cannot be the "constant remaining" (*ständiger Verbleib*; *SZ* 96) of present-at-hand Things. Rather, since phenomena are temporal in character, his insistence on spelling out the phenomenal basis of formal relationships rules out a (platonist) picture of language as a detemporalized, static realm in which various semantic or inferential relations hold. Heidegger is thus committed to thinking about language in terms of *processes* rather than as a domain of fixed structures—or, in Humboldt's terms, to viewing language as an activity rather than a (finished) work.[4]

Heidegger's presentation of language in the broader context of the phenomenon of disclosure thus reflects a pervasive tendency that can be summed up by saying it is fundamental to his approach that language, just as Dasein, is to be understood in terms of its embeddedness in the world, so to speak, as *language-in-the-world*. These features, as I have indicated, are palpably motivated by Heidegger's commitment to the need for philosophical views to reflect lived experience accurately—that is, to phenomenological accountability. With this in mind, however, they can be taken as bringing out a pattern of general commitments that should characterize any phenomenological conception of language: namely, that language has no inside and outside, is embedded in or distributed over the world, and should be understood nonreductively, without formalism, and as processual rather than static.

Suppose it were agreed that these features adumbrate the general shape a phenomenological conception of language ought to have. What of it? How might such a conception contribute to philosophical understanding of language? A first answer is that at this general level Heidegger is simply proposing a certain picture or a better way of seeing language. And since the way we "picture" some phenomenon is simply shorthand for the overall way we conceive of it, to urge a particular picture in this way is clearly of some critical interest. Indeed, even at this general level Heidegger's approach can be taken to contest directly some of the more strident abstractions found in philosophy of language—for example, forthright claims that "a language" is a set of paired sentences and meanings, an "abstract entity," or essentially an internal representation conforming to "Universal Grammar."[5] It is plausible to contend, as Heidegger does, that such claims either neglect to examine the ontology required for their realization or disregard the actual or full phenomenology of language, leading to a conception of language that is at best inadequate and at worst highly misleading.

Matters are often not so clear-cut, however, as can be illustrated by considering how Heidegger might respond to a conception of language based on predicate calculus of the kind originally due to Frege. This kind of approach, which I will call the "semantics approach," can reasonably be considered standard in contemporary analytic philosophy of language. Nonetheless, it is fairly clear that Heidegger would object that in its outset the semantics approach ignores the phenomenology of language and that the conception of language it yields is a variant of formalism.[6]

What is the substance and force of such objections? Heidegger's claim is in effect that a formal or algorithmic conception of language is superficial. Suppose the means of predicate calculus could be honed to represent convincingly and comprehensively the semantics of natural languages. While perhaps allowing this to be an extraordinary technical and intellectual achievement, a Heideggerian antireductionist would maintain that it fails to tell us anything about the phenomena and processes in which the use of language is embedded. However, it is not obvious that this is necessarily a problem for the semantics approach, which might simply acknowledge the need Heidegger identifies and view his approach as complementary.[7] I am not suggesting here that that would be the proper response, but simply that this possibility brings out a genuine limitation of picture-level engagement in that one needs to establish not merely that pictures or approaches appear different, or present their common subject matter in a different way, but that the views they yield genuinely contradict or conflict with each other. For, failing this, two apparently dissimilar approaches or conceptions may simply be providing complementary and equally correct accounts of disjunct aspects of the same (e.g., linguistic) phenomena. To put it another way, picture-level engagement is not sufficiently specific to allow a properly informed judgment as to the precise relationship between, in this case, a phenomenological approach modeled on Heidegger and the semantics approach.

If the risk of degeneration into a polemical, or simply indifferent, stand-off is to be averted, a more specific kind of engagement is therefore required. The requisite focus for this, I suggest, is hinted at by Heidegger's treatment of predicative Statements as "derivative" in virtue of their foundation in purposive understanding. Examining what this claim amounts to in more detail, as I will now do, will bring out both what is distinctive about Heidegger's position and an identifiable tension between this position and the semantics approach.

2 The Idea of Prepredicative Founding

The philosophical poignancy of Heidegger's view of predicative Statements (*Aussagen*) in *SZ*—discussed above in chapter 1, section 4—can be further brought into focus by considering the parallel, more extensive discussion in his 1925–1926 lectures entitled "Logic." In an exposition of Aristotle's *On Interpretation* Heidegger there highlights the importance of "combination and division," or synthesis and diairesis, as a precondition for the truth/falsity of propositions.[8] As he puts it, a "proposition can only be true at all ... because qua proposition it moves *a priori within the 'as'*" (*Logik* 135; italics added). In Heidegger's view, Aristotle had correctly recognized the need for the phenomenon of combination/division but failed to explicate it.[9] As in his discussion of signs, Heidegger allows that the "formal structures of 'combining' and 'dividing'" can become the object of a kind of calculus or "'calculation.'" But here too he insists on the need to spell out the phenomenal basis of such a "superficial 'theory of judgment'" (*SZ* 159). Accordingly, to understand the "basic structure of the λόγος qua Statement," Heidegger thinks it crucial to explicate how this a priori "*as*-ness"—what it is for an object to be (as) that object—is constituted.[10] And this of course is exactly what the Preparatory Analytic of Dasein in *SZ* claims to do in highlighting the founding of the "apophantic *as*" in the "hermeneutic *as*," or of predicative in purposive awareness.

The first thing to notice here is that Heidegger's approach to the content of predication centers on an underlying "prepredicative character of the *as* structure" (*Logik* 144). In this respect his position takes up a characteristic theme of the phenomenological tradition, motivated by its general commitment to antireductionism and antiformalism. For Husserl's aim had always been to investigate the overall constitution of intentionality, and in so doing to relate logical-categorial acts to their foundation in underlying phenomena, in particular perceptual acts. And Husserl himself, most notably in *Experience and Judgment*, offers an extended account of "prepredicative" structures, the aims of which Heidegger's closely resemble.[11] Heidegger's view differs markedly from Husserl's in claiming that the prepredicative should be understood in terms of purposiveness rather than perceptual states. Thus he not only talks of "prepredicative ... seeing" of the ready-to-hand (*SZ* 149), but suggests that "so-called straightforward having-there and grasping" of an entity is "not a direct grasping at all" since this presupposes having "already dealt with it previously," understanding it "in terms of that for which it serves."[12] The question this section will focus on, however, is not the substance of such claims—that is, what the prepredicative is taken to consist of (since this is addressed by subsequent chapters)—but what they are to be interpreted as claiming: What does it mean to say that something is "prepredicative"?

In answering this the starting point is to observe that if propositional content is taken to be based on predication, then "prepredicative" factors will be concerned with the way the content of propositions is constituted. In saying this, it should be borne in mind that Heidegger thinks such "content" is arrived at only by way of abstraction, or "formalization," and that it encompasses heterogeneous phenomena. In other words, what looks homogeneous at the level of propositional content may be eliding important differences in the way entities are understood. This raises the possibility that there might be several different kinds of content-constitution process that underlie meaningful sentences. But whether there are several such processes or just one, the core idea is that prepredicative factors are to effect the constitution of propositional content.

To be more precise, however, and to allow the relationship between a Heideggerian and the semantics approach to be explored more fully, I want to suggest that Heidegger should be understood as claiming that prepredicative factors have the three following defining features.

(i) The Prepredicative as Subpropositional

Heidegger's basic claim that propositional content is formed by prepredicative factors has two sides. On the one hand, he is clearly denying explanatory priority to the ("derivative") propositional level. On the other hand, he is claiming that the existence and functioning of prepredicative factors are required for, and are hence implicit in, contentful use of sentences. A further clue as to what prepredicative factors are assumed to be is provided by the fact the genetic phenomenological approach Heidegger takes to the articulation of Significance leads him to focus on words—that is, on the structural components of sentences. Thus his concern is with factors that underlie propositional content both functionally and structurally. In this dual sense I will describe the prepredicative factors Heidegger is concerned with as "subpropositional."

(ii) The Prepredicative as Subinferential

A second defining feature of the prepredicative can be drawn from an aspect of the transition to predicative awareness highlighted by Robert Brandom. As he puts it, the distinctive feature of predicates relied on by Heidegger's treatment of Statements is that "predicates come in inferential families," serving to "codify" the "inferential significances" already grasped purposively (Brandom 1983, 402). To be sure, this is not how Heidegger himself describes the transition, but Brandom's suggestion—that predicates encode inferential properties—is at least consistent with Heidegger's stated view. Given that inferential properties (or "significances") concern the interrelations between propositional contents, and that Heidegger understands propositional content in terms of predicates, it seems right to conclude that predicates are systematically connected with inferential properties. It should also be noted that the shift to an awareness of inferential properties brings with it commitment to the notion of semantic truth and the ability to engage in reasoning, both of which are internally related to the systematic interconnectedness of propositional contents that predicates "encode." In other words, recognizing its connection with inferential properties makes clear that predicative awareness corresponds to various abilities that are often taken to be the hallmarks of full-fledged rationality.

Recognizing the link between predication and inference also sheds further light on the transition to properties-based awareness, which was interpreted above (chapter 1, section 4) as involving a shift from purpose-specific to purpose-independent individuation of entities. This interpretation might be challenged by arguing that purposive understanding already entails an awareness of some properties. In Heidegger's paradigm example

Dasein must know what a hammer looks like, at least in the sense of being able to recognize hammers from nonhammers. Or in the case of language, of principal interest here, a speaker must surely have a corresponding recognitional grasp of the syntactic properties of words—their distinctness from one another and grammatical properties—in order to make use of them. What Brandom's observation adds is a constraint clarifying what it is to be aware of something in a predicative manner, or to individuate it in terms of objective properties: namely, that one be able to involve it in making inferences, so that the "properties" involved in predicative awareness are inferential properties.

This might initially seem to modify Heidegger's claim by restricting his talk of "properties" to some subset "inferential properties." However, I think this would be to misconstrue the constraint Brandom identifies. After all, in principle any property can be involved in inferences, if an agent is able to introduce them into reasoning processes. Conversely, if an agent is unable to involve some feature in inferential use, then it must be at least doubtful that he or she is aware of that feature as such.[13] Brandom's constraint thus bears not on the properties that an entity as such might have, but on the kind of grasp agents have of their properties. Predicative awareness, on this view, involves being aware of some entity or feature in such a way that it is immediately available for use in making inferences, whereas prepredicative awareness lacks this quality. This has the important implication that prepredicative awareness is in some sense "prior" to the ability to reason. However, it is equally important to keep in mind that this obviously does not mean it is "irrational" in the sense of lacking any link with inferential processes and reasoning. Rather the whole point of talking of a *transition* from prepredicative to predicative awareness is that, this modification notwithstanding, inferential properties are somehow shaped or determined by prepredicative factors.

The second defining feature is therefore that the prepredicative can be described as *subinferential* in a dual sense: (a) grasping an entity or feature prepredicatively (in Heidegger's picture, as being "for ...") does not entail the ability to reason about that entity or its uses—that is, does not entail awareness of how these relate to other beliefs or propositionally expressible commitments; (b) inferential properties are based on, and somehow shaped by, prepredicative factors.

(iii) The Functional Discontinuity of the Prepredicative

A third defining feature of the prepredicative can be approached by considering the relationship between disclosure and truth. In *SZ* Heidegger famously advocates an unorthodox view according to which "truth" refers not specifically to a property of sentences or propositions but to the broader phenomenon of disclosure, which—in antireductionist spirit—is intended to encompass all the various aspects of how the world is revealed or open to us. Heidegger sees this broader notion of disclosure as reflecting the proper meaning of the Greek term (ἀλήθεια), standardly translated as "truth" (*SZ* 220–221, 219).

This unorthodox view was no doubt partly motivated by Heidegger's reading of book VI of the *Nicomachean Ethics*, which distinguishes five different ways of attaining truth, including practical wisdom and science.[14] But apart from historical considerations, Heidegger also had a philosophical motive for incorporating propositional truth within a broader phenomenon: denying a basic subject-object dichotomy in favor of the notion of being-in-the-world meant the idea of being radically out of touch with the world, or with truth, makes no sense. Dasein is always "in the truth" (*SZ* 221) in the sense that it is in genuine, unbreached contact with its surroundings. The difficulty, as Heidegger sees it, is instead to understand how truth is founded in the world, or how "the roots of propositional truth reach down into the disclosedness of [purposive] Understanding" (*SZ* 223). It is not my aim here to consider the objections that have been or might be made to this view as a theory of truth.[15] For present purposes, however, it will be convenient to allow, in the spirit of the traditional view rejected in *SZ* §44a, that the label "truth" should apply only to propositions and sentences. Although it was no doubt part of Heidegger's rhetorical strategy to incorporate the propositional level in a more comprehensive notion of truth qua disclosure, respecting this terminological convention makes it easier to articulate an important feature of Heidegger's position. In contrast to "truth"—as an attribute of sentences/propositions—Heidegger's term *disclosure* will therefore be taken in the following to refer to the broader phenomenon of openness to and contact with the world, and as encompassing subpropositional, propositional, and suprapropositional levels.[16]

With this distinction between truth and disclosure in place, it becomes easier to see prepredicative and predicative factors as distinct. For as long as both prepredicative and predicative awareness are described in the same terms, whether related to "disclosure" or "truth," it might seem that their functioning is governed by the same considerations. Yet one of the principal theses of the Analytic of Dasein (as seen in chapter 1) is that this is not the case: readiness-to-hand and presence-at-hand, purposive and predicative awareness, are to be distinct categories, even though the latter are somehow to depend on the former. This ambivalent relationship of both distinctness and dependency is underwritten by the thought that prepredicative and predicative awareness are *functionally discontinuous*—that is, function in different ways that are to be described in terms of different factors.[17] This third feature of Heidegger's view of the prepredicative not only underlies the assumption that the predicative and prepredicative are mutually irreducible, but also gives genuine substance to the idea of foundation by allowing that subpropositional elements play a role that is neither arbitrary nor reducible to their contribution to propositional truth/falsity.

Whether or not one thinks Heidegger's account satisfactory, it is clearly helpful to preserve this feature of his foundational claims by characterizing the functioning of prepredicative factors in some way other than its relation to propositional truth. In the following I will therefore talk of the prepredicative as being governed by considerations of *appropriateness*.[18] In addition to distinguishing prepredicative functionality from propositional truth, this has two advantages. First, it avoids the a priori assumption of bivalency that is usually taken to apply to propositional truth, along with any other limitation to an arbitrarily chosen number of discrete states. This allows the possibility—as is arguably more plausible—that appropriateness comes in degrees: the tone of my comments can be more or less appropriate to the situation, a tool better or worse suited to the job at hand, and so on. At the same time, second, it suggests that conditions or considerations of felicity (e.g., relative to other means, expressions) are relevant—that is, that there is something to be said about what constitutes appropriateness or a lack thereof.

An implication of recognizing appropriateness as distinct from (propositional) truth is that a conception of the overall phenomenon of disclosure should respect both modes of functioning. An instructive attempt to do this is made by Martin Seel, who sees "pragmatic appropriateness" and "propositional truth" as complementary aspects in "disclosure of the world." According to Seel, appropriateness—or, as he calls it, "rightness"—governs our access to the world and constitutes "domain-opening understanding of things" in practice, whereas propositional truth functions normatively, as a corrective influence on rightness, which is not itself "a dimension of validity" (Seel 2002, 50, 61). It seems to me that Seel is right to discern both mutual irreducibility and functional complementarity between appropriateness and truth. However, despite the obvious philosophical appeal of emphasizing the corrective potential of propositional truth, his approach seems a little one-sided. For although not necessarily directed to epistemic goals or restricted in its functioning to a bivalent scheme of evaluation, pragmatic appropriateness is clearly a kind of (nonepistemic) normative requirement governing disclosure. To emphasize the corrective role of propositional truth alone would therefore somewhat obscure Heidegger's point that Equipmental access is not accountable to epistemic standards. Indeed the lesson of Heidegger's analysis in this respect, it seems to me, is rather that disclosure, particularly linguistic disclosure, should be thought of as genuinely multifactorial, governed by the interplay and tensions between both pragmatic and cognitive criteria.

Having distinguished these three defining features of Heidegger's notion of the prepredicative, one further concern is that his prepredicative founding thesis might seem to be compounded and complicated once language is in the picture. For it might be thought necessary to distinguish between prepredicative (prepredicative$_s$) awareness of linguistic signs and prepredicative (prepredicative$_o$) awareness of the objects referred to using these signs, inevitably prompting the question as to the relationship between the two. In particular it would then be tempting to wonder about inconvenient permutations such as (a) prepredicative$_s$ awareness of signs effecting predicative awareness of objects, or (b) prepredicative$_o$ awareness of objects being mediated by predicative awareness of language. In fact, either of these possibilities is problematic: (a) implausibly suggests one could be aware of the inferentially relevant properties of an object without knowing the inferentially relevant properties of the term being used to refer to it; equally unconvincingly, (b) suggests one could be aware of the inferentially relevant properties of signs without being similarly aware of the corresponding properties of the objects to which they refer.

The fact the two possibilities just mentioned are both problematic can be seen as symptomatic of an intimate link between predicative awareness of objects and that of signs, one which ensures that prepredicative$_s$ and prepredicative$_o$ are actually correlative. For on the one hand, the use of signs is involved in predicative thinking, since symbolic representation is required for generalized cross-purpose awareness.[19] Yet on the other hand, the whole point of representation is that the inferential properties signs have *qua signs* are due to what they are taken to stand for. In other words, predicative awareness of an object is possible only by using a sign that by definition shares its inferential properties. Conversely without predicative awareness of its supposed object, a sign cannot be considered to have inferential properties. As a result, our awareness of both signs *and* their supposed referents is either predicative or prepredicative, so that a single distinction between prepredicative and predicative awareness (applying correlatively to both) is all that is needed. This result, which reflects the involvement of signs in predicative awareness, is perhaps not surprising, given Heidegger's view of language as embedded in the world more generally. Indeed, the temptation to think there is a philosophically important distinction between the grasp we have of language and the grasp of the world we arrive at through the use of language is perhaps due to residual dualist habits of the kind Heidegger was seeking to overcome.

One complication that does arise on Heidegger's picture is that whereas he suggests the use of signs is necessary for predicative awareness, he clearly does not think it sufficient (see chapter 1, section 4). This complication is implicit in Heidegger's view—again seen in his discussion of Statements—that the distinction between purposive and predicative awareness falls within language use, rather than coinciding with the linguistic/nonlinguistic distinction. But it also points the way to a distinctive and interesting feature of Heidegger's position in allowing that at least some language use is "prepredicative" in character. In the terms distinguished here, this should be interpreted as the claim that certain kinds of language use are to be understood in terms of subpropositional features that play a subinferential role that cannot be reduced to the functional logic apparently governing language at the propositional level.

This hints at two questions to be addressed in the following chapters. First, having here clarified what claims to be prepredicative amount to, there is a need to fill out the substance of this claim: What is supposed to be "prepredicative" about language? In other words, which factors are being appealed to in such cases, and what does prepredicative language use amount to in practice? Second, even if one accepts Heidegger's characterization of prepredicative and predicative awareness as individuating entities, and so functioning, in different ways, he seems to go no further in describing their relationship than to characterize the "modification" of one to the other in somewhat abstract functional terms. So how is this "modification" to be understood in practice? How exactly is it supposed to feature in phenomena familiar to us as language users?

3 The Disclosive Function of Linguistic Signs

The last two chapters have focused on what I described as the more general features of the Heideggerian framework. While the first chapter set out where language fits into Heidegger's overall view of disclosure, the preceding chapter has identified various features of his conception of language that distinguish it as a phenomenological one. This chapter shifts to a more specific level by considering Heidegger's view of the disclosive function of linguistic signs—that is, the role these play in disclosing the world to agents. To this end, the aim here will be to build up a more detailed understanding of his view of what linguistic signs are used to do and how they reveal our surroundings to us in an articulate way.

The most general determination of this disclosive function, and hence the function of language, in *SZ* is found in two basic functions that Heidegger attributes to Articulacy. The first of these was met in chapter 1 with Heidegger's discussion of Articulacy as the ontological foundation of language. In its definition as the "articulation of Significance" and its four-point structural characterization that discussion highlighted the *articulatory function* of Articulacy, its role in structuring a determinate understanding of the world. The second basic function of Articulacy, its *demonstrative function,* is introduced in Heidegger's definition of phenomenology as a method in §7B through a distinction between two kinds of truth. The more basic or primordial "Greek" sense of truth is taken to be that of directly seeing or intuiting how things are, via the senses (*aisthesis*) or intellection (*noein*): the "purest and most primordial" sense of truth, Heidegger says, is "direct inspection, by looking, of the simplest determinations of Being of that which is as such."[1] By contrast there are truths that we are not immediately aware of and that therefore need pointing out to us. Such "pointing out" is to be the role of phenomenological discourse and the logos, or Articulacy (*Rede*). Accordingly, the demonstrative function of Articulacy, as Heidegger explains it, is that of "pointing to" what is talked about so as to allow it to be "directly seen" or "inspected."[2]

Heidegger's talk of "direct seeing" might appear obscure and/or metaphorical, but it is akin to common uses that are not obviously problematic. It should be noted first that his use of the term *see* is a deliberately broad or "formalized" one, referring to any act of apprehending features of the world that has the immediacy of seeing something visually (*SZ* 147). In this respect it refers to a familiar basic feature of phenomenological method, corresponding closely to Husserl's reliance on the idea of "intuition" or "evidence."[3] However, a similarly broad use of "seeing" is also familiar beyond the confines of phenomenological philosophy from idiomatic—and presumably "literal"—phrases such as "I see what ... means," "He doesn't see it that way," and so on.

All the same, Heidegger's talk of "direct" or straightforward (*schlicht*) apprehension of the world, especially where the use of signs is involved, might seem paradoxical. Part of its point is to signal commitment to the descriptive method of phenomenological study (*SZ* 35), which—as explained in the introduction—involves a contrast with inference (i.e., consciously moving from given data to corresponding conclusions). In this respect "direct" awareness means noninferential awareness.[4] Yet for Heidegger "direct" awareness also implies unsevered contact with the world, having an awareness of entities or features of the world as such, rather than an awareness in which some interposed medium (e.g., language) continues to play a constitutive and potentially distortive role.[5] Thus Heidegger's thought is that truth in the basic "Greek" sense will be direct in these two respects and that such truth can be realized through the demonstrative feats or "pointing out" of Articulacy.

The central focus in this chapter is on understanding this demonstrative function of Articulacy and language. There are several clues that this pointing out or showing of linguistic signs must be a more complicated operation than one might initially suspect. This follows first from the fact—as shown in chapter 1—that Articulacy manifests itself in different modes, corresponding to the contrast between purposive and predicative articulation of Significance and between embedded showing and overtly demonstrative practices. The demonstrative function of Articulacy must therefore be assumed to generalize over these cases. However, a further complication is suggested by *SZ*'s ambivalent view of the role of language in everyday understanding, which clearly implies that in many cases the showing operation does not succeed in the way it might. My approach in this chapter will be to explore this ambivalence, attempting to understand its basis and in this way to build up a more complete picture of Heidegger's view of the functioning of linguistic signs.

The chapter begins by reviewing this ambivalence about everyday language use, before canvassing and rejecting a couple of initially plausible explanations. To understand what underlies this ambivalence, the second section looks at Heidegger's earlier notion of "formal indications" as a paradigm of the specific (demonstrative) function he saw signs having in phenomenological inquiry. The third section of the chapter shows how these earlier ideas reappear in *SZ* and how, together with the foundational role of purposive understanding discussed in chapter 1, they can account for *SZ*'s ambivalent attitude toward language. The final section sets out how the disparate factors underlying Heidegger's ambivalence about language are brought together in *SZ*'s conception of signs and how the limitations of Heidegger's discussion define the task for subsequent chapters.

1 Heidegger's Ambivalence about Language

An important source of insight into Heidegger's view of language in *SZ* is provided by its discussion of what he calls *Gerede* and defines as the "everyday way of being of Articulacy" (*SZ* 167–168). In contrast to Heidegger's examples of hearing or remaining silent, this everyday mode of Articulacy—as "spoken-out-ness," "communication" (*Mitteilung*)—is concerned with the public life of sign use, and in particular with the role of Language in relation to Dasein's everyday, "inauthentic" understanding of being. As such it constitutes Heidegger's view of most language use. The term *Gerede* means "rumor" or "gossip" and is usually translated as "idle talk." However, since Heidegger denies that this term is intended to be pejorative or "detractive" (*herabziehend*; *SZ* 167)—a "denial" that would be unnecessary, were the term's negative connotations not so manifest—I will render it here more neutrally as "routine" language use. Note that routine language use should not be confused with the use of "everyday language," if the latter is taken to refer to the use of a particular vocabulary, comprising commonplace linguistic words and concepts as opposed to highbrow or technical ones. Rather, routine use is a *manner* or mode of language use, potentially applying to everyday and technical terms in the same way.

What characterizes this routine mode of language use? Given that Dasein is "thrown" into a historically conditioned intentional situation, the need arises to account for the constitutive role of language in setting up such a situation. Heidegger recognizes this in emphasizing that routine use is a "positive phenomenon that constitutes everyday Dasein's Understanding and Setting-out" (*SZ* 167). It corresponds to the amorphous agency of everyday life that articulates Significance.[6] Language, as normally manifested in spoken and written use, "harbors in it a way of Setting-out [*Ausgelegtheit*] Dasein's understanding," and so too "already a developed conceptuality [*Begrifflichkeit*]" (*SZ* 167, 157). Accordingly, Dasein and Language complement one another in their historicity: Language is a product of "the discoveredness respectively attained and handed down" that "preserves, in the whole of its divided-up contexts of signification, an understanding of the disclosed world," of self and others. Hence, for thrown Dasein, it conditions "the available possibilities and horizons for new approaches by Setting-out and conceptual articulation" (*SZ* 168).

So far, so good. But the pejorative air of Heidegger's term *Gerede* is no coincidence. Rather it rhetorically accentuates a supposed negative aspect of routine language use. Routine use, Heidegger says, is the medium of the "public way of Setting-out" (*Ausgelegtheit*; *SZ* 169), which contents itself with an average kind of understanding. What is said can "be largely understood without the hearer's bringing himself into an original understanding being toward what is being talked about … one means *the same*, because one understands what is said in the same averageness" (*SZ* 168). Words themselves, what is said, gain a "life" of their own and can be passed from speaker to speaker as empty tokens in processes of "hearsay."[7] Without an "original appropriation" of what is being spoken about, the "initial lack of firm grounding [*Bodenständigkeit*]" of such freely circulating linguistic signs ascends to "full groundlessness" (*SZ* 168). In this sense, the disclosive function of routine language use culminates in a countertendency to "close off," a phenomenon that remains largely unnoticed because its "indifferent intelligibility" leaves speakers with the impression that they understand everything (*SZ* 169).

Heidegger's view of routine language use is thus conspicuously ambivalent: while recognizing its constitutive role in everyday understanding of the world, he also clearly *insinuates* that the understanding it constitutes lacks something important. Understanding and explaining this ambivalence will, I hope to show, prove instructive with regard to Heidegger's view of the disclosive function of linguistic signs.

From what Heidegger says about routine language use as a "positive phenomenon," three important commitments can be elicited that help provide a clearer picture of the task. The first of these, which Heidegger explicitly recognizes, is that everyday understanding, as mediated by routine language use, is a necessary precondition for attaining any kind of understanding. As he puts it, Dasein can "never escape" everyday understanding, in which, out of which, and against which "all genuine Understanding, Setting-out and communication, rediscovery and new appropriation" occur (*SZ* 169). Heidegger does not think this everyday understanding is the only mode of understanding of being, and contrasts it with "authentic" understanding. Nonetheless, he maintains that *all* genuine understanding, not least "authentic" understanding, "remains dependent on One [*das Man*] and its world" (*SZ* 299): "*authentic* existence is not something that floats above falling everydayness," it is merely a "modified grasping"; "*authentic* disclosedness" is still "guided by the concernful lostness in One," but "modifies" its understanding of being without a change "'in its content'" (*SZ* 179, 297).

A second commitment explicitly recognized by Heidegger is that routine use of language in attaining everyday understanding commits us to a "developed conceptuality" linked with a certain way of interpreting being. While it might sound immediately plausible, this claim is in considerable need of elucidation and will turn out to be a complex commitment. For in what way does routine language use embody a "conceptuality" and in what way does it bind us? One issue, again recalling chapter 1, is whether the "conceptuality" involved in routine language is supposed to purposive, predicative, or both. Another is due to the fact that the German term *Begriff* is commonly used in two senses to mean either term/expression or concept. Accordingly, Heidegger's talk of a "Begrifflichkeit" could mean simply that routine language use commits us to a particular stock of words, the inherited vocabulary, rather than a particular articulation of conceptual understanding.

To appreciate further why these issues are nontrivial in Heidegger's framework a third commitment should be considered. For his claim that much of what we learn in life never exceeds the "average understanding" of "One" clearly implies that such understanding, as a rule, is perfectly adequate to the purposes of everyday life (*SZ* 169). Language must, so to speak, be failing despite its continual and pervasive successes. So whatever supposed deficiencies underlie Heidegger's ambivalence toward routine language use, they must accommodate this triple commitment to its *indispensability*, its role in *conceptual conditioning*, and its *pragmatic adequacy* to the majority of everyday needs.

In considering the source of the negative aspect of Heidegger's ambivalence toward routine language use, there are two obvious candidates. The first is suggested by his talk of the "averageness" of the "public interpretedness" of being, which might seem to align his view with the familiar idea of a "division of labor" in the use of language. According to the latter, speakers ordinarily have only schematic or "stereotypical" awareness of concepts' intensional properties, but are thereby enabled to use words in their "social meaning," the full intensional and extensional character of which is determined by respective experts.[8] However, disregarding the fact that Heidegger does not explicitly discuss such a distribution of competence, a difficulty remains with this proposal in that speaking this way about concepts treats words as objectlike bearers of properties—in Heidegger's terms as present-at-hand Things. And while it is reasonable to require an account of the kind of limitation implied by Heidegger's talk of "averageness," his unequivocally dismissive claim that analysis in terms of "word Things" "shatters" language (*SZ* 161) signals that his reservations about routine language use can hardly be due to limited knowledge of intensional properties.

The second obvious candidate for explaining Heidegger's reservations about routine language use is suggested by his emphasis on the foundational function of purposive understanding. This makes it tempting to suppose that Heidegger thinks of everyday understanding of being as just purposive rather than predicative, that routine use of language mediates an awareness of what environing entities are for, but not of their objective properties. Disregarding again the fact that Heidegger says nothing explicit to this effect, there are two reasons for rejecting this proposal. First, it would conflict with his argument that traditional and "vulgar" concepts engender a misguided ontological picture that conceals the foundational role of purposive awareness by presenting the world to us in an ontologically indifferent way as a nexus of present-at-hand Things (*SZ* 225). For this argument implies that the language we (routinely) speak presents the world to us as predicatively structured. Admittedly, given Heidegger's view that using sentences with a predicative form is not a sufficient condition for predicative awareness (cf. chapter 1, section 4), assuming that established language is predicatively structured does not entail that speakers be aware of the predicative implications of the sentences they utter. Nonetheless, the argument does reveal something about the mode of conceptual conditioning that Heidegger is assuming routine language use to effect: namely, that this cannot be construed in terms of purposive factors alone.

The proposal that Heidegger's everyday understanding of the world is simply purposive in kind should also, second, be rejected because it is intrinsically—and obviously—implausible to suggest that such understanding could exclude predicative awareness completely, as though this were simply a subsequent theoretical superstructure. For as Heidegger himself emphasizes—in the passage (*SZ* 158) discussed in chapter 1, section 4—theoretical predication is one extreme kind of speech act performed with sentences of a predicative structure. While Heidegger there stresses that it is unreasonable to assimilate all use of assertoric sentences to theoretical predication, it would be equally wrong to claim that properties' attribution takes place only in "theoretical" contexts: what something weighs, what color it is, whether it is damaged by water, and so on, are properties that are often involved in everyday activities but that are quite independent of specific functional expediency. So, given Heidegger's distinction between purposive and predicative awareness and his suggestion that these are two extremes on a graduated continuum, everyday understanding of the world—and with this the conceptual conditioning of routine language use—must be assumed to encompass a blend of the two. Whatever deficiency Heidegger discerns in routine language use cannot therefore be captured straightforwardly using the purposive/predicative distinction.

Having discounted these two initial candidates, the negative side of Heidegger's ambivalent attitude must be explained in some other way. To this end there is one useful, though rather cryptic, clue as to what he thinks routine language use lacks. Drawing on his characterization of the structure of Articulacy, Heidegger identifies a specific sense in which routine use is a limited actualization of its possibilities: in routine use—*Gerede*—"hearing and understanding" cling to "what is said [*das Geredete*] as such," "Being with one another moves in talking with one another and concern with what is said," "What is said as such ... takes on authoritative character."[9] So routine use focuses on "what is said," rather than on what is being talked about. Stated in this elliptic way, as it is in *SZ*, it is clear neither what this distinction amounts to, nor why it should be considered to identify a deficiency—after all, what could be more appropriate to language than attending to what is said? To get a clearer sense of both this claim and the overall perspective at work in Heidegger's ambivalence about language, it will be revealing to look back several years prior to *SZ*. For, as will become clear, the conception of language found in *SZ* is shaped by Heidegger's earlier view of signs, so that reviewing the earlier model allows some important features to emerge more clearly than in *SZ*'s dense and sometimes occlusive narrative.

2 Phenomenological Concepts as Formal Indication

From 1918 through to the early 1920s Heidegger intended his philosophy to be a radicalization of Husserl's phenomenology. This project, as he construed it, involved understanding the way presuppositions, in particular its "entire conceptual material," are made by phenomenological-philosophical inquiry. Accordingly, the "problem of concept formation"—in the form of "a phenomenology of intuition and expression" took on "central importance" in Heidegger's project of radicalizing phenomenology.[10] His response to this problem was a general characterization of the way signs or symbols function in phenomenological inquiry that he refers to as "formal indication." Unfortunately, although the term pervades his work around 1920–1922, Heidegger never provided a systematic or comprehensive account of formal indication, so that reconstruction of his view must rely on interspersed comments in his writings.[11] Nevertheless, what he does say provides valuable clarification of what Heidegger took to be the proper function of concepts in the context of phenomenological inquiry.

In considering Heidegger's notion of formal indication it is necessary to keep in mind two distinctions from the first of Husserl's *Logical Investigations*. The first, between "indication" and "expression," concerns the way signs function as meaningful. Something functions as an "indication" when its existence is taken by an agent to motivate belief in the existence of "certain other objects or states of affairs" (i.e., of that which is indicated) (Husserl 1992b, 31–32). This "motivation" is due to there being some link between the indication and what's indicated, so that it makes sense to describe them together. The links underlying such a "descriptive unity" might be either causal links or the intentional use of arbitrary signs to stand for something, but Husserl excludes connections that are immediately obvious, intuitive, or "insightive" (*einsichtig*; Husserl 1992b, 32, 31; cf. 35). Thus an indicating sign suggests, but does not make it intuitively obvious, that whatever is indicated exists.

Husserl contrasts indications with "expressions" as—so to speak, properly—"meaningful" signs (Husserl 1992b, 37), or the sound of a word "animated by meaning."[12] The meaningfulness of expressions, he tells us, is due to certain "meaning-conferring acts" or "meaning intentions" that comprise the "phenomenological characteristic of the expression as opposed to the empty sound of a word."[13] Although Husserl initially contrasts the two, indication and expression are not intended to be mutually exclusive functions. In spoken communication, for example, both occur: spoken words are expressions, since "the speaker generates them with the intention of thus 'speaking out about something'" and so "conferring meaning on them"; yet at the same time in "communicative talk," "all expressions ... function as indications" by standing for just those "meaning-giving acts."[14]

The second distinction of Husserl's relevant to formal indication is that between meaningful intentions "empty of intuition" and those that are "fulfilled." This distinction concerns the difference between meaning and knowledge and in this case the contrast is exclusive. Husserl defines fulfillment as the conscious "realization" or "actualization" of the "objectual relationship" signaled by a meaningful expression. Fulfillment, he explains, is inessential (*außerwesentlich*) to expressions as such, but can bring out their "appropriateness" by "confirming, enhancing, [or] illustrating" them (Husserl 1992b, 44). Fulfillment, then, is a conscious experience in which—otherwise "empty"—expressions are validated.

In his attempt to take up Husserl's project Heidegger develops a modified view of the role of signs in phenomenological inquiry. He starts by making a distinction between "philosophical concepts" and concepts in the individual sciences—which he took to be defined in relation to a corresponding domain of entities and would later call "ontic" or "positive" sciences.[15] Heidegger supposed the basic function of concepts in individual sciences to be classification or taxonomy of the entities concerned, with these being apprehended in terms of properties (i.e., as present-at-hand Things). Whether or not this claim is plausible, it indicates an important delimitation in that Heidegger sees the role of concepts in philosophy as differing from the essentially classificatory function of properties-based understanding. In a move characteristic of his reorientation of phenomenology to lived experience—one which would later facilitate his incorporation of existentialist ideas into a phenomenological framework—Heidegger sees the key distinction of philosophical concepts as being due to the *performed* or *actualized character* of philosophical understanding.[16] Correspondingly, the "peculiarity of philosophical concepts" is to be understood in terms of the role they play in "the manner of philosophical experience and ... the manner in which philosophical experience makes itself explicit" (*PAA* 171).

The basic point of reference for Heidegger's "philosophical concepts" is a kind of apprehending experience, from which a definition or concepts result and to which they remain accountable. Such "grounding experience" (*Grunderfahrung*), as he calls it, is the "specific performed context" of understanding, "the situation of evidence" or "experience in which the object properly [*eigentlich*] gives itself as that which it is and how it is" (*PIA* 35–36). Characterized in this way, grounding experience clearly corresponds both to Husserl's acts of fulfillment, in which the objectual relationship is actualized in consciousness, and to Heidegger's own later definition of a phenomenon as "that which shows itself in itself" (*SZ* 31). As this lineage suggests, the notion of "experience" relied on here is intended to be broad, encompassing not only experience via sense perception, but any validating experience in the realm of thought or intentionality—in which sense Heidegger was later to credit Husserl with having rediscovered "the sense of all genuine philosophical 'empiricism' [*Empirie*]" (*SZ* 50n). The most important point to note, however, is that this is the basis of Heidegger's talk of "originality": grounding experience or the "concrete situation" are the "sense genesis" or "original contexts of sense"; these comprise the "proper original sphere."[17] In other words, when Heidegger talks of originality—or "primordiality," as *Ursprünglichkeit* is often rendered—he has in mind a connection with an origin in a particular type of understanding experience.

Heidegger defines the function of philosophical concepts, as formal indications, in relation to such grounding experience. The task of "phenomenological definition"—Heidegger standardly treats definition as paradigmatic of formal indication[18]—is to indicate the "grounding experience" such that the way "back" to this is clear; accordingly, Heidegger demands, the concepts describing or defining an object should be created so as to reflect "the way *in which the object becomes originally accessible*" (*PIA* 20).

Corresponding to this task, the term *formal indication* has both a negative and a positive connotation. As with Husserl, the negative side is that formal indications are merely (unfulfilled or empty) *indications*. In Heidegger's words: "indicating" definitions "precisely do not give the object to be determined fully and properly [*eigentlich*]"; "indication of something by another" means not "to show it in itself, but to represent [*darstellen*] it indirectly, mediately, symbolically."[19] This constitutive shortfall—that signs present things that do not themselves appear—is marked by the term *indication*.

Conversely, the qualification as "formal" indication is to signal a positive characteristic (*PIA* 32–33). A formal indication is one in which, although itself "empty," the "'formal' provides the 'character of the approach' to performing ... the original fulfillment of what is indicated" (*PIA* 33). The idea here is that the link between formally indicating signs and what they stand for is to be tighter, or more determinate, than with arbitrary or unstructured indications. So whereas a high temperature can be a symptom, or indicative of indefinitely many kinds of illness, the link between signs and what they conceptualize is in some way to be revealed by the form of signs. As Heidegger somewhat enigmatically describes it, formal indication is a use of signs that shows "the 'way,' in its 'outset.' Undetermined in content, a determinate bind on performance is given in advance." The "'formal' is such content as refers the indication in the [right] direction, prefigures the way."[20] Thus in broad outline the function of philosophical concepts on Heidegger's account is as follows: formal indication is a use of signs the content of which is indeterminate, but that comprises a constraint serving to bring about the state of understanding from which they arise. In this sense, formal indications are to function as a pointer, a signpost, or a path "back" to the corresponding grounding experience.

When compared with Husserl's conception of signs, it is conspicuous that Heidegger's discussions of formal indication make no mention of meaning-conferring acts. Along with his general eschewal of subjective psychological concepts, this can be explained by recognizing a further influence on Heidegger's view. For although the idea of formal indication adopts Husserl's indication-fulfillment distinction, it replaces Husserl's notion of expression with that of Wilhelm Dilthey, which centers on the mutual interdependence of experience, understanding, and expression.[21] According to Dilthey, the lifeworld is permeated with ordered objective structures and practices that result from, and continue to shape, human purposes and understanding. Such "objectifications of spirit" or "expressions of life" range from "morals, law, the state, religion, art, science and philosophy" through to "every square planted with trees, every room in which chairs are arranged," and not surprisingly include linguistic expressions (Dilthey 1990, 252, 177, 256). Because they result from processes of understanding and determinate possibilities of experience, and are hence literally expressions of these, Dilthey reasoned that such objectifications allow the same possibilities to be relived or reactualized.[22] So on this view—which I am suggesting Heidegger adopts—the expressiveness, or meaningfulness, of signs is due not to meaning-conferring intentions, but to their embodying and standing in a determinate relationship to specific situations of actualized understanding. Moreover, forms of expression that emerge from acts of understanding are taken to be intrinsically conducive to reattaining that same understanding, and it is in this sense that they "point to" their grounding experience.[23]

Heidegger's notion of formal indication goes beyond Dilthey's view, however, by intimating which means expressions deploy in pointing to their grounding experiences. As hinted above, the qualification "formal" means for Heidegger that the "sense structure" of the "empty" content "provides the direction of performance" (*PIA* 33). In other words, the structuredness and interrelatedness of the signs used are taken to "point" the way to, and so enable, experiences of understanding. The determining sense structure of linguistic forms thus comprises a clue to the link between the indication and indicated that distinguishes formal indication from arbitrary or unstructured indications. Nevertheless, to preserve the distinction between indication and performed apprehension—corresponding to Husserl's stipulation that indications are not "insightive"—the clue provided by the sense structure cannot suffice to induce immediate understanding.[24] Rather, Heidegger explains, in "order to grasp the complete sense" of a formal indication, to follow where it points, "radical interpretation of the 'formal' itself is required" (*PIA* 33). This interpretation of "empty" symbolic forms is a means to the end of "performed understanding of the formally indicating definition"; it is part of a process of "working forth to the situation" in which formally indicating "characteristics" become "deformalized" by "receiving the concrete factual categorial determination from the respective direction of experience and interpretation."[25] So put simply, "formal" or "empty" signification and actualized or performed understanding are two extremes between which interpretation mediates. In fully actualized understanding, one would appreciate how each of the various features of one's symbolic representation is motivated by corresponding features of the phenomenon in question.

Against this background Heidegger's view of concepts can be seen to epitomize commitment to phenomenological accountability in being guided by the idea that concepts are somehow proper to their respective phenomenal origin, the grounding experience in which they arise and to which they remain internally related. Accordingly, formal indications are to be characterized by being both receptive to and expressive of phenomena. The first of these characteristics requires that concepts be literally shaped by phenomena: "the conceptuality of the object in the respective definitional determination must be drawn from the way *in which the object becomes originally accessible*" (*PIA* 20). In this way "the function of 'meaning-differentiation by forms' proceeds from the [phenomenal] 'material'" (Heidegger 1996a, 13), such that definitions and concepts (i.e., use of signs) are literally to receive their articulation from the grounding experience, the intensity of sense or determinacy of a phenomenon in its proper context. The second characteristic, expressiveness, is the correlate of the first: it implies that in being used "significantly" signs capture to some degree the sense manifested in phenomenal contexts, so that formally structured signs quite literally express specific features of phenomena. As Heidegger puts it, the meanings of "linguistic expressions" should be drawn from "the phenomenal context and its categorial tendencies" and should themselves "spring into" that context.[26] In other words, by capturing, and so being expressive of, certain phenomenal features, signs in turn acquire the peculiar capacity to evoke the original sense context. The efficacy of signs, on this picture, is due to their formally articulated expressiveness, which makes them conducive to actualizing the understanding in which they are based.[27]

It is not difficult to see that the idea of formal indication is based on a fairly straightforward representational model, according to which linguistic articulations correspond to phenomenal features. It is significant, however, that formal indication is assumed to be approximative or schematic in character, and so to lack constitutively the intensity or concretion of sense in actual phenomena. It has, Heidegger emphasizes, a "necessarily restricted mode of performance" since, although its content "directs [us] to the manner of proper encounter," it remains inconclusive in leaving "the genuine phenomena to become determinately decisive" (*PIA* 74, 33–34). This constitutive inconclusiveness is significant because it underlies Heidegger's view of the way phenomenology makes (conceptual) presuppositions—that is, of how formal indication solves the problem to which it was addressed. The "ineluctable significance" of formal indication, he claims, lies in its being "genuinely motivated by the concrete and factual ... as nonprejudicial, but also nondecisive, prefigurative touching on the factual" (*PAA* 85). In other words, formal indication is a use of signs through which phenomena are addressed in a provisional, tentative, and hence corrigible manner. Understanding the functioning of signs in this way means that phenomenological inquiry can be understood as an ongoing cycle of interpreting phenomena—with signs formally indicating phenomena and phenomena prefiguring the forms of symbolic indication.[28] Formal indication is accordingly the "method of outset of phenomenological interpretation in each stage of its execution," with "the interpretation's preconception" each time "stem[ming] from the respective stage of appropriation" (*PIA* 141, 87).

Before returning to consider its relevance to *SZ*, I want to highlight briefly three features of Heidegger's notion of formal indication relevant to assessing not only this notion, but also a phenomenological conception of language more generally. The first is that Heidegger's discussions seem to invoke an ideal of full actual presentation of sense, the kind of "*presence* of sense to a full and original intention" that Derrida (1993, 3) has portrayed as a dubious metaphysical assumption underlying Husserl's phenomenology. For in grounding experiences, Heidegger tells us, the basis of "concrete work," the "(ultimate) structural sense of the full object" is to be possessed in its full determination; a "phenomenon" is simply the "being present of an object" (*PIA* 28; Heidegger 1988, 69). However, whether or not Heidegger was assuming such an idealized presence of sense, it is important to realize that his view of the function of formal indication could survive without it. For in thinking of their function in an approximative or schematic manner, he avoids suggesting that formally indicating signs could, even in principle, capture any full presence of sense. Although his picture undeniably invokes some kind of actualization of understanding, that role might be understood in relation to aspectual or partial presentations such as Husserl's "adumbrations" or Merleau-Ponty's corresponding view of the indirectness of intentional objects (cf. chapter 4, section 4).

The second feature to highlight is that on Heidegger's model the function of signs is what might be called "simple preservation": actual or performed understanding is laid down in language such that it can later be reactualized in just the same way. Formal indications thus function, or fail to function, as deposition and reactivation, in much the same way as (say) a compact disc allows us to listen to music just as it was recorded. Underlying this simple preservation model is Heidegger's assumption that there is an internal link between formal indications and determinate grounding experiences, between signs and their "origins." This implies that the relationship between expressions and their sense-genetic origins is essentially atemporal. Actualized understanding may deteriorate into "empty" use of indications and so stand in need of "reappropriation," but no allowance is made for the ravages of time or the internal workings of language eroding the internal link between expressions and the grounding experiences proper to them. Rather, on this model, language would be a pseudotemporal phenomenon, a "process" with no intrinsic temporality.[29]

Finally, a major source of obscurity in Heidegger's idea of formal indication is the idea of "form" it relies on. In being contrasted with "deformalizing" situations of "evidence," it is clearly supposed to function as one pole of something like a form-content distinction, or more precisely a form-fulfillment distinction. Unfortunately, the one passage where Heidegger undertakes to explain the sense of *formal* in formal indication is of limited use. He there compares and contrasts his use of the term *formal* with Husserl's distinction between (eidetic) formalization and (classificatory) generalization (cf. Husserl 1992c, 31 [§13]). As Heidegger sees it, generalization is ordering within a framework of genres and species that entails reference to the subject matter being classified, whereas formalization is to be subject-matter independent and pertain to "formal ontological" categories such as thing, experience, object (Heidegger 1995, 58–59). He attempts to link this with formal indication by claiming that the latter should "indicate the relation to the phenomenon in advance" or what he calls "relational sense" (Heidegger 1995, 63). This claim relies on a distinction Heidegger standardly made during his early phenomenological period between three "directions of sense" found in any phenomenon: (a) content sense (*Gehaltssinn*): "what" is "originally experienced in it"; (b) relational sense (*Bezugssinn*): the "original manner in which it is experienced"; and (c) performative sense (*Vollzugssinn*): the "original manner in which the relational sense is performed" or actualized.[30] However, it seems to me that Heidegger is here in a tangle: First, he fails to offer any reason as to why the "formal" should not convey all three kinds of sense he distinguishes. Further, it is far from clear that a formal indication could serve its purpose without some determination of "what" is involved. Finally, the restriction to relational sense is inconsistent with Heidegger's own subsequent discussions of formal indication, which routinely treat such indications as having "content."[31]

Nonetheless, quite independently of the question of what exactly it conveys, Heidegger does not explicate what he means by "form." To the extent that language is to be construed as formal indication—which is clearly implied by Heidegger's discussions (e.g., by construing definition as formal indications)—the relevant notion of form can be taken to be that of linguistic form at all syntactic levels (i.e., definitions, sentences, words, etc.). But this leaves open important questions as to which aspects of linguistic form are relevant in directing us to features of the world: Are the inferential properties of sentences or expressions "formal" in Heidegger's sense? Do letters, words, and sentences all "point" to phenomena in the same way? Are we supposed to agree with the view halfheartedly proposed by Socrates that individual letters have their own expressive or representational properties?[32] And, finally, how might Heidegger have responded to the obvious difficulty posed by the "arbitrariness" of linguistic signs? Answering such questions clearly requires a more detailed view about the "formal" qualities in virtue of which linguistic signs "point out" features of the world.

3 Heidegger's Ambivalence Explained

The motivation for examining Heidegger's earlier views about the role of philosophical concepts as formal indications was to explain his ambivalence about the role of routine language use in *SZ*. The task, as set out in this chapter's first section, is to explain Heidegger's apparently negative attitude in *SZ* toward routine language use despite his recognition of its indispensability, its role in conceptual conditioning, and its pragmatic adequacy. This section will consider how the notion of formal indication feeds into *SZ* and how it allows Heidegger's ambivalence toward routine language use to be understood.

As *SZ* contains no account of formal indication and references to it are sparse and seemingly incidental, some evidence should perhaps be mentioned for assuming its relevance to the later work.[33] To begin with there is testimonial evidence in that Heidegger himself continued to emphasize the importance of formal indication both at and beyond the time of *SZ*'s publication. For example, in a letter to Karl Löwith in August 1927, he writes: "Formal indication, critique of the customary doctrine of the *a priori*, formalization and the like, all of that is still for me there [in *SZ*] even though I do not talk about them now" (quoted from Kisiel 1993, 19). And in his 1929–1930 lectures formal indication is still described as the basic or pervasive character of philosophical concepts (Heidegger 1983, 425, 430).

More important, perhaps, are several terminological and structural parallels between formal indication and views expounded in *SZ*. Most obviously, the emptiness-fulfillment distinction on which the formal-indication model is built suggests a basis for Heidegger's qualified disqualification of routine language use. In particular the negative characterization of routine language use in *SZ*, as lacking actualized understanding and grounding (experience), clearly corresponds to the "emptiness" of formal indications.

Particularly telling, however, is the fact that the same two regulative concepts are relied on in discussing both formal indication and Dasein's disclosedness in *SZ*. Thus the notion of "originality"—or "primordiality" (*Ursprünglichkeit*)—is linked with grounding experience, as the source of sense, not only in the early conception, but also in *SZ*. "Originality," as Heidegger puts it here, requires the securing of "a phenomenally suitable hermeneutic situation," and this involves finding an appropriate provisional interpretation of a "grounding experience" (*Grunderfahrung*).[34]

Similarly, the distinction between formal emptiness and fulfillment in grounding experience is cast in terms of *Eigentlichkeit* and *Uneigentlichkeit*. These are the centrally important terms in *SZ* that are usually rendered as "authenticity" and "inauthenticity," but which were discussed in the previous section in terms of indications being proper/improper. Thus indications, being empty, are "improper," whereas apprehension, or understanding, comes to be "proper" when fulfilled, actualized, or performed.[35] To be sure, in the context of *SZ* these terms are quite naturally taken to have an existentialist connotation, bearing on Dasein's self-realization and appropriation of its hermeneutic situation. But in light of Heidegger's phenomenological conception of signs it should be noted that "proper" apprehension (also) stands for an epistemic distinction. The problem with the "situation of understanding in its outset," Heidegger explains, is that "the object" does not offer "itself fully and properly [*eigentlich*]" (*PIA* 34). By contrast, grounding experience is to be the "situation of evidence" in which a decision is made according to "experience in which the object properly [*eigentlich*] gives itself as that which it is and how it is" (*PIA* 35). Hence there are two sides to the "propriety" of grounding experience: it is not simply that an agent "appropriates" objects, making them cognitively its own, but also that the object of understanding simultaneously shows itself ("properly") as it in itself is.[36]

Finally, the idea that signs should be shaped by phenomena also recurs in *SZ*. For example, in his discussion of the development of Understanding Heidegger emphasizes that Setting-out can either "create the conceptuality belonging to the entity ... from this itself," or "force [this entity] into concepts" that conflict with its "manner of being" (*SZ* 150). Or again, as he puts it in introducing Articulacy—in precisely the terms used to characterize the deficiency of routine language use—"*what* is said" should be drawn or created (*geschöpft*) from "what is talked about" and so make this manifest and accessible to others.[37]

Given these parallels, we can now consider how the notion of formal indication provides various components required to explain Heidegger's ambivalence about routine language use in *SZ*. As the parallels above suggest, the principal importance of the idea of formal indication is its positive aspect—that is, what it tells us about the way Heidegger thought signs could, or ideally should, function as phenomenological concepts. The ideal supposedly underpinning the proper functioning of concepts is concisely stated in his summer 1924 lectures. When "meaning and the use of a word" function optimally, he explains there, the object addressed is broken down into the "proper 'respectiveness' comprising such an object ..., so that I see it in its proper articulatedness" (Heidegger 2002, 37–38). Heidegger's ideal—that of *transparent presentation*—is thus as follows: that the function of signs is to allow features of the world to be seen as articulated in the way linguistic forms present them to us. Significantly, the same ideal is also found in *SZ*'s definition of what I called the demonstrative function of Articulacy as "pointing to" what is talked about so as to allow it to be "directly seen" or "inspected" (*SZ* 32, 34). The proper function of Articulacy is thus the same as that of formal indication, namely to see the world directly in the way it is presented (by language) as being. The core function of signs, according to this ideal, is to present features of the world in a certain way, which can then prove to be appropriate or otherwise.

It seems to me that the function just described is not only a central feature of Heidegger's conception of language, but must in some sense also be right. Indeed, a central feature of our experience of language is that it often makes us aware of how the world (itself) is. For example, typically if someone tells me that the tram is ten minutes late, or that it is raining in Brunswick, then I learn something straightaway about the trams or the weather in Melbourne—without the latter themselves being affected by that linguistic act. The use of language thus mediates an awareness of the world that is direct in the dual sense Heidegger attributes to "Greek" truth. Further, I think Heidegger is right to suggest that the capacity of language to direct our awareness to (or "point out") specific features in this way is intimately connected with the linguistic forms used to present the world in a certain "articulation," or more precisely with some kind of meaning or sense that is bound up with those forms. Thus while I am not suggesting that Heidegger's view of this pointing function or its connection with linguistic forms is completely satisfactory, I think it correctly identifies part of what it is for linguistic signs both to disclose the world and to be meaningful. In the following, I therefore want to hold on to this important feature of Heidegger's position by saying that linguistic expressions have a *presentational sense,* a kind of sense that underlies their capacity to present features of the world in an articulate manner and that is due specifically to the form of linguistic expressions.[38]

Of course, if the distinction between empty and fulfilled, or improper and proper, sign use is to be upheld phenomenologically, it must be possible for such transparent presentation to fail to take place—indeed *SZ*'s discussion of routine language use suggests this is usually the case. This possibility is accommodated by the role Heidegger attributes to interpretation and the approximative or schematic character of formal indications. For, as seen in the preceding section, interpretation has the role—so-called deformalization—of cashing out formally indicating features of signs in terms of phenomenal features of the world. The need to do this arises from the schematic character of formal indications, which allows that it may not be immediately obvious how they relate to phenomena.[39] In other words, the formal-indication model allows us to have "empty" awareness, as opposed to awareness of the phenomenal "fulfillment" underlying the forms of language we use—that is, to be in possession of linguistic forms without being able to map them interpretatively onto the features of reality (assuming there to be such) that make them appropriate.

With regard to "empty" understanding, however, an important limitation of the formal-indication model emerges, because this is conceived of almost entirely negatively in relation to grounding experiences. Although the significance of signs is taken to be anchored in form, it is conceived of only as the sketchy imprint of actual phenomena. Through this one-sidedness, however, the formal-indication model seems to suggest there is nothing in which empty intending itself consists.[40] Whether or not this possibility is coherent in considering the function of signs in the rarefied context of phenomenological inquiry, to which the notion of formal indication was tailored, it would be obviously inadequate in the framework of a general account of language. However, by integrating the understanding of linguistic entities within its general account of the foundational function of purposive awareness, *SZ* has the means to fill this lacuna in Heidegger's earlier conception of signs. And indeed, as seen in chapter 1, Heidegger extends his view of the foundational function of purposive understanding to language. The result is a pervasive instrumentalism in *SZ*'s characterizations of language: "Statements" are "ready-to-hand" entities, as are words, the totality of which—as the "spoken-out-ness" of Articulacy—comprises language (*SZ* 161, 224, 161). This instrumentalism emerges most clearly, as one might expect, in Heidegger's conception of signs in §17, in which signs are characterized as ready-to-hand Equipment, the use of which is embedded in and serves to make manifest instrumental relations.[41] Thus by the time of *SZ* Heidegger's view is that our basic grasp of language, of linguistic entities, is of a kind with the purposive understanding of other tools, with words and sentences being grasped in terms of "what they are for" and not as objectlike entities bearing certain ("semantic") properties. Hence it can be allowed that the understanding we have in using language is of this instrumental kind even if it is "empty" in the sense of not meeting the positive ideal of formal indication.

At this point some clarification is necessary. What has just been said might be thought to suggest that the distinction between emptiness and fulfillment coincides with that between purposive and predicative understanding. But that would be a mistake for two reasons. First, it would imply that routine language use is purposive *as opposed to* predicative, a view rejected in section 1 above, partly on the grounds that everyday understanding of the world cannot plausibly exclude predicative awareness completely. Second, the idea of transparent presentation ("fulfillment"), which I am claiming is the source of Heidegger's negative assessments of routine language use, is indifferent to and applies over the purposive-predicative distinction. To see this, recall that Heidegger's distinction between purposive and predicative modes of Setting-out implies that one can have or fail to have an articulate grasp of the world in two different ways. The ability to interpret ("Set Out") entities purposively amounts to knowing how to use them in particular contexts, while predicative Setting-out involves the ability to grasp context-independent properties, make inferences, and reason. Both these modes of Setting-out involve an awareness of how language is connecting us to features of the world, an awareness we can presumably have to a greater or lesser degree, which suggests that both modes can be thought of in terms of emptiness and fulfillment. While this complicates matters, as the purposive-predicative distinction can no longer be equated with the emptiness-fulfillment distinction, it provides a more plausible basis for Heidegger's claims about routine language use. For Heidegger can be understood as claiming that an average amount of *both* purposive and predicative awareness will be enough to get by in everyday life, which is of course compatible with emphasizing a failure to attain full awareness of either kind. In more concrete terms this means that routine language use involves an average grasp of the contexts in which and to what ends a given word might be used, along with an average grasp of its context-independent implications. Heidegger's notion of routine language use can thus be thought of as involving both purposive and predicative understanding of the world (of "what is talked about") that suffices for everyday activities. In this way it can plausibly claim to meet the constraint of pragmatic adequacy, while simultaneously exhibiting inadequacies that motivate Heidegger's ambivalence about its disclosive efficacy.

We are now in a position to summarize Heidegger's ambivalence toward routine language use. Overall the conception of the disclosive function of linguistic signs in *SZ* is the result of trying to fuse the ideal underlying formal indication with the story Heidegger tells about the foundational role of purposive understanding. Heidegger's ambivalent attitude results from this fusion. On the one hand, the presentational function of formal indications identifies a positive ideal for the functioning of phenomenological concepts, that of actualizing awareness of how what we are talking about is presented to us by language, by which routine language use is held to fail. This failure amounts to being unable to interpret features of the language we use in terms of correlative phenomenal features, and thus in not properly appreciating the presentational sense of the expressions we are using. But the negative assessment this suggests must be tempered, on pain of absurdity, since the failure to actualize understanding in routine use cannot be complete. Such absurdity is avoided by the claim that language (what is said) is grasped instrumentally in conjunction with the thought that in being so grasped language mediates an understanding of the world (what is talked about) that suffices for everyday purposes.

Against this background, it is possible to appreciate better the difference between Heidegger's talk of "averageness" of understanding and the "stereotypical" awareness of words' intensional properties linked with a "division of labor" in language use (section 1 above). To begin with, the latter view seems phenomenologically unconvincing, since we are not usually conscious of the supposed properties of words mediating understanding. This fact can be explained by Heidegger's view, which suggests a different conceptual priority: it is not that (intensional) properties of words mediate awareness of features of the world, but that what it is for a word to have certain (intensional) properties must be spelled out by interpreting it in terms of features of the world. Heidegger's view thus offers a direct account of the routine functioning of language, whereas the stereotype thesis introduces intensional properties as an initially obscure, and presumably superfluous, intermediary. On this more direct account the "averageness" of everyday understanding refers to a limited awareness of how linguistic features are connected with and motivated by the world—that is, to a shortfall relative to Heidegger's ideal of transparent presentation.

What of Heidegger's otherwise cryptic characterization of routine use as being directed to "what is said," rather than "what is talked about"? Although this characterization clearly intends to suggest that "what is talked about" is understood only superficially or minimally, it is less clear how to construe Heidegger's formulaic "what is said" (*das Geredete*). Does this have the honorific sense, "what is *said,*" referring to successfully expressed propositional content? Or is it a slighting reference to "what is said," as the words uttered with a quality bordering on that of mere noise? Though Heidegger does not address this matter directly,[42] there are various clues that he means the latter. To begin with it seems highly unlikely that Heidegger has in mind the honorific sense, given his view of the derivative nature of propositional content (cf. chapter 1, section 4). This conclusion is supported by considering Heidegger's view that different degrees of understanding can be brought to bear on the same utterance, according to how well it can be interpreted phenomenally. For on this view "what is said" remains constant between superficial and more penetrating ("deformalizing") interpretations of an utterance's phenomenal implications in a way that would be unintelligible if "what is said" were taken to mean "propositional content." It is therefore most plausible to read "what is said" as referring literally just to the words uttered, such that the superficial understanding of routine use borders on the merely verbal, of knowing merely which linguistic forms to use in which circumstances.[43] In this case, however, it becomes clearer why an awareness of "what is said" might be thought superficial or minimal. What routine language use lacks, on Heidegger's picture, is an awareness of how features of the language we are using are connected with features of the world (of "what is talked about"). Heidegger's negative comments about it reflect the absence of transparent presentation, the fact that we are merely using language rather than seeing the world in the way language presents it to us.

4 Linguistic Signs as Compound Instruments

If the preceding reading of Heidegger's views is correct, it faces an obvious challenge. Knowing which words to utter in which circumstances, it may be objected, is perfectly adequate as a characterization of what understanding language consists in. By contrast, Heidegger's ideal of transparent presentation might be dismissed as irrelevant to the functioning of language as a public phenomenon, and as reminiscent of the simplistic hypostatization of linguistic imagery mocked by the later Wittgenstein. Now in a sense this objection is redundant. Far from denying that knowledge of this kind suffices as knowledge of language, that is precisely Heidegger's point. It is despite this that he discerns language's failure to convey fully interpretative awareness of phenomena. Nevertheless, the objection does pick out something important. For Heidegger's negative comments about routine language use arise from modeling the function of signs in terms of presentational feats, *even though* this model seems, by his own account, to be largely irrelevant to the practical functioning of language. This reliance on two different kinds of standard seems to some extent incongruous, and might be thought indicative of a tension, because the standard of presentational sense is not only extraneous to practical concerns but appears to conflict with Heidegger's rhetorical emphasis on the foundational importance of purposive understanding.

In fact there is no such conflict, as both kinds of standard are built into Heidegger's view of the disclosive function of signs. This can be seen in the conception of signs presented in *SZ* §17. Summarizing his view, Heidegger there explains that the "sign is an ontic ready-to-hand [entity] [*ontisch Zuhandenes*] that … *simultaneously* … indicates the ontological structure of readiness-to-hand, the totality of instrumental relations and worldliness" (*SZ* 82; italics added). It is crucial to appreciate that this definition is twofold. In a basic sense—as "ontic[ally] ready-to-hand"—signs are instruments with a specific function. So, like other instruments, they are "constituted by instrumental relating" and are a means to getting things done within practical contexts (*SZ* 78). Clearly in this sense signs can have many different functions, according to the various tasks in which they are made use of as Equipment.

However, in addition—"simultaneously ..."—Heidegger attributes to signs a "*distinguished* use" that sets them apart from other instruments.[44] Thus, unlike other instruments, a sign "explicitly raises an instrumental whole into circumspection"; signs "allow the ready-to-hand to [be] encounter[ed]," or "more exactly, a context" of such readiness-to-hand, they "show primarily that 'in which' one lives" (*SZ* 80, 79–80). In this sense, all signs are to have the same function, the generic function of drawing attention to a context of action. Heidegger emphasizes this "peculiar instrumental character of signs" with regard to the creation or "institution of signs": because the worldly context usually remains inconspicuous, an instrument is needed to do the "'work' of *allowing* the ready-to-hand to *become conspicuous*" (*SZ* 80). Thus, independently of the various specific uses to which they may be put, this dual conception also identifies a generic standard of propriety for sign instruments, insofar as these can be more or less well suited to the task of inducing awareness of their context.

In this characterization of signs' "distinction" or "peculiarity" it is not difficult to recognize again the influence of formal indication's presentational function, in particular the idea that signs should be shaped by and expressive of phenomena. Similarly, this view of signs' generic use recognizably corresponds to the demonstrative function of Articulacy (its "pointing-out" function). What §17 makes clear, however, is that in *SZ* the presentational sense signs have due to their form is conjoined with what I will call their *pragmatic sense*, the meaning or sense they have due to the respective roles they have as instruments in established practices. In other words, *SZ* has a twofold conception of signs tailored to meet the demands of both pragmatic sense, since signs are instruments, and presentational sense, since these instruments are formed so as to be good at "pointing to" phenomena.

Heidegger himself does nothing to emphasize this conjunction of the pragmatic life of signs with their more "distinctive" capacity to point out features of the world, either in *SZ* or in later works that accentuate the importance of language. Nonetheless, I want to claim that this is a key feature that makes *SZ*'s conception of signs particularly interesting for a phenomenological approach to language. We might note in passing that it makes Heidegger's reliance on two kinds of standard less surprising. For although the ideal of transparent presentation is perhaps extraneous to pragmatic concerns, both kinds of aim—so to speak, the foundational and the aspirational—are combined in his conception of a sign. However, what makes *SZ*'s conception of signs particularly interesting is that it brings together two distinct factors in the articulation of lived meaning—the structure of linguistic forms and the role of expressions in human practices—each of which clearly features in our prereflective experience of language and has implications for the inferential properties that linguistic expressions manifest in sentential contexts.[45] To put it another way, thinking of linguistic signs as bearers of both presentational and pragmatic sense allows them to be seen as the locus of a correspondingly complex disclosive function, as entities that can reveal features of the world to us through either a proper appreciation of their presentational implications or a grasp of their pragmatic utility.

How is the relation between presentational and pragmatic sense to be understood? Are both needed in order to understand the function of linguistic signs in disclosing the world, or could one of the two be eliminated or explained in terms of the other? One obvious reason for thinking of them as genuinely distinct is that identical practices can be carried out in different languages, such that words that are in some respects pragmatically equivalent can nonetheless have quite different implications in virtue of syntactic differences. To be more precise about their relation a more detailed conception of each kind of sense would be required. But this is effectively where the Heideggerian path comes to an end. For although Heidegger places great emphasis on the foundational importance of purposive understanding in *SZ*, he offers no more detailed account of the way conceptual commitments—or alternatively, the structure of disclosure—are inherent in the practices in the midst of which Dasein is "thrown." Further, despite orienting his whole account of signs toward the goal of presenting the world in an articulate manner, Heidegger never provides anything more than passing hints as to what comprises the "formal" aspect of language (or concepts). This problem was highlighted above in discussing formal indication (section 2) and it persists into *SZ*. While insisting on the need to liberate "grammar" from "logic," Heidegger there postpones consideration of the form of Articulacy, claiming only—as pointed out in chapter 1—that settling "the basic forms" of an "articulation of what can be understood altogether" presupposes his analytic of Dasein (*SZ* 165–166). However, as long as this question is not answered it remains unclear, for example, how the presentational feats of language are to be understood and how he might respond to the arbitrariness of linguistic signs. Heidegger's twofold instrumentalist conception of signs thus ultimately offers only a schematic framework, an outline of the disclosive function of linguistic signs that again stands in need of filling out. An important task for the following chapters will therefore be to give substance to the notions of presentational and pragmatic sense.

In doing this an additional constraint will need to be considered. For since, as chapter 1 showed, a distinctive feature of Heidegger's position is that it allows prepredicative uses of language, both pragmatic and presentational sense should be explicable in a way consistent with their involvement in prepredicative use. This means, according to the view proposed in chapter 2, that both should be understood in terms of factors that are subpropositional, subinferential, and irreducible to the predicative functioning of propositions. This does not, of course, imply that presentational and pragmatic sense should be isolated from predicative uses—for they are supposed to provide the (subinferential) foundation for such "derivative" uses. Indeed, part of the rationale for describing presentational or pragmatic sense as "sense" is precisely that they can be exploited in these different—prepredicative or predicative—ways. That is, the term *sense* in the following is intended to refer to factors that structurally influence the linguistic articulation of lived meaning, with the distinction between these two types of sense corresponding to two different kinds of structural influence. Hence the task for the following two parts of the book will be to draw on the ideas of Merleau-Ponty and Wittgenstein to set out, in a phenomenologically plausible way, how linguistic form and practical factors are of structural relevance to both prepredicative and predicative uses of language.

Before moving on to this larger task, note that it is now possible to see how Heidegger's position respects the three commitments identified in section 1 of this chapter. The need for pragmatic adequacy, first, is most obviously addressed by Heidegger's integration of the foundational role of purposive awareness into his conception of signs. However, the acknowledgment that routine everyday understanding of the world involves some predicative awareness and that fully transparent presentation is not required are both features of Heidegger's view that reflect the need to account plausibly for both the achievements and the limitations inherent in everyday practice. Second, the general shape of Heidegger's answer to the question of how routine language use embodies a "conceptuality" is twofold, with conceptual commitments being shaped by both the requirements of various practices and the constraints imposed by operating with a certain set of linguistic forms. Third, the indispensability of signs to both empty and fulfilled, or improper and proper, understanding can be appreciated. Signs are required by fulfilled understanding, since this presupposes being presented (or "indicated") in some articulation by linguistic forms. But they are also required for empty understanding since—even in the absence of "fulfillment" or transparent presentation—routine language use is involved in the constitution of the instrumental nexus of Significance.

Finally, describing the function of linguistic signs as "instrumental" introduces a potential misunderstanding due to an accident of historical usage. Locke, for example, describes language as "the great Instrument" and words as instruments that people use arbitrarily "as marks for the Ideas within [their] own Mind" (Locke 1975, 402 [III.1.§1; cf. §5]). On this usage, language is merely a means (an "instrument") for communicating thoughts that is *extrinsic* to—that is, not essentially involved in—meaning constitution. Against this background it is sometimes suggested that a conception of language that is "instrumentalist" *as such* contrasts with any view which recognizes language as having a role in constituting intentionality.[46] Yet there is no reason to assume that these two determinations necessarily exclude one another, such that conceiving of language in instrumental terms rules out its having a role in constituting meaning. Moreover, although some of Heidegger's comments in *SZ* might be seen as aligning him with an extrinsic view of instrumentality, such a view is neither directly entailed by his characterization of language as ready-to-hand Equipment nor obviously reconcilable with the role he attributes to routine language use in conceptual conditioning. For these reasons I will interpret Heidegger's views as relying on a notion of linguistic instrumentality that allows a more intimate link between language use and meaning constitution.

II Merleau-Ponty: The Presentational Aspect of Language

4 Language as the Expression of Lived Sense

So far I have been looking at Heidegger's work to draw out the basic framework for a phenomenological conception of language. I have tried to show that at a very general level Heidegger provides a coherent and phenomenologically plausible overall view of the role of language in the broader phenomenon of disclosure. At a more specific level I have tried to show he also offers a distinctive view of the functioning of linguistic signs, as complex instruments characterized by two kinds of sense that I labeled presentational and pragmatic sense. The aim of the next two parts of the book is to fill out this framework, at both levels, into a more detailed phenomenological conception of language. To do this I will draw on the work of Merleau-Ponty and Wittgenstein in turn, using these to complement and complete Heidegger's general picture of the role of language and to develop a more detailed conception of each aspect of linguistic signs' disclosive function.

Merleau-Ponty's approach to language is guided by his conviction that perception is a paradigmatic phenomenon that "teaches us ... the true conditions of objectivity" relevant to "the present and living being" of cultural and historical phenomena such as language (Merleau-Ponty 1996a, 67–68). The influence of this paradigm is discernible in two defining characteristics of his conception of language. The first is that his views of perception and language both focus on the role of embodied finite agents. Merleau-Ponty does not, as Heidegger does, arrive at this focus through arguments against the coherence of the idea of transcendental subjectivity or its compatibility with phenomenology. Rather, it was an implicit assumption from his earliest works onward that a philosophical conception of subjectivity should reflect the findings of empirical biology and Gestalt psychology. However, this led him to claims about perception and language that clearly correspond to Heidegger's rejection of transcendental subjectivity as the locus of fully constituted eidetic relations. Thus Merleau-Ponty rejects not only the "prejudice of the world," the idea of an "exact world, entirely determined" that is "posited first" by "objective thinking" (*PdP* 11, 39), but also the idea of a fully explicit ideal language, or a "language before language" (*PdM* 10). Hence the second defining characteristic of Merleau-Ponty's approach is the rejection of what I will call *final determinacy*—that is, any state of determinacy that functions as the supposed telos or end point in terms of which intentionality is to be understood. Eschewing this idea, the task for a philosophical conception of subjectivity or language becomes precisely to explicate how structures bearing determinate meaning are generated in actual experience and so to be a "phenomenology of the genesis" of such structures (*PdP* xiii).

Although Merleau-Ponty's overall view of language remained largely constant throughout his work, my discussion here will accentuate some decisive refinements to his view that came with his appropriation of Saussurean ideas. In particular, I want to show how the notion of "indirect sense" he then developed can be used to explicate the idea of presentational sense. That task is undertaken in the next chapter, which concentrates on texts from the early 1950s that develop in greater detail Merleau-Ponty's view of signs as bearers of indirect sense. By contrast, this chapter relies primarily on Merleau-Ponty's earlier works *The Structure of Behavior* (*SdC*) and *The Phenomenology of Perception* (*PdP*) in setting out some of the main features of his conception of language. In doing this the aim is both to show how these features fit into the Heideggerian framework's overall picture of language and to provide the requisite background for understanding the notion of indirect sense and its role.[1]

This chapter begins by outlining how Merleau-Ponty's focus on "lived sense" responds to the need he sees for a conception of language to reflect the intimate connection between language and the formation of thought. The first section sets out the biological and existential basis of such lived sense, while the second looks at its experiential side, particularly Merleau-Ponty's assimilation of language to gestures and the role of "sedimentation" in distinguishing linguistic expression. I also identify here two problems with Merleau-Ponty's early views that would be solved later by his assimilation of Saussure. In the third section I turn to Merleau-Ponty's emphasis on creative expression, arguing that, although this sheds light on certain kinds of language use, it should not be considered explanatorily primary in the way Merleau-Ponty supposes. The fourth section introduces Merleau-Ponty's notion of indirect sense—which will be the main focus of the next chapter—by considering which characteristics of "directness" it is taken to contrast with. The chapter concludes with a brief review of how

Merleau-Ponty's views, as set out here, parallel and complement the general picture of language provided by Heidegger.

1 The Efficacy of Language

Much of Merleau-Ponty's discussion of language is geared to the requirement that the use of linguistic signs should be inseparable from what makes them meaningful. In *PdP* he characteristically introduces this requirement by criticizing views he attributes to "empiricism" and "intellectualism" respectively. The "empiricist" view thinks of language in terms of the causal efficacy (*existence effective*) of "verbal images," or the "traces" of previous language use (*PdP* 203). Because this affords no role to the speaking subject, Merleau-Ponty sees the "intellectualist" position, according to which the meaning of linguistic utterances is constituted by the categorial operations of a pure thinking subject, as a modest improvement. Nevertheless, he objects that this—essentially Kantian—picture fails to account for the connection between thought and language. Although subjectivity is now involved, it is as the "thinking subject," not the "speaking subject"; thought has meaning while "the word remains an empty envelope," lacking "an efficacy of its own [*efficacité propre*]" and functioning merely as an "exterior accompaniment" of thought (*PdP* 206). Merleau-Ponty therefore concludes, somewhat enigmatically, that one surpasses both these positions through the "simple remark that the word has a sense [*sens*]," which is to imply that "the word, far from being the simple sign of objects and of meanings, inhabits things and is the vehicle of meanings [*véhicle les significations*]" (*PdP* 206, 207). This is of course a general point: Merleau-Ponty's objection applies not only to the idea of a pure thinking subject, but to any postulated mode of meaning constitution that stands in a merely external relation to actual linguistic processes—that is, to any view of language as an instrument in the extrinsic sense mentioned earlier, including attempts to construe linguistic signs merely as indications in the Husserlian sense (*PdP* 193, 211; cf. *PdM* 24).

Merleau-Ponty also characterizes the more intimate link he discerns by observing that "thought tends toward expression as toward its completion," such that "thought and expression are therefore constituted simultaneously" (*PdP* 206, 213f.). While there are some experiences (e.g., literary writing) of which this Kleistian thought is perhaps literally true, such formulations are clearly challenging.[2] Is one supposed to be incapable of having a thought without vocalizing it? Is it not obviously wrong to suggest, as Merleau-Ponty does, that "the thinking subject himself is in a sort of ignorance about his thoughts as long as he has not formulated them for himself or even said or written them"?[3] These claims are best interpreted not as simply reporting experience but in a criterial sense, as saying that what a thought is can be understood only in terms of its being expressed. The "tendency" toward expression can then be explained by the fact that the ability to express it in language is the best, and often the only, criterion of an agent's having a particular thought. In this sense, to cite two of Merleau-Ponty's own examples, the way thought "tends toward" language is analogous to the way abilities to type or play a musical instrument are directed toward performance (*SdC* 131). In all these cases, just as Merleau-Ponty suggests, it is a matter of culmination or accomplishment, such that a thought is nothing more precisely and more properly than its expression.

As an elucidation of the link between language and thought the view just outlined is somewhat limited. To begin with, it seems simply to paraphrase the claim that linguistic expression is the realization of thought with the tautology that linguistic behavior wouldn't be what it is without the use of linguistic expressions. Nor is it clear that it improves on intellectualism's failure to account for the "efficacité propre" of words. For the claim that thought tends toward expression leaves the linguistic articulation of a thought looking simply like a final cause and does nothing to establish a constitutive role for language in the *formation* of thoughts. To understand the distinctiveness and potential appeal of Merleau-Ponty's position it is therefore necessary to get clearer about the two features that he has suggested distinguish words from "simple signs." First, what is it for words to "inhabit" things? Second, in what way are words the "vehicle" of significance?

Merleau-Ponty's answer to these questions centers on the idea that sense is generated by embodied agents living their lives. Such *lived sense* has two sides or aspects, to be discussed in this section and the next. The first is an underlying biological and existential aspect, according to which language is a kind of "incarnate sense." Thus the realization of sense through the body, Merleau-Ponty says, comprises "a primordial operation of meaning [*signification*] where the expressed does not exist apart from the expression," this being "the central phenomenon of which body and mind, sign and meaning are abstract aspects" (*PdP* 193). The emphasis in *PdP* on the body's explanatory importance, together with a suggestive reliance on subjectivist terminology, might give the impression that Merleau-Ponty's preferred view fails to meet his own requirements any better than the intellectualist.[4] In particular, his descriptions of the body as "a power of natural expression" that we see "secreting in itself a 'sense'" that it "project[s] onto its material surroundings and communicates to others" (*PdP* 211, 230) might seem to imply that sense is constituted by the body—so to speak, as a transcendental medium—and contingently externalized via linguistic signs.

That this is not Merleau-Ponty's view can be seen by keeping in mind his earlier work in *The Structure of Behavior* (*SdC*), which underlies many of his claims in *PdP*.[5] Comparing biological and physical systems, Merleau-Ponty talks there of "sense" as something implicit in the "structure" of the interaction between a system and its surroundings (*SdC* 112–113). Rather than straightforwardly tending toward physical equilibrium with the outside, organisms are distinguished as having a complex "structure" that enables them interactively to "constitute an environment of their own," as comprising a "unity of meaning" with behavior coordinated by "*sense*" rather than physical "laws" (*SdC* 157, 169). So despite his suggestive formulations in *PdP*, Merleau-Ponty's is not the subjectivist claim that the world acquires meaning ("extrinsically") in virtue of bodily sense constitution projected outward by expressive acts. Rather, his idea is that sense is constituted by the *interaction* between an organism and its surroundings, that *behavior* is so to speak the locus of sense, so that it is in virtue of being bodily behavior that meaning is immanent in language.[6] It is this view of sense as a biological feature manifested in behavior that underlies *PdP*'s talk of "incarnate sense" and its claim that linguistic expression, as "living sense," "presents, or rather ... *is* the subject's taking up a position in the world of its meanings" (*PdP* 225).

With regard to language Merleau-Ponty makes an important distinction between two forms of biologically lived sense. As he explains it in *PdP*, "below the conceptual meaning of acts of speech [*paroles*]" there is an "existential meaning" that they do not simply translate, but that "inhabits" and "is inseparable" from them (*PdP* 212). In making this distinction Merleau-Ponty relies on empirical findings from Gestalt psychology. Thus in arguing the need for an "existential theory" of aphasia—the loss of linguistic ability due to brain damage—he discusses patients who are unable to sort a series of samples into color groups. Because for them red samples fail to stand out as forming a group, these patients are unable to subsume particular cases under the color concept. In more general terms, characteristic of Gestalt psychology, in failing to discern general patterns or saliences within experience in the way normal subjects do, such patients lose the ability to apply the "categorial attitude" to experience and remain stuck in the "concrete attitude" (*PdP* 204f., 222f.). Despite this loss, however, such subjects reportedly retain the ability to use words where they are of "affective and vital interest," leading Merleau-Ponty to distinguish between "the word as an instrument of action and as a means of disinterested denomination" (*PdP* 204).

The simple absence of classificatory abilities contrasts with cases in which the categorial attitude is somehow preserved but fails to engage appropriately with its existential foundation. Merleau-Ponty illustrates this kind of failure using the case of Schneider, a brain-damaged patient studied in detail by the Gestalt psychologists Gelb and Goldstein. Schneider's impairment was such that his intellectual or judgmental faculties were at some level left intact: he showed "no sign of a weakening of general intelligence" and retained the ability to explicitly subsume particulars under general concepts (*PdP* 228, 148). His difficulty was that this general intelligence no longer seemed attuned to human relevance. For instance, although motorically capable of "concrete" movements—such as those habitually acquired, or reflex responses to pain—and able to understand instructions concerning "abstract" movements simply to move his arm to a certain position, Schneider was unable to carry out the latter or, generally speaking, spontaneously to translate intellectual understanding into motoric action. Similarly, his behavior failed to manifest any sense of fictive or "virtual" scenarios, or of involvement in sexual or conversational interactions.[7] With regard to language he had difficulty understanding simple analogies and metaphors, even where these deployed familiar words—familiarity that sometimes enabled him to explicitly reason out the analogy via their literal meanings (*PdP* 148–149).

An obvious concern one might have about Merleau-Ponty's discussions of Schneider is whether the force of his claims depends on their reliability as empirical hypotheses, which would potentially expose him to the objection that his views are scientifically anachronistic, or simply that the single case of Schneider is an inadequate inductive basis for his confident extrapolations.[8] However, this would be to misconstrue the point of Merleau-Ponty's discussion, which uses Schneider as a paradigm highlighting the contrast with certain competences tacitly at work in normally engaged subjectivity. As such, the force of his claims rests on the use of this case to exemplify certain deficiencies—a role that might be equally well fulfilled by considering a fictive patient. Thus in discussing various possibilities of disintegration Merleau-Ponty aims to show that to be (causally) effective in the sensible world intellectual capacities require "orientation"—that is, they must connect with practically and emotionally significant features of lived experience. In terms of the distinction between the concrete and categorial attitudes, this involves the distinct feats of picking out such practically and emotionally significant features as figure-ground saliences in the lived world, and organizing these saliences into intellectual categories.[9] What makes Schneider's case of particular interest to Merleau-Ponty is that, despite embodying intellectualist philosophical assumptions about the nature of subjectivity, he remains recognizably dysfunctional as a human agent in lacking the overall ability to connect the categorial attitude properly with the concrete attitude. This powerfully illustrates his claims that such "orientation" is needed and that it must be understood as a bodily feat. The body "is the condition of possibility ... for all expressive operations" since—to generalize Merleau-Ponty's comments about the notion of space—talk of the "body schema [*schéma corporel*]" is "a way of expressing that my body is in the world ... one's own body is the third term, always implied, of the figure-ground structure."[10]

The distinction between the concrete and categorial, or existential and conceptual meaning, can be illuminated further by the corresponding distinction between "detachable" (*amovible*) and "symbolic" forms of behavior discussed in *SdC* (115–133). Detachable forms of behavior are already distinguished there from straightforward "syncretic forms"—that is, responses to stimuli that remain bound to "the material of certain concrete situations," by being "relatively independent of the materials in which they realize themselves" (*SdC* 115). Detachable behaviors thus involve a degree of flexibility in responding purposively to given situations; examples include chickens responding to differences between color stimuli (rather than to specific hues), rats choosing a shortest path without being confused by their initial direction, or chimpanzees exploiting the "instrumental value of an object" (*SdC* 124). While such behaviors manifest some awareness of situations' structure, the detachability of this awareness from specific situation types is limited. Hence, as Merleau-Ponty explains, "the animal cannot adopt a point of view chosen at discretion with regard to objects" and so "cannot recognize a same *thing* in different perspectives" (*SdC* 127f.). In sum, detachable forms are "behavior adapted to the immediate and not to the virtual, to functional values and not to things" (*SdC* 130). By contrast "the 'thing structure'" and "the sense of the virtual" are made possible by and characterize "symbolic behavior" (*SdC* 128, 130). Here a "symbol" is used to allow "substitutions of different points of view," thus freeing "'stimuli' from actual relations" or the "functional values" assigned to them by "the defined needs of the species" (*SdC* 133). This type of feat, Merleau-Ponty further suggests, is explained by the "systematic principle" inherent in a sign system along with the idea of a "structural correspondence": "The true sign represents the signified ... as long as its relation to other signs is the same as the relation of the object signified by it to other objects" (*SdC* 132–133).

We can now begin to see how Merleau-Ponty would respond to the two questions identified at the start of this section. Both forms of behavior he distinguishes—detachable and symbolic behaviors—manifest living or "biological sense" (*SdC* 19), and this explains his use of the term *inhabitation* to characterize the intimacy of bodily presence in the world.[11] In addition, both forms of behavior can involve the use of linguistic signs, since—as mentioned above—the use of words "as an instrument of action" is already possible in detachable or concrete behaviors. However, it is clearly symbolic behavior that Merleau-Ponty has in mind in talking of the feat specific to words—their "efficacité propre"—and that provides the basis for describing words as the "vehicle" of significance. For whereas in detachable behaviors the "vocal sign does not mediate any reaction to the general meaning of the stimulus" (*SdC* 131), in symbolic behavior the use of an expression is essential to cross-situational organization. Moreover, because the articulation of thought is linked with the "systematic principle" underlying a sign system, the means of mediation Merleau-Ponty suggests—although in itself still in need of explication—provides a basis for explaining thought's "tendency" toward expression. Thus, on Merleau-Ponty's view, it is only in symbolic behavior that linguistic expressions mediate sense and so function as its vehicle; such behavior does not "have," but "is" itself meaning (*signification*).[12]

2 The Phenomenology of Lived Sense

The preceding discussion of the biological-existential side of lived sense is, however, only part of the story for Merleau-Ponty. He also emphasizes that "the theory of language must make its way through to the experience of speaking subjects," so that the notion of lived sense also has a visible aspect: it is not simply lived through but experienced, phenomenological as well as biological.[13] Correspondingly, several features of Merleau-Ponty's conception of language seek to characterize our experience of language. In this spirit he highlights the transparent feel that language has in carrying our attention to the world, rather than opaquely fixing it on the signs used. The "marvel of language," he says, is that we "forget" it, that successful "expression erases itself before what is expressed."[14] In fact, Merleau-Ponty considers this "transparency" to be largely an effect of habitual familiarity, one we often misinterpret as "absolute" or "direct" awareness of the world.[15]

A major feature of Merleau-Ponty's characterization of our experience of language is his suggestion that an act of speech (*la parole*) is "a true gesture" containing its "sense" in the same way as other gestures.[16] This assimilation of language to gestures has several implications for Merleau-Ponty. One is active uptake: "The sense of gestures is not given but understood, that is, seized again by an act of the spectator" (*PdP* 215). This is a rejection of the behaviorist (or empiricist) idea that gestures might be learned passively by conditioning processes, but the phenomenology of such learning further implies that grasping a gesture, though spontaneous, is a nonintellectual and context-specific act: "One day I 'caught' the word [e.g.] 'sleet' as one imitates a gesture ..."; the word was "never inspected, analysed, known, or constituted, but grabbed and taken on by a speaking power"; the "sense of a word" is learned like "the use of a tool, by seeing it employed in the context of a certain situation" (*PdP* 461–462; cf. 216). Finally, gestures are grasped in relation to a form of life. As Merleau-Ponty points out, as humans we naturally have no sense of the "sexual mimicry" of other species, yet if the sense of a gesture "were given to me as a thing" it would not be clear "why my comprehension of gestures is for the most part limited to human gestures" (*PdP* 215).

These phenomenological claims about the gestural character of language use reflect Merleau-Ponty's biological-existential view that the sense inherent in embodied behavior is due to the way the world is presented in the optic of the organism's vitality: "Behavioral gestures ... do not pick out [*ne visent pas*] the true world or pure being, but being for the animal ... a certain manner of dealing with the world, of 'being in the world.'"[17] However, because even sexual or otherwise affective (e.g., anger) behaviors are not transcultural constants, the form of life to which gestures relate is also modulated by cultural factors. As Merleau-Ponty puts it, such behaviors are "invented" just as words are and "transcend" the body's "simply biological being" to "create meanings that are transcendent with regard to the anatomical arrangement and yet immanent in behavior" (*PdP* 220–221). Thus it is in relation to a human blend of nature and culture that the "phonetic gesture ... realizes for the speaking subject and for those listening a certain structuration of experience" (*PdP* 225).

Two features of this assimilation of language to gestural communication should be highlighted. The first—reflected in the expression "to gesture at" something—is that gestures are more a matter of intimation than fully determinate presentation.[18] Extended to language, this means that "once 'acquired'" a word's sense is "just as precise and just as little definable as the sense of a gesture. ... The consciousness that conditions language is merely a global and inarticulate grasp of the world" (*PdP* 462–463). The second feature is closely connected: although a gesture only "outlines [*dessine en pointillé*] an intentional object," this "object becomes actual and is fully understood when my body's powers adjust to and recover it" (*PdP* 215–216). Thus, in keeping with *PdP*'s overall focus, the second feature is that gestures are fundamentally bodily behaviors. In this respect Merleau-Ponty likens gestures to perceptible things, the identity of which "through perceptual experience is merely another aspect of the identity of one's own body," and which comprise "a system of equivalences … under examination by a bodily presence" (*PdP* 216).

There is, however, one feature that Merleau-Ponty sees as distinguishing language from other kinds of gesture: "speech is capable of becoming sedimented and constituting an intersubjective acquisition"—"in the case of speech the expressive operation can be iterated indefinitely" (*PdP* 221–222). Because expressed concepts or thoughts can be reused or taken up (*repris*), one is able to "overcome the temporal dispersion of the phases of thought," meaning that "speech is precisely the act through which thought eternalizes itself as truth" (*PdP* 441, 445). This interplay of speech and sedimentation is intended to address a central requirement of Merleau-Ponty's overall metaphysical picture by accounting for the genesis of the "mental landscape" in lived experience—that is, by indicating how the mode of being of explicit ideas is constitutively linked with acts performed by embodied subjects.[19]

Merleau-Ponty also appeals to the phenomenon of sedimentation in distinguishing language from other forms of expression, such as music or painting. The difference, as presented in *PdP*, is that whereas acts of speech consciously take up and are founded on previous acts of linguistic expression, music and painting do not depend on antecedent acts of expression, so that "each artist takes up his task from the beginning" (*PdP* 221). An obvious objection to this—that an artist's work is similarly conditioned by the personal, historical, and discursive situation—is accommodated by his later concession that painting too is *capable of* sedimentation (*PdM* 139ff.). Refining his earlier claim, Merleau-Ponty now suggests a distinction between sedimentation that "accumulates" and sedimentation that "integrates." Unlike painting, which "accumulates" or juxtaposes its productions over history, language is to "integrate" (i.e., preserve and develop) its past products and so to establish the realm of "integral" truths (*PdM* 142–143). This distinction is motivated by Merleau-Ponty's conviction that the development of language is animated by the aim of establishing stable bodies of truth, and that this requires "integration," with the result that whereas sedimentation is *possible* for painting, it *constitutively* shapes linguistic expression.

Against this background it is possible to identify two basic problems with Merleau-Ponty's views around the time of *PdP* that would be improved on by his subsequent appropriation of Saussurean ideas (to be discussed in the next chapter). The first is that although the notion of biologically and phenomenologically lived sense explains in some detail how words are "inhabited" by sense, Merleau-Ponty's early view of how they are the "vehicle" of sense is less developed. To be sure, in telling us that expression realizes thought and that language builds up a stable matrix of "sedimented" relations Merleau-Ponty hints at how words are the bearer—and in this sense the vehicle—of meaning. Furthermore, in *SdC* he had appealed to the "systematic principle" underlying relations between signs to explain their mediative role in symbolic behavior. But his early attempts to explicate further how words are involved in the articulation of thought—what exactly their "efficacité *propre*" consists in—are less than convincing. This can be seen in a passage evincing *PdP*'s view of how syntactic and lexical forms bear their meaning. Arguing against the idea that linguistic signs are conventional, Merleau-Ponty picturesquely suggests that if one takes into account their "gestural sense," "words, vowels, and phonemes are so many ways of singing the world" (*PdP* 218). He expands on this with the hypothetical suggestion that, if it were possible to eliminate all "mechanical" phonetic rules, foreign-language influence, and grammatical rationalizations from a language, the result would be an "emotional essence," a "somewhat reduced system of expression" in which, say, "it is not arbitrary to call light 'light' if one calls night 'night'" (*PdP* 218). In the analogy with singing and language's supposed emotionally expressive core it is not difficult to discern echoes of Condillac and Rousseau's, and even Herder's, speculations concerning the origin of language.[20] Nonetheless, it is tempting to doubt the coherence of Merleau-Ponty's hypothetical construction and to object that if stripped of grammatical and phonetic regularities a language would not be a *system* of expression at all; further that by lacking systematicity it would also lack the means to be genuinely expressive.

Yet, even if Merleau-Ponty's hypothetical scenario is assumed to be coherent, it is difficult to see the claim that pairs of words, such as *lumière* and *nuit*, stand in some deep-seated *emotional* contrast as anything more than an article of faith. First, the fact that light and dark are distinguished and referred to with different general terms seems to be a perceptive or cognitive response to the world rather than an "emotional" one. And while Merleau-Ponty might have defended his claim in the case of light/dark—perhaps we have a deep-seated fear of darkness, night, and so on—it seems wrong to suggest that in general the distinctions marked by language are emotional rather than cognitive. Second, even if we suppose the *distinctions* marked by language to be shaped by something like Merleau-Ponty's existential meaning, and in this extended sense "emotional," this is surely of no relevance to the labels used to mark those distinctions. Yet Merleau-Ponty seems to be suggesting that the *words* themselves stand in a nonarbitrary emotional contrast. Whatever the attractions of this view might be, it is hardly a convincing characterization of linguistic representation.

Merleau-Ponty's motivation for making these claims can nonetheless be understood and is revealing. As his discussion of the relationship between "existential" and "conceptual" meaning makes clear, he is attempting to identify preconceptual aspects of language that would parallel his critique of objective thinking about perception. His aim, here as there, is to describe the "prepredicative" level of "operant intentionality" (*PdP* xiii). However, lacking the conceptual means to do this convincingly, in *PdP* he resorts to obscure and clearly inadequate metaphors—"singing the world," "emotional gesticulation" (*PdP* 219)—suggested by his overall somatocentric perspective and vague analogies with artistic expression. As a result *PdP* ultimately fails in its aspiration to articulate what it is for linguistic signs to be a "vehicle" of meaning.

The second basic problem with *PdP*'s view is that the assimilation of language to other bodily behaviors obscures its specific ontological character. This can be seen by considering a passage describing the persistence of linguistic entities (i.e., words) over time that compares "the presence ... of the words I know" with that of perceptible objects and perceptive horizons. For any given word to endure, Merleau-Ponty suggests, it "suffices that I possess its articulatory and acoustic essence as one of the modulations, one of the possible uses of my body" (*PdP* 210). However, since words are routinely longer lived than the "fleeting hold" (*prise glissante*; *PdP* 462) that mortal subjects have on them, this clearly cannot be the whole story. Merleau-Ponty recognizes this in criticizing Descartes's failure to consider the importance of acquired language—the "spoken cogito"—and in underlining the "power of language" to make things "exist" and to open up "new dimensions, new landscapes to thought." However, he then brushes aside the idea, as he puts it, that language "envelops us," suggesting that this would be to "forget half the truth" (*PdP* 460). Merleau-Ponty himself focuses one-sidedly on this supposedly forgotten half of the truth, that presupposed by explicit language use, which he enigmatically refers to as "a tacit cogito," "a silence of consciousness that envelops the speaking world." He characterizes this silence as "a motoric presence," such that a word's "generality is not that of the idea, but that of a style of behavior which my body 'understands'" (*PdP* 461–462).

Here a tension in Merleau-Ponty's account emerges. On the one hand, if the tacit cogito's motoric presence is individuated so as to correspond to individual bodies, then its existence is equally "fleeting" or transient and a more sophisticated account of linguistic entities' persistence is required—that is, of how the horizonal structure of language is distributed over many speakers and equally many tacit cogitos. On the other hand, there are some passages in which Merleau-Ponty stresses the distributed, or transindividual, character of language: "The sense of a word ... is above all the aspect it takes on in a human experience .... It is a meeting of the human and inhuman, it is like a behavior of the world" (*PdP* 462). If, as such passages seem to suggest, the tacit cogito's motoric presence is an aspect of the world-as-agent, it surely would provide a context sufficiently comprehensive to account for the horizonal structure of language. But the price would be to obscure and perhaps destroy the force of Merleau-Ponty's methodological and explanatory reliance on lived experience: for it would then be unclear in what relation the tacit cogito's motoric presence stands to that of individual bodies, and indeed in what sense the tacit cogito is a "cogito." This tension is part of a general problem with *PdP*'s exclusive focus on the body as a transcendental explanans. Presumably it is such tensions that eventually led Merleau-Ponty to seek an ontology more suited to his views and so to a decentered metaphysical idiom.[21] Yet it is particularly acute with regard to language. For the attempted reduction of linguistic entities to motoric presence is latently subjectivist, fails to acknowledge that the former's mode of persistence differs from that (say) of physical objects, and fails to distinguish linguistic sense from that of other embodied behaviors. These defects all indicate the ontological evanescence of language in *PdP*, which fails not only to characterize the linguistic horizon specifically, but even to acknowledge that language makes up a distinct kind of horizon with its own ontological import.

3 Creative and Established Expression

It might seem surprising that I have so far not focused on Merleau-Ponty's likening of linguistic to artistic expression, which is often thought to be the most important feature of his view of language. The use of artistic expression as a model is clearly important to Merleau-Ponty. It is already found in *PdP* (213), where painting and music are appealed to as uncontroversial examples of expression in which sense is inherently realized, and is central to later discussions of language, particularly his extensive discussions of literary expression and painting in *The Prose of the World* (*PdM*).

The focus on artistic expression is closely connected with what Merleau-Ponty sees as a basic distinction between two different modes of expression. In presenting this distinction, as applied to acts of speech (*parole*), Merleau-Ponty's terminology varies: he refers to "authentic" and "constituted" speech, speech that is "original" or "secondary," "speaking" or (already) "spoken," and finally "transcendental" or "empirical" speech. However, rather than hinting at a rich typology of different modes of speech, in each of these pairings the first term refers to acts in which "new sense" is generated as opposed to reusing previously available "sedimented" sense.[22] This tendency is perhaps at its clearest in the distinction between "original" speech, which Merleau-Ponty exalts as "the primordial function of expression," and "secondary" speech (*PdP* 446). Nonetheless, the message common to all these terminological variants is that a certain mode of language use is philosophically primary: to *say* something is, properly speaking, to say something *new*; that to speak "authentically" is to create new sense; that this is the "transcendental" mode of expression, or the paradigmatic act, in relation to which all speech must be understood. To avoid both terminological confusion and implicit acceptance of Merleau-Ponty's primacy claims, the following discussion will, however, refer to the two modes of speech distinguished by the use of old and new sense as *established* and *creative* expression respectively.[23]

Merleau-Ponty describes these two modes of expression as differing phenomenologically, reflecting his claim that familiar uses of language have a transparent feel. Whereas already "spoken language" "disappears before the sense of which it has become the bearer," "speaking language" is to lead one "from the signs to the sense"—that is, makes one aware of signs' mediation (*PdM* 17). To illustrate what this amounts to he concentrates on literary experience as a model. Although this phenomenon relies on the use of established words, Merleau-Ponty likens communally shared language to an "anonymous corporality" that conveys others to us only schematically or "in general."[24] At the same time in "speaking" speech it is susceptible to small modulations, a "coherent deformation" of established language, that Merleau-Ponty characterizes using the notion of "style."[25] In literary experience, according to Merleau-Ponty, we see that the words used by a writer, such as Stendhal, are to have been "subjected to a secret torsion" investing them with "new sense" (*PdM* 19). The "moment of expression," he says, occurs when the "book takes possession of the reader," who responds by acquiring, for example, "Stendhal's language" (*PdM* 20).

Now it might be disputed that this is a representative description of linguistic, or even literary, experience. However, whatever else the experience involves, perhaps it is plausible that in literature we are exposed to and become to some extent familiar with an author's idiolect. Yet, even if this is allowed, what is supposed to distinguish creative expression as philosophically primary? Merleau-Ponty sometimes relies on the idea of genetic priority, such that creative expression's primacy lies simply in that established uses of language must at some time have been initiated: first use is original, established use is derivative. Indeed, he often suggests that reusing linguistic expressions in a previously established sense is not really to say anything at all, and that really, truly, or properly ("speakingly") to say something means forging new meaning.[26] This sometimes translates into a pejorative attitude toward noncreative expression, reminiscent of Heidegger's treatment of routine language use (*Gerede*), with Merleau-Ponty claiming that the previously "constituted speech" of "everyday life" presupposes "expression's decisive step," and that established ways of speaking are "banal," "already formed meanings" that "excite in us only secondary thoughts" (*PdP* 214).

Merleau-Ponty also relies on a more important, transcendental line of thought, which I will call the *new sense argument*. Thus he assumes (quite reasonably) that unless the phenomenon of communication is to be an illusion, speakers must be able to learn something new from utterances. Explaining this fact without appealing to an antecedently constituted realm of meanings such as transcendental subjectivity or a "pure language" requires, in Merleau-Ponty's view, an account of how "new sense" is constituted (*PdP* 208; *PdM* 12–13). However, because the conditions of intelligibility of speech are effectively the same from the first- and third-personal perspectives—literary experience reminds us that "to speak to and to be spoken to" is "the same thing" (*PdM* 197)—the new sense argument must be taken to apply not only to communication, but also to the expression of one's own thoughts.[27] Reflecting this thought, that both communication and thinking rely on creative expression, Merleau-Ponty proposes that "the fundamental fact of expression" is that the signifier makes possible reference to a signified that "exceeds" or "transcends" it.[28] Accordingly, he describes speech as the "paradoxical operation" of using "already available meanings" to realize "an intention that on principle goes beyond and modifies, in the final analysis itself fixes, the sense of the words by which it is translated" (*PdP* 445–446).

As arguments for the philosophical primacy of creative expression these are unconvincing. First, the importance of genetic priority is counterbalanced in the—integrating, sedimentary—case of language by the fact, as Merleau-Ponty freely concedes, that creative expression itself relies on taking up "already available meanings, the result of previous acts of expression."[29] However, if established language is just as essential to new sense constitution as vice versa, their relationship is clearly one of interdependence rather than one-sided priority. In addition, the new sense argument, as Merleau-Ponty presents it, is faulted. To begin with, it trades on an ambiguity as to whether learning "something new" means new facts or new ways of saying things. Yet once disambiguated it is obvious that learning new facts does not entail learning new ways of using words, leaving it far from clear that new sense constitution is needed to explain, say, the phenomenon of communication.

Despite these arguments' inadequacies, it is true that Merleau-Ponty's overall position structurally requires an account of new sense constitution, because his rejection of final determinacy commits him to explicating how determinate meaning is generated in actual experience. This structural need, which both *SdC* and *PdP* seek to address, can be seen to underlie Merleau-Ponty's emphasis on both the inherent realization of sense in expression and creative expression. Indeed, it also explains some of the more obscure features of Merleau-Ponty's views, such as his interest in the emergence of the "first word" from the "silence" underlying language—again echoing the Condillac-Rousseau-Herder tradition's speculations about the origin of language—and his somewhat mischievous suggestion that philosophical reflection on language should foster awareness of the "paradox of expression" and the "mystery" of language.[30]

From the viewpoint of exegetic accuracy, it is clearly important to recognize that Merleau-Ponty focused on creative/artistic expression and to examine his reasons for doing so. However, it seems to me that he provides no good reason to think of creative expression as philosophically primary. This is not to deny that his discussions of creative expression are of interest. But, rather than thinking of them as identifying a transcendental, original, or authentic function of language, we should understand them simply as shedding light on specific kinds of language use. To understand how they do this and how they can be used to explicate the notion of presentational sense, it is necessary to consider the distinction between "direct" and "indirect"—or "lateral"—sense that Merleau-Ponty links with established and creative expression respectively.

4 The Aspectual Presence of Language

The idea of "indirect" sense is central to Merleau-Ponty's discussions of language after *PdP*. It is introduced as contrasting with the "direct sense" of "already acquired expressions," which "corresponds point-for-point to the turns, the forms, of instituted words." It is the "sense of expressions in the course of making themselves [*en train de se faire*]," resulting "from the commerce of words themselves," that is "lateral" or "indirect" (*PdM* 64–65; cf. *S* 53). Although this direct-indirect distinction clearly corresponds to that between established and creative expression, what exactly is meant by "indirect" sense is difficult to pin down. This is perhaps partly because some of the relevant texts were incomplete and only published posthumously, but it is also because Merleau-Ponty clearly uses the label to mark several distinct theses. Thus the "indirectness" of linguistic meaning variously refers to the fact that signs play a role in mediating awareness of the world, alludes to the Saussurean emphasis on differences between signs, including its holism, and hints at the unspoken, implicit content of linguistic acts. The next chapter will have the task of exploring the positive aspects of this view in some detail. However, to understand the demarcation intended by the label "indirect," it will be helpful to begin by focusing on what he is rejecting—that is, by asking: What does Merleau-Ponty mean by "direct" sense?

This question is again complicated by equivocation in Merleau-Ponty's use of the term *direct*. In the initial definition just quoted, *directness* simply denotes conformity to established or "instituted" patterns of language use. Merleau-Ponty takes such conformity to characterize both everyday language use and mathematical or "algorithmic" thinking (*PdP* 214; *PdM* 180). In contrast to what he considers the "fundamental fact of expression," both of these are "secondary" phenomena because they fail to exploit the expressive potential of terms and leave signs functioning as "simple indices of univocal thought" (*PdP* 446). Nevertheless, despite attempting to downplay its importance for a philosophical conception of language, Merleau-Ponty clearly accepts that linguistic sense can be "direct" in this way.

The same cannot be said of a second, highly consequential, demarcation that Merleau-Ponty refers to with the term *direct*. This concerns the implications of the paradigm of perception, specifically Merleau-Ponty's rejection of final determinacy, on the mode of givenness of intentional objects.[31]

The intended demarcation is most succinctly expressed in the statement that all perception is "indirect or even inverted in relation to an ideal of adequation that it presumes, but that it does not look at face on" (Merleau-Ponty 1968, 12). In this context the term *adequation* does not mean (mere) adequacy, or just-about-enough-ness, but refers to the *ideal* of adequation defined by Husserl as the sum total of all possible experiential "evidences" in relation to the perceived object.[32] By thus adding or integrating over all possible perceptions, this ideal is supposed to eliminate perspectival properties and become something like the "full" (or "absolute") presence of the object. In Merleau-Ponty's view, remaining true to lived experience means rejecting this kind of idealization, so that perception "can only be grasped via certain of its parts or certain of its aspects" (Merleau-Ponty 1996a, 49). And while this clearly echoes Husserl's analysis of perception in terms of "adumbrations" (*Abschattungen*), Merleau-Ponty differs decisively in seeing no role for a fully present and fully determinate object as its telos. Instead, he attempts to conceive of the functioning of language without reference to such an ideal, underlining that "it is necessary to see ... that [the act of] meaning [*la signification*] does not transcend the *factual presence* of signs" (*PdM* 148–149). Merleau-Ponty's approach therefore takes a "*decentering as the foundation of sense*," such that "the thing is not frontal," but "given in an indirect grasp" (*PdM* 63n). This approach to language mirrors his view of the "perceived thing" as "an open totality," "an indefinite number of perspective views that match up according to a certain style, a style that defines the object concerned" (Merleau-Ponty 1996a, 49). Indeed, such emphasis on the openness, or ongoing processuality, of intentional phenomena is a hallmark of Merleau-Ponty's thinking overall: perception is an "open field," human experience an "open totality whose synthesis cannot be achieved," the world even an "open unity" or "incomplete work."[33] The second sense of *indirectness* is therefore to preclude reliance on the idea of idealized "fully present" intentional objects.

It might be objected that Merleau-Ponty's analysis of perception is faulted and so cannot be generalized in the way just outlined. After all, in the case of perception of a physically present object the goal of adequation seems underwritten by the existence of a material substrate. So although an object is only ever perceived perspectivally (in "adumbrations"), it seems reasonable, necessary even, to posit a kind of direct (physical) presence that underwrites the idea of adequation. Hence one might object that Merleau-Ponty's view fails to acknowledge a key feature of perception: that it is a *pars pro toto* operation in which experienced adumbrations relate to a determinate whole.

Whatever the case may be with the perception of physical objects, there is a disanalogy that, I suggest, makes his conclusions more convincing in the case of language. First, it is not clear what could correspond here to the physical object as a determinate whole, given that the intentional objects referenced by language (unlike perception) might be fictive. Moreover, even if it were assumed—in platonist manner—that some equivalent of the physical object could be appealed to, there is no reason to suppose this would be relevant to linguistic representation. For while it might appear plausible to suggest that sensory perception aspires to represent its objects in rich detail, this is clearly not the aim in the case of language. Hence, whether or not one agrees with his analysis of sensory perception, Merleau-Ponty seems right to highlight that the intentional objects of linguistic discourse cannot generally be thought of as having a direct, fully determinate, presence in the world. This has the profound consequence that on Merleau-Ponty's view there is no such thing as a fully "adequate" saying of what is meant. Or rather, the term *adequate* no longer makes sense: because the presence of thoughts in the world is realized in linguistic expression, the implication is that thought itself is constitutively indirect, aspectual, or perspectival.[34] Fully representing "content" in language is impossible—"expression is never total"—precisely because there is no such thing as full content.[35]

Many of Merleau-Ponty's claims about language respond to the need to reinterpret the "representational" function of language entailed by this repudiation of direct statement of thoughts. One example is his rejection of the idea of a language "of things" themselves (cf. *PdM* 7–8, 92–93). The thought behind this is presumably that a full direct presentation of objects *would* make sense of language's accountability to its objects, and enable it "in principle" to represent them fully. But Merleau-Ponty's disavowal of the adequation ideal implies that there is no aperspectival object of accountability, so that *"the very idea of adequate expression, that of a signifier that comes to cover exactly the signified"* becomes "inconsistent" (*PdM* 42). Linguistically expressed thoughts, on this model, constitute not a presence but an absence, a "determinate void," such that "the meanings in speech are always ideas in the Kantian sense, the poles of a certain number of convergent acts of expression that magnetize discourse without for their part being properly given" (*S* 112). As a consequence, precision of expression must be understood without recourse to an ideal standard. Linguistic expression can no longer be thought of as gravitating toward unequivocally determinate thoughts or structures. Instead it serves to narrow down indeterminacy, constraining possible interpretations sufficiently for equivocation, to all intents and purposes, to be eliminated: "the very idea of *accomplished expression* is chimerical: what we call [accomplished expression] is successful communication"; "all expression is perfect to the degree that it is *understood* unequivocally."[36] Thus the ideal of fully determinate presence of intentional objects yields to the pragmatic criterion of successful communication as a standard of univocity.

Merleau-Ponty seems to have thought of the two senses of directness distinguished here as connected. In cases "where it *seems* to us that what is expressed is itself attained ... it is simply that the gesture is habitual, that our take up is immediate, and that it does not demand from us any reorganization of our ordinary operations" (*PdM* 43; italics added). It is fairly simple to see how this association might arise: when talking in established ways communication generally succeeds, so that it can seem that we are somehow speaking the language of things themselves.[37] Yet even if they suggest one another in this way, it is important to be clear that the two notions of directness are quite distinct: one concerns the relation to established regularities, the other to an idealized presence of thoughts. At the same time, there is a common feature in that both hint at a standard of determinacy, or univocity. In the first case direct sense appears to be unequivocally laid out in ("corresponds point for point to") established patterns of language use in which there are no "gaps" (*lacunes*); in the second case direct sense lies in conformity to the fully determinate presence of (intentional) objects—an "eidetics of language ... of univocal relations susceptible, in their structure as in their functioning, to total explication" (*S* 105–106). Despite their differences, then, in both cases "directness" also implies a lack of equivocation or ambiguity. From this it becomes clear that Merleau-Ponty's talk of "indirectness" has the further point of signaling the absence of an objective standard of determinacy in language use.[38]

In summary, then, Merleau-Ponty's talk of the "indirectness" of linguistic sense stands for three (independent) theses, each marking the demarcation from a certain sense of "directness": "indirect" sense is not explicable (a) in terms of established rules of language use; (b) in terms of an ideal of representational adequation; (c) in relation to a standard of univocity.

5 The Heideggerian Framework Revisited

Before moving on to examine the positive aspects of indirect sense in more detail, I want to highlight briefly how the views set out in this chapter fit into or augment the Heideggerian framework's general picture of language. Heidegger's and Merleau-Ponty's approaches to language obviously have a lot in common. As outlined in the introduction, both take concrete, finite agency—"In-der-Welt-sein" and "être-au-monde" respectively—and lived experience as their starting point, and in so doing depart from Husserl's emphasis on pure experience and transcendental subjectivity.

One reason for potentially questioning their overall compatibility is that Merleau-Ponty's position is presented as a theory of "expression." As such it might be thought to be concerned primarily with the externalization of emotions or other subjective states and so to conflict with Heidegger's talk of "disclosure" of the world. However, despite *PdP*'s emphasis on "emotional" and bodily significance, Merleau-Ponty always conceived of expression as a complex function, occurring at the subject-object interface of behavior and shaped by the "system 'me-others-things'" (*PdP* 69; cf. *PdM* 95). So although this emerges clearly only in later attempts to develop a decentered ontological picture, it is *at least* an *implication* of Merleau-Ponty's approach—from *SdC* onward—that expression has multiple inputs and is not specifically concerned with the externalization of subjective states.[39]

Another concern one might have, suggested by certain uses of the term in aesthetics, is that Merleau-Ponty's talk of "expression" marks a contrast with representation, cognition, or the literal. This concern is dispelled by recalling the emphasis he puts on linguistic expression as the realization of thought.[40] Understood in this way, Merleau-Ponty's view differs not only from the simplistic view of expression as flowing from a person's "feelings or his character," but also from more refined views that distinguish expressive properties, say, as "those properties of artworks (or natural objects) whose names also designate intentional states of persons," as contrasting with a work's "representational properties," or as "metaphorically exemplified" in contrast to those properties (e.g., consisting of certain colors, shapes, etc.) a work literally exemplifies.[41] For insofar as linguistic expression effects the *formulation* or embodiment of thoughts, it obviously cannot be opposed to the cognitively relevant and hence precedes distinctions between the literal and the metaphorical, or representation and expression. Merleau-Ponty's conception of expression is thus of a very general kind, corresponding to that found in Dilthey and underlying Heidegger's formal indication (cf. chapter 3, section 2). The concern in all these cases is with what Charles Taylor (1985, 238) describes as "expressive power, the power to make things manifest," where "what is made manifest is not exclusively, not even mainly, the self, but a world."

Apart from the terminological difference, there therefore seems to be no significant barrier to assimilating Merleau-Ponty's conception of expression to Heidegger's notion of disclosure and to his overall view of language as language-in-the-world. A further convergence worth noting is that Merleau-Ponty's rejection of final determinacy and his corresponding emphasis on the generation of sense means that, like Heidegger, he conceives of language as a process rather than a static or atemporal structure. Merleau-Ponty also clearly shares Heidegger's antireductionist and antiformalist attitude, as can be seen, for example, in his emphasis on the need to understand how conceptual meaning is founded or in his association of algorithms with direct sense. "Formalization," as Merleau-Ponty puts it, is "always retrospective" and "only ever appears complete" while remaining dependent on underlying processes of "intuitive thinking" (*PdP* 441; cf. 118, 145ff.).

In addition to these general similarities in outlook, Merleau-Ponty and Heidegger both see predicative awareness as founded in something prepredicative on which their respective analyses focus: just as present-at-hand Things are founded in ready-to-hand Equipment for Heidegger, for Merleau-Ponty conceptual meaning is founded in existential meaning. This structural similarity comes to the fore in several features of their respective positions. For example, Merleau-Ponty's description of the grasp we have of word gestures as spontaneous, but nonintellectual and context specific closely parallels Heidegger's characterization of our purposive grasp of Equipment. Similarly, the contrast between the concrete and categorial attitudes, or between detachable and symbolic forms of behavior, in Merleau-Ponty's discussion parallels that between purposive and predicative awareness—or the hermeneutic *as* and the apophantic *as*—underlying Heidegger's discussion of Statements. Schneider's inability to connect the two attitudes fluently in his behavior thus graphically illustrates the need for both attitudes in everyday life—underlining my earlier emphasis on this need in interpreting Heidegger's ambivalence about routine language use (see chapter 3, section 1).

Against the backdrop of this compatibility in approach, Merleau-Ponty's reliance on the model of perception brings two important refinements to the previously outlined Heideggerian picture. First, Merleau-Ponty makes clear that embodiment is an essential aspect of finite agency, reminding us, so to speak, that Dasein is flesh and blood. Second, his repudiation of final determinacy leads to a view of intentional objects as indirect presences and constitutively open horizons unified by a certain style, rather than determinate full presences. The distinctiveness of this view can be highlighted by contrasting it with Heidegger's conception of formal indication. To recall, Heidegger had conceived of linguistic signs as indeterminate/determinate pointers ("indications") to the performance of originary understanding in acts of "grounding experience" (see chapter 3, section 2). On this picture, apparently as with Merleau-Ponty, linguistic signs function by gesturing at or toward thoughts rather than fully embodying them, by intimating rather than stating outright. There is a fundamental difference, however, in that Merleau-Ponty rejects the idea of any such fully and finally determinate presentation of sense in "grounding experience." So whereas Heidegger conceives of formal indication as constitutively inadequate gesturing toward original understanding, Merleau-Ponty rejects the very idea of adequation and accepts that the actual reality of understanding is perspectival, partial, or "indirect." True to this commitment, as will emerge in the following, Merleau-Ponty's conception of language provides a way of understanding the presentational function of language freed of reliance on the idea of a full and original "presence of sense" that Derrida critically describes as "the foundational concept of phenomenology as metaphysics" (Derrida 1993, 3, 111).

5 The Art and Science of Indirect Sense

In the last chapter I outlined Merleau-Ponty's approach to language in *PdP* and attempted to show that it is both broadly compatible with, and in some respects complements, the general Heideggerian picture of language set out in chapter 1. However, the Heideggerian framework set out in part I of this book also incorporates the more specific proposal that the disclosive function of linguistic signs relies on two different kinds of sense, which I labeled presentational sense and pragmatic sense. In this chapter the aim will be to show how Merleau-Ponty's views about the function of linguistic signs can be used to develop the Heideggerian framework at this more specific level by explicating the notion of presentational sense.

The basic idea I want to develop is that Merleau-Ponty's discussions of "indirect sense" do this in three ways. First, they provide the means for understanding what kind of form underlies the presentational feats of language, and so define the *structure* of presentational sense. Second, they provide a model of how embodied agents exploit such form to point out features of the world, and so define the *function* of presentational sense. Third, by focusing on various features of both language and painting they identify a *mode of intelligibility* that characterizes presentational sense in its attunement to embodied finite agency. Developing these thoughts requires a shift in exegetic focus to works from the early 1950s—particularly *The Prose of the World* (*PdM*)—in which Merleau-Ponty assimilated Saussurean ideas into his general view of language as an expressive behavior of embodied agents. In keeping with his treatment of creative expression as a paradigm, Merleau-Ponty thought of indirect sense as having a privileged role in understanding all language use. However, having argued in the last chapter that he overestimates the importance of creative expression, my more limited claim here is that his discussions of indirect sense provide a partial account of the disclosive function of linguistic signs, offering valuable insight into the features of language that are relied on in certain kinds of situation (e.g., creative language use) but not others (for many of which the notion of pragmatic sense will be required).

An interesting feature of Merleau-Ponty's conception of indirect sense is that it draws on relevant considerations from both science and art. This chapter will look at each in turn. The first two sections focus on Merleau-Ponty's appropriation of Saussurean linguistics. The first clarifies how the structure of indirect sense should be understood, and how this solves the two problems previously identified with his earlier views, while the second examines the mode of intelligibility or rational character of indirect sense that is suggested by Saussure's conception of signs. The following three sections discuss Merleau-Ponty's analogy with modern painting, showing how this analogy sheds light on the presentational function of indirect sense and how it converges with Saussurean views. Thus the third section of the chapter discusses how painted forms serve to present or point out features of the world in an articulate but allusive manner, the fourth considers how the activity of painting provides a model of embodied deliberation, while the fifth considers how the notion of style connects painting and language. The sixth and final section sets out how Merleau-Ponty's indirect sense fills out and improves Heidegger's view of the disclosive function of linguistic signs in a phenomenologically convincing way.

1 The Differential Structure of Indirect Sense

Merleau-Ponty's reliance on the idea of "indirect" sense coincided with important developments in his views on language due to his adoption of Saussurean ideas from 1947 onward.[1] Accordingly, it is to Saussure's conception of linguistic signs that one must look to understand Merleau-Ponty's enigmatic claims that "language expresses as much through what is *between* words as through words themselves, and through what it does not say [as much] as through what it says" (*PdM* 61–62).

Merleau-Ponty's interest in Saussure, as economically stated in his 1953 inaugural lecture "In Praise of Philosophy," centers on the idea that his conception of linguistic signs suggests "a theory of historical sense" and hence "a new philosophy of history" (Merleau-Ponty 1960, 56). The problem Merleau-Ponty saw for such a theory is to reconcile the roles of contingency and dialectic necessity in historical development, conflicting demands that are exemplified in the "paradox" that the history of language is "made of too much chance to admit a logical development," yet "produces nothing that is not motivated" (*PdM* 32). Saussure's virtue, in Merleau-Ponty's view, was to offer an account of the "living present" or the "synchronic sense" of language freed from etymological considerations, while reconciling the influences of reason and chance, or necessity and contingency, in "a new conception of reason."[2] Merleau-Ponty interpreted these moves as a turn to "speech [*la parole*], to experienced language" and as such took them to constitute a phenomenological view (*PdM* 36; cf. Merleau-Ponty 1996b, 107).

Before going any further, several exegetic difficulties should be mentioned briefly. First, although there are controversies about the posthumous editing of the *Course in General Linguistics*, these can be neglected for the present purposes and, since this was the source available to Merleau-Ponty, Saussure's views taken to be those of the published text.[3] Even so, there are notorious difficulties in reconciling Merleau-Ponty's presentation of Saussure with the Saussure of that text.[4] Foremost among these, and potentially bearing on the compatibility of the two views, are Merleau-Ponty's references to Saussure's key distinction between *parole* and *langue*. For example, in his 1951 lecture "On the Phenomenology of Language" Merleau-Ponty suggests that Saussure distinguishes "synchronic linguistics" of *parole* from "diachronic linguistics" of *langue* (*S* 107; cf. also *PdM* 33). As it stands, this seems straightforwardly mistaken, since Saussure's own distinction was between *parole* as "the execution" of speech by individuals, and *langue* as the conventional, social, and above all *synchronic* system of language (*langage*).[5] Furthermore, it might seem that, due to his own emphasis on acts of speech (*parole*), Merleau-Ponty is oblivious to the fact that Saussure (1972, 36–37) thought *langue* the proper object of linguistic theory, with *parole* being systematically derivative.

Despite the impression given by such passages, it would be an oversimplification to think that Merleau-Ponty simply fails to understand (what is after all) a fairly basic distinction of Saussure's.[6] The discrepancies are, I suggest, primarily terminological, and arise because rather than straightforwardly adopting Saussure's standpoint, Merleau-Ponty is *incorporating* certain of Saussure's views into his own position. Moreover, given Merleau-Ponty's focus on "parole," this incorporation naturally creates the impression that Saussure's views are being misrepresented. In this respect, though it is surprising that Merleau-Ponty saw no need to clarify such terminological shifts, it is in fact of secondary importance whether he is being true to Saussure. Nonetheless, terminological issues aside, Merleau-Ponty's view of *parole* as the living present of language does respect key Saussurean motifs, and his talk of a "return to living spoken language" is clearly justified given Saussure's emphasis on synchronic meaning and the primacy of spoken language.[7] This fidelity, for what it is worth, should emerge more clearly below.

Returning to the focus that Merleau-Ponty found attractive in Saussure: What constitutes the "synchronic sense" of living speech? The view Merleau-Ponty develops in response to this question centers on a distinction between the "categories" of "official grammar" and an underlying "expressive system." Despite the fact that, in his use, "grammatical" rules attribute "meanings" (*significations*) to signs, Merleau-Ponty sees them as lacking explanatory importance (*PdM* 40). This is partly because he takes post-Saussurean linguistics to have established that "grammatical categories" are merely approximative, a "retrospective and inessential expression of our proper power to speak" (*PdM* 39, 38). The underlying thought, however, is that grammatical—that is, for Merleau-Ponty, both semantic and syntactic—categories are not autonomous, but are to be accounted for in terms of something more basic: "Before language bears its meanings," he explains, its "internal arrangement" must "secrete ... a certain original sense from which meanings are taken" (*PdM* 44). In keeping with this conviction, Merleau-Ponty's discussion focuses exclusively on the functioning of the underlying "expressive system," just as his views on perception focus on the preobjective.

It is in characterizing this "expressive system" that Merleau-Ponty relies on Saussurean ideas. To begin with, he identifies phonemes and morphemes as the elements of which words, and hence language, are comprised.[8] The internal organization and interaction of these *sublexical elements*—as I will call them—are to account for the "effective function of speech" (*PdM* 44), which Merleau-Ponty sees as a holistic process and describes using two Saussurean notions. First, he emphasizes above all the importance of differences: "each element of the 'verbal chain' ... only signifies its difference with regard to the others"; "The most exact characteristic of a word is to be 'what the others are not'" (*S* 110; Merleau-Ponty 2001, 83). Indeed, Merleau-Ponty goes so far as to suggest that "in language there are only differences in meaning [*signification*]."[9] Second, Merleau-Ponty adopts Saussure's idea of value, describing language as a "collection of linguistic gestures ..., each of which is defined less by a meaning than by a use value"; similarly, all morphemes are "'linguistic tools' that have less a meaning than a *use value*."[10]

Although Merleau-Ponty never discusses this in detail, it is worth reviewing more precisely the role these two notions—difference and value—play in Saussure's conception of linguistic signs in order to understand properly the implications of the view Merleau-Ponty is proposing. Saussure (1972, 99) defines a sign as a whole formed by the pairing of "signifier" and "signified." This is not to be mistaken for a pairing of sign and referent, or any similar modeling of linguistic signs—which Saussure calls "nomenclature"—on the operation of naming independently individuated objects or ideas. Indeed, his very point in introducing this terminology was to avoid confusion with views that picture linguistic signs as the pairing of antecedently individuated—conceptual and acoustic—elements. The relationship Saussure intended is far more intimate, with signifier and signified comprising two aspects, "two faces," of the linguistic sign as a "psychic entity" (Saussure 1972, 99). Saussure offers several ingenious illustrations of this intimacy—for example, language is to "serve as the intermediary between thought and sound"; it resembles a sheet of paper, the two sides of which are thought and sound; linguistic signs are like chemical compounds, such as H_2O, which are formed from structured elements but acquire new properties in combination (Saussure 1972, 156, 157, 145). Perhaps the most challenging feature of this picture is that the two aspects are supposed to individuate one another, without a basis of primitive or "atomic" units. Thought and sounds, as Saussure (1972, 155, 157) puts it, are simply "amorphous and indistinct" masses, such that each acquires discrete form only through their "combination" in linguistic signs. It is "their union" that "leads necessarily to the reciprocal delimitations of units" (Saussure 1972, 156), so that signs are, so to speak, the intersections to which the articulation of both sound and thought are due.

It is in describing the signifier-signified relationship that Saussure deploys his technical notions of difference and value. With regard to the structure *within* each of the sign's conceptual and material (phonetic, morphemic) aspects he speaks of "differences"; to discuss relationships *between* two (such) "orders," as he calls them, Saussure instead talks of "value." The characteristic feature, on this view, of something with value is therefore to stand in two kinds of relationship: to things "similar," of the same kind ("order"), with which it can be "compared"; and to things "dissimilar," or differing in kind, with which it can be "exchanged." Thus in the case of money, which Saussure uses to illustrate his view, a $10 bill can be exchanged either for other bills or coins (similar) or for goods (dissimilar) to the same value. Analogously, he is suggesting, a word can be "compared" with other words (similar) or exchanged for an "idea" (dissimilar).[11]

To appreciate the point of distinguishing difference and value, it is helpful to consider another important, though often neglected, distinction made by Saussure. This immediately follows a passage that is prone to misuse in which Saussure (1972, 166) states "in language [*langue*] there are only differences." Crucially, however, in saying this he is talking about "all that precedes," about signifier and signified respectively, so as to distinguish these from the sign as a whole. Nonetheless, with "the sign in its totality," as a signifier-signified pairing, "one finds oneself in the presence of a thing [that is] positive in its order," and in relation to this, Saussure states, "one cannot speak of difference." Signs, he explains, are characterized not by "difference," but by "distinctness" and "opposition."[12] The key phrase here is "in its order": as a "positive," Saussure is claiming, the sign is something that can be considered discrete, delimited, or individuated without reference to something different in kind. In this it differs structurally from the aspects of signifier and signified, which, Saussure has argued, are not composed of independently articulated units.

At first glance this might look like mere terminological posturing. It might, for example, be objected that where there are differences there is identity, or discrete units. I think this rather "intuitive" objection is clearly wrong, as is shown by looking at a map of any coastline: here we find a form that undeniably has structure, shape, yet with no natural "units" of which it is comprised. In Saussure's terms the coastline's features are "different" but not "distinct."[13] Although both terms imply contrast, they do not denote the same mode of structural contrast: distinctness is a relation between individuated units, whereas difference can lie on a continuum. Thus whereas coins are distinct, the prodigious range of sounds the human vocal system is capable of producing or an autumn forest's rich spectacle of colors are examples of differential structure.

Alternatively, it might be thought Saussure is suggesting that the prior units of which linguistic signs are composed are morphemes and sememes—that is, the minimal syntactic features that make a semantic difference and (putatively) basic semantic elements. Yet this is precisely the kind of suggestion against which Saussure's analysis is directed, as expressed in his description of language as having no "simples," as being "an algebra possessing only complex terms" or a "form" rather than a "substance" (Saussure 1972, 168–169; cf. 157). The point of these descriptions is precisely that the constitution of sublexical structures is such that they cannot function as "units" "prior to" the sign system. One reason for this is that they lack the right kind of determinacy—namely, discrete individuation or "distinctness" in their own "order"—to be considered "units." Instead, on Saussure's view, morphemes and sememes are semideterminate structures, representing (orthogonal) cross-sections through the system of signs, such that in focusing on one dimension (e.g., syntactic), the other (semantic) loses all definition.[14] A second reason is that sublexical structures do not precede the sign system. As Saussure (1972, 157) puts it, one must start with the "interdependent whole" in order to "obtain by analysis the elements that it contains." His point is not merely epistemological or methodological, but rather that "elements" such as morphemes or phonemes would not exist but for the role they play in a system of constituted signs.

The purpose of Saussure's terminology is thus to characterize precisely qualitative structural differences between signs and their constituent aspects (i.e., signifiers and signifieds). The result is a hierarchy of structural levels defined in terms of the notions of difference, value, and distinctness respectively. At the lowest level are two "orders," referred to by Saussure as those of "acoustic images" and "ideas," characterized by inhomogeneity, variation, or "difference," but not by individuated units. Difference is discernible, but not in systematic arrangement. At the next level are structures, such as Saussure's signifier and signified, characterized by "value." Value concerns the relationship between two structural axes (or "orders"). Since, by definition, it is a principle of equivalence that regulates the exchangeability or comparison of all relevant structures, value is necessarily holistic. Morphemes and sememes, though not full-fledged signs in Saussure's sense, are good examples of structures with value. Simultaneous semantic and syntactic focus comes only at the level of what Saussure calls "signs." Signs are discrete units, individuable as such (i.e., in their own "order"), which stand in relations of "distinctness" to one another and bear "characteristic properties" (*caractères*; Saussure 1972, 168). Saussure's scheme of concepts can thus be summarized as in table 5.1.

One final point requires emphasis: Although these features are structurally distinguishable, serving to identify particular aspects and relationships of the organization embodied by a system of signs, Saussure does not consider them independent. Rather, as he puts it, value is something

Table 5.1
Saussure's structural distinctions

Structural characteristic	Examples	Distinct units ("positives")?	Holistic?
Difference	Acoustic images Ideas	No	No
Value	Signifier, signified Morphemes	No	Yes
Distinctness	Signs Words	Yes	Yes

"emanating from the system," "it is difference that makes the character [of signs], as it makes [their] value and unity" (Saussure 1972, 162, 168). It is for this reason that Saussure might be—and often is—interpreted as providing a reductive account of the functioning of signs in terms of difference.

Against this background, and notwithstanding any terminological laxness in some of Merleau-Ponty's references to Saussure, it is not difficult to discern both convergence and complementarity in their views. Clearly, in addition to their (as Merleau-Ponty supposed) shared interest in the "living present" of language, both share the view that thought is realized in language. There is also an important parallel between Saussure's rejection of antecedent simples, the view that language is a "form" rather than a "substance," and Merleau-Ponty's view that expressed thoughts are a "determinate void" (*S* 112). In this respect both seek to account for the functioning of language without recourse to language-independent entities, the presence of which might originate or ground linguistic meaning.

Most striking, however, is the structural parallel between Saussure's positive and differential levels and Merleau-Ponty's interest in finding a foundation for objective or conceptual awareness.[15] This parallel allows Merleau-Ponty to see Saussure as providing an approach to language corresponding to his own treatment of perception, linked by the claim that differentiation is a process underlying distinct units: just as gestalts are preobjective structurations in the realm of perception, so too is linguistic differentiation preobjective and preconceptual. Conversely, Merleau-Ponty's gestalt-psychological approach to the formation of categories can be seen as bolstering a weak point in Saussure's approach. For Saussure relies on the thought that both the phonetic and conceptual domains are fundamentally differential. Yet whereas it is reasonably clear how this applies to the linear "chain" of phonetic "acoustic images," Saussure does not explain how the conceptual domain, or as he also calls it the conceptual "chain," is to be thought of as a process of differentiation (Saussure 1972, 146). It is this gap that Merleau-Ponty's view of gestalt formation—the "differentiation of excitations that appears to be the essential function of the nervous system" (*PdP* 88)—can at least claim to be filling.

Given these parallels Merleau-Ponty was able to integrate Saussure's diacritical conception of signs quite smoothly within the framework of his earlier views. Thus he continues to treat language as gestural, links Saussurean differentiation with his earlier metaphors of the melody or style of articulation, and re-rehearses his argument that language is underlain by an original language of expressive contrasts—except that whereas these were formerly "emotional," they now comprise a system of "coherent motivations."[16] Similarly, as in *PdP* our grasp of language is still to be bodily, gestural, preintellectual, preobjective, and nonconceptual. It belongs to the "perceptive order," is "quasi-sensory," is something we "sense [*sentir*]."[17] But his characterization of this basic linguistic sensitivity now takes on a Saussurean flavor: "speaking is not having a certain number of signs at one's disposal, but possessing the language as a principle of distinction" (*PdM* 46).

At the same time, the appropriation of Saussure's ideas brings significant changes that address the two basic problems with *PdP*'s position identified at the end of the second section of chapter 4. One of these problems was that Merleau-Ponty's earlier view seemed unable to account properly for the ontological character of language. The subsequent improvement is signaled by the fact that Saussure's conception of signs is not straightforwardly, if at all, amenable to somatocentric interpretation. Whereas in *PdP* Merleau-Ponty had construed the tacit underpinnings of language simply as a "tacit cogito," or a "motoric presence," this "silence" now encompasses both a somatic and a linguistic dimension, or "horizon." As a result he now claims that phenomenology of language has an "ontological bearing," and somewhat awkwardly concludes that language itself is "something like a being."[18] In other words, Merleau-Ponty now recognizes that language has a temporally extended, materially anchored manner of existing that is not straightforwardly correlated with individual bodies. Accordingly, *PdM* draws on *two* sciences, psychology *and linguistics*, to show how one "can renounce timeless philosophy without falling into irrationalism."[19] The appropriation of Saussure's views thus brings what might be described as Merleau-Ponty's personal version of the linguistic turn, the recognition that the being of language—as an intersubjective, sedimented acquisition—cannot be straightforwardly assimilated to the behavior of a single embodied cogito but requires a "new conception of the being of language" (*S* 110).

The second problem with Merleau-Ponty's earlier position—bearing directly on the disclosive function of linguistic signs—was its inadequate characterizations of what it is for signs to be the "vehicle" of sense and thought constitution—that is, the "efficacité propre" of signs. The diacritical conception of signs provides Merleau-Ponty with a much more satisfactory approach to this issue. In Saussurean terms the functioning of the "expressive system" is now explained as the interaction of phonetic-morphemic and conceptual differentiation, and it is to this that signifier-signified pairings, and hence linguistic "categories," are to be due. This provides Merleau-Ponty with an articulate account of the constitution of conceptual meaning or categories in terms of "a more original differentiation activity ... an inexhaustible power to differentiate one linguistic gesture from another" (*PdM* 47). Language, as Merleau Ponty now puts it, becomes "a profusion of gestures, all occupied with differentiating themselves from each other and intersecting ... this sublinguistic life, the entire industry of which is to differentiate and systematize signs" (*PdM* 161). Thus Saussure provides Merleau-Ponty with a way of detailing the specifically linguistic structurations that underlie and shape the conceptual meaning of linguistic expressions. This means that, rather than relying on misleading claims about emotionality or vague metaphors of singing the world, Merleau-Ponty is now able to draw on an articulate and sober (indeed scientifically respectable) conception of the prepredicative workings of language.

2 The Inchoate Rationality of Indirect Sense

Another feature Merleau-Ponty draws from Saussure's conception of signs is to conceive linguistic form as imbued with a kind of sense, as having a specific mode of intelligibility that distinguishes it from the products of both mechanical causation and fully developed rationality. This feature is closely linked with Merleau-Ponty's view that Saussure's conception of signs leads to a "historical theory of sense" and a reconciliation of contingency and necessity in a "new conception of reason." The key to understanding these claims is found in the following passage: "chance is the basis of all the restructurations of language; in this sense one can say that language is the domain of the *relatively motivated*: nothing rational is found there that does not derive from taking up and elaborating some piece of chance by means of systematic expression by the community of speaking subjects" (Merleau-Ponty 2001, 85).

The expression "relatively motivated" again stems from Saussure, who had used it to mark a limitation of his thesis of the "arbitrariness" of the linguistic sign—that is, the thesis that there is no "natural link" between signifier and signified.[20] Notwithstanding this "fundamental principle of the arbitrariness of the sign," Saussure distinguished features of signs that are "relatively motivated" from those that are absolutely, or "radically arbitrary" (i.e., "unmotivated") (Saussure 1972, 180–181). As an example, he explains that in French whereas the use of the term *dix* to mean 10 or *neuf* to mean 9 is unmotivated, use of *dix-neuf* to mean 19 is relatively motivated and correspondingly less arbitrary. Thus relatively motivated features are what might be called intelligible features of signs, syntactic traits with conceptual implications, or as Saussure puts it they imply a "syntagmatic relation" and an "associative relation." In doing this, they effect a "limitation of the arbitrary" and are "nothing other than the mechanism in virtue of which any term whatever lends itself to the expression of an idea" (Saussure 1972, 182). This—for Saussure clearly essential—aspect of the functioning of language highlights that his thesis of the ultimate, absolute arbitrariness of the sign does not commit him to the claim that either the *distinctions* marked by linguistic signs or relatively motivated semiotic features themselves are arbitrary. Indeed, as Saussure (1972, 167) himself argues, on pain of becoming lost or confused "every difference" of ideas "perceived by the mind seeks to express itself by distinct signifiers."

Merleau-Ponty's point in the passage cited above is that relatively motivated features testify to the meeting of contingency and rational order in acts of speech. On the one hand, language deploys contingent (phonetic) material to do its intentional work (cf. Saussure 1972, 211ff.); yet, on the other hand, traces of such intentional use become inscribed ("sedimented") in the sublexical structure of linguistic signs. "What Saussure saw," Merleau-Ponty emphasizes, "is precisely this meshing of chance and order, this taking-up of the rational, [and] of chance" (Merleau-Ponty 2001, 86). This interweaving of contingency and reason, though already implicit in Saussure's pairing of phonetic signifier and conceptual signified, is epitomized by Merleau-Ponty's notion of *parole*. For it is acts of speech, understood to be both contingently conditioned and intentional, that mediate between chance and logic, yielding a "constituted" expressive system shot through with relatively motivated features. Acts of speech are, so to speak, the stitches that weave the fabric of historical and intersubjective sense.[21]

Two clarifications are required here. The first concerns the rational status of relative motivations. To take Saussure's example, it does not seem to be completely contingent that *dix-neuf* means 19 in French. This compound label, we might say, makes sense, but equally there is no compelling *reason* for adopting this label, rather (say) than a simple one such as *seize* for 16. It is precisely this intermediacy between causes and reason that Merleau-Ponty expresses with the term *motivation*: "One phenomenon triggers another, not by an objective efficacy, like that which links the events of nature, but through the sense it offers—there is a reason for being that orients the flow of phenomena without being explicitly posited in any of them."[22] Motives, on this view, are inchoately rational—that is, meaningful in a way that sets them apart from simple causation yet falls short of fully explicit patterns of reasoning that might be based on them. This inchoate rationality of motivation has two features. First, a motive serves as a basis for reasons that it cannot be considered to embody antecedently. This important asymmetry, which I will call *nonretrojectability*, means that whatever reasons we are led to find by motives cannot be projected back into the motivating structure itself. This thought becomes clearer and more plausible when one bears in mind, second, Merleau-Ponty's view that "*lived logic*" does not "possess the full determination of its objects" (*PdP* 61). Nonretrojectability is not simply a stipulation, but is due to the fact that motives are *semideterminate*: they constrain and guide the formulation of reasons, but in principle remain "ambiguous" (in the terminology of *PdP*) and could always motivate different reasons. Hence, on this view a motive, or a motivated feature, is something with sense that enters into intentional activity, and shapes or "orients" this without fully predetermining it.

It is in the technical sense just outlined, I suggest, that Merleau-Ponty adopts Saussure's talk of the "relatively motivated." His claim is that, despite their ultimately contingent phonetic basis, the sublexical structure of words conveys something about the way their referents are to be thought of, that relative to other features of the lexical system morphemes can signal aspects of similarity and difference. A simple example is provided by the four English words *telephone, microphone, telescope,* and *microscope,* which are clearly cognate and each tell us something about the function of their referents (e.g., hinting at the link with speaking or seeing at a distance). And whereas the various objects to which each term refers might have acquired a different label in the language, the established terms could not be interchanged without failing to make sense in a way particularly appropriate to their referents. It is this kind of loose connection between the structure of words and what they can be used to mean that makes language, as Merleau-Ponty puts it, "the domain of the relatively motivated." The form of individual expressions constrains, though it does not fully determine, what they can be used to say. So while morphemic features might suggest reasons for preferring one word over others, their relative character implies that not all such features are thus interpretable, and their motivational character means that such reasons as we find cannot be projected back into language—as its inherent underlying rationality or suchlike.

The second clarification concerns the scope of Merleau-Ponty's claims. It might look as though he has overstated the case in describing language, without further qualification, as the domain of relative motivations. Isn't

Saussure's point to distinguish some signs from others? Doesn't he imply that relatively motivated features are found only in "compound" words formed of several morphemes? Well, compound terms, such as *dix-neuf*, no doubt provide clear paradigms of the principle of relative motivation. Nevertheless, Saussure's own claim extends further. Relative motivation, he says, is the "mechanism in virtue of which *any term whatever* lends itself to the expression of an idea" (Saussure 1972, 182; italics added).

Two thoughts can help in understanding the broad scope of this thesis. The first is that words belong to an articulate system, and that as such each word means what it does only in relation to others, and ultimately to all others. In Saussure's terms this is reflected in the notion of value, with its concomitant holism, and the distinctness of linguistic signs from one another. So even where there is no manifest structure to words, they nonetheless form a distinct node within the network of phonetic and conceptual differences. As Merleau-Ponty puts it, the "expressive power of a sign depends on its being part of a system and coexisting with other signs"; accordingly, language is "a system of signs that only have sense relative to one another and of which each is recognized according to a certain use value that comes to it in the whole of the language" (*PdM* 52; Merleau-Ponty 1960, 56–57). The second thought is the Austinian idea that the mere existence of an established word should serve as prima facie evidence for its having a distinct function.[23] This idea is based on the evolutionary heuristic that words are more likely to establish themselves in language, rather than disappearing from circulation, if they are found to have a specific use. What both these thoughts assume is that there is some kind of reason for the existence of a particular word, as distinct from all others. Or rather, since this assumption is presumably defeasible—perhaps there are genuine synonyms and/or redundancy in use—and because any claim to identify "the" reason for any such differences is bound to look suspect, the intelligibility of words' distinctness is presumably somewhat weaker. This weaker mode of intelligibility, I am suggesting, can be understood, with Merleau-Ponty, by saying that the existence of a particular word is motivated relative to all others (i.e., is relatively motivated).

Before leaving the subject, it should be noted that there is one subtle but significant difference between Merleau-Ponty's and Saussure's view of signs. Saussure, despite describing them as "emanating" from the underlying patterns of phonetic and conceptual differentiation, allows that signs as distinct, "positive" linguistic units do result. Merleau-Ponty, however, adopts a more reductive attitude. He takes Saussurean linguistics to show that "grammatical categories" are only ever approximate and, so to speak, epiphenomenal, so that they play no proper role in accounting for the functioning of language (cf. *S* 49 and section 1 of this chapter). Although perhaps making explicit Saussure's tendency in talking of "emanation," Merleau-Ponty's claim goes further in suggesting that the macrosemantics of conceptual relations can be accounted for reductively by the microsemantics of the prepredicative expressive system.

A reductive approach of this kind would, however, exacerbate a problem inherited from Saussure. For perhaps the most serious deficiency of the latter's conception of signs is that relations to the extralinguistic world (i.e., reference) play no role in it, which makes it susceptible to the charge of linguistic idealism, of viewing language as an abstract medium disconnected from the world.[24] To some degree Merleau-Ponty's overall picture of language suggests a corrective to this tendency. After all, his view of language as embodied expressive behavior rules out many of the ontological assumptions Saussure encourages—for example, viewing *langue* as a platonist abstract object or a closed system of semiotic relations, or as somehow licensing the structuralist tradition's subject-free picture of linguistic phenomena.[25] But the ontological embeddedness of Merleau-Ponty's overall picture is not carried over into his conception of how signs function. For he not only fails to take any account of the use of signs to refer to objects, but effectively eliminates that aspect of Saussure's picture—signs as distinct units—to which an account of language's referential function might be annexed.

Three lines of argument can be found in Merleau-Ponty in support of his reductionist move. The first is that our use of language does not seem to involve specific awareness of the conceptual—or "categorial"—properties of linguistic expressions.[26] Although plausible as a phenomenological observation, this hardly supports Merleau-Ponty's reductive claim, since in speaking we are no more specifically aware of the values of sublexical elements than we are of grammatical categories.

The second argument is that "it is essential to language that the logic of its construction is never one of those that can be put in concepts." Instead, Merleau-Ponty continues, language is an expressive activity driven by the "blurred logic of a system of expression that bears the traces of another past and the germs of another future" (*PdP* 52–53). This argument, that the expressive power of language depends on sublexical factors that are irreducible to conceptual relations, is stronger. Nonetheless, it too is inconclusive, for the fact that linguistic functioning is not reducible to conceptual relations does not entail that conceptual relations are not also important, such that linguistic functioning can be explained reductively in terms of nonconceptual relations, as Merleau-Ponty seems to think.

The third argument is that Merleau-Ponty seems to think of himself as more consistently assimilating the insights of Saussurean linguistics, hinting that the latter remains too close to "positivism" (*PdM* 55). Perhaps his thought is that, having dispensed with primitive or atomic entities at the start of his account, Saussure lapses into a corresponding commitment to (final) determinacy by reintroducing the notion of distinct signs. However, unless it is shown that emergent distinctness relies on a problematic notion of full presence, it is difficult to see what the problem might be, since the rejection of presupposed determinacy in no way implies that there should be anything problematic about emanating determinacy. Alternatively, perhaps Merleau-Ponty's thought is that if distinct linguistic units simply "emanate" from underlying differential processes, they have no proper function and are hence explanatorily vacuous. This is quite true of course, but begs the present question about the legitimacy of a reduction in simply assuming that distinct units have no functional importance.

The requisite conclusion is that Merleau-Ponty fails to provide convincing reasons for his apparent view that conceptual meaning can be completely accounted for reductively in terms of underlying differential processes. If the referential use of signs is to be incorporated in an overall picture of how language functions, as it no doubt should be, linguistic signs should in this respect be understood as closer to Saussure's own model. Thus one can accept that sublexical differentiation has a role in explaining the functioning of language, yet allow that talk of determinate conceptual meanings (or "positives") is not otiose. This leaves open the possibility that *something like* Saussurean positives comprise a higher level of linguistic organization, characterized by irreducible level-specific (emergent) features of a kind suggested by Saussure's (1972, 145) analogy with H_2O. At such a "positive" level signs might behave more or less as they would if characterized only by determinate conceptual meanings, as they perhaps appear to be doing when used in well-established ways.

The latter suggestion is in fact less difficult to reconcile with Merleau-Ponty's approach than it might seem. For the key requirement is that *something* functions as a distinct sign, which is unproblematic provided that this is not conceived of in terms of a full objective presence. Thus on Merleau-Ponty's picture we need only replace the talk of something "positive" with that of a "determinate void" (*S* 112) behaving, to all intents and purposes, as though it were unambiguously determinate.

3 The Presentational Function of Style

A second major theme of Merleau-Ponty's discussion of indirect sense is an analogy with painting, motivated by the thought that both this and language are forms of "creative expression" (*S* 59). This section and the following two focus on this analogy, the development of which is clearly intended both to provide a complementary perspective on processes of linguistic disclosure and to correspond to Merleau-Ponty's appropriation of Saussure. I begin here by showing how this analogy provides a model for the function of indirect sense, specifically the way it allows features of the world to be presented or "pointed out" in an allusive manner. The following two sections set out how it lends support to thinking of language as having the inchoately rational mode of intelligibility suggested by Saussure's conception of signs.

There is of course nothing new in comparing the representational feats of language with those of pictures or images. What is distinctive about Merleau-Ponty's comparison is that it centers on modern painting, which might initially seem surprising, given the commonplace thought that this departs from traditional painting's concern with accurately representing the natural world—a task plausibly shared with language. In keeping with his very general notion of expression, Merleau-Ponty rejects the idea that modern painting can be understood as a move away from objective representation to subjective self-expression. In doing so he foregrounds the objection that a subjectivist notion of self-expression cannot account for acceptance of incompleteness (the *inachevé*) in modern painting.[27] This objection is telling. It is not simply the banal point that it became acceptable for artists to leave gaps on a finished canvas, but attests that modern painting is no longer directed toward the ideal of complete, perfect, or "adequate" representation.[28] Somewhat paradoxically therefore, it is precisely because it does not attempt to represent nature adequately that Merleau-Ponty is drawn to modern art. Nonetheless, it is significant that, while breaking with traditional academic techniques such as central or aerial perspective, the work of the early modern painters Merleau-Ponty has in mind—in particular Cézanne and Matisse—is generally figurative or representational, rather than purely abstract, in character. This fact, as will be seen, enables him to treat painting as a paradigm of indirect encounters with the world that deploy an "allusive logic" rather than supposedly "sufficient signs" of the visual (*PdM* 91, 71).

Merleau-Ponty faces an obvious question: If not as "adequate" representation of the world, and not as subjective self-expression, how is the task of modern painting to be understood? His response is: "What the painter puts into the painting is not his immediate self ..., it is his style" (*S* 65). This appeal to style is finely balanced. On the one hand, style is the hallmark of embodied action and is therefore to be understood as founded in our body as "a general power of motoric formulation capable of transpositions that effect the constancy of style."[29] On the other hand, it is fundamental to Merleau-Ponty's view that artistic output is the product of intentional, deliberate expressive acts (cf. *PdM* 82–84; Merleau-Ponty 1996b, 23ff.). This is what allows him, despite emphasizing the intimate link between an individual's life and art, to distance himself from both a crude expressivism—such that, for example, sad painters automatically paint sad pictures—and the inadequacies of psychoanalytic theories of art.[30] Indeed, although he does not use these terms, and though an artist will surely have both, Merleau-Ponty clearly intends to distinguish *artistic style*, as features characterizing works themselves, from *personal style*, as characteristics of an embodied agent's overall way of being.

Merleau-Ponty defines (artistic) style as a mode of "formulation" that can be characterized for "each painter [as] the system of equivalences that he constitutes for this work of manifestation"; "what makes 'a Vermeer' for us, or that that it speaks Vermeer's language, is ... that it observes the particular system of equivalences."[31] Although this definition allows for style to be highly individual, in several ways it is an impersonal feature. For example, an artist is not assumed to have privileged knowledge of it: "it is a mode of formulation that is just as recognizable for others, as little visible for him as his silhouette or his everyday gestures" (*PdM* 82). Indeed, while immersed in the task of creative production the artist may not be conscious of style as such—it may develop, so to speak, without him knowing (*à son insu*; *S* 67). Moreover, there is no reason to think of it one-sidedly as manifestation of an artist's self: "When style is at work the painter knows nothing of the antithesis of the human and the world"; "the painter himself [cannot] ever say ... what is due to him and what is due to things" (*PdM* 83, 95). Style instead characterizes an artist's representational technique as such, a particular way of depicting the world. Vermeer, to take Merleau-Ponty's example, has artistic significance not as an "empirical figure," but as the inaugurator of "an essence, a structure, a Vermeer sense" (*PdM* 100). In this perspective style is akin to scientific or technological discoveries—intellectual acquisitions that no one thinks of as inextricably linked with their creator. As a characteristic way of doing something, a technique, style is clearly impersonal or "objective," so that there is no reason to suppose that the inventor of a technique, an artist perhaps, is in a privileged position to understand the significance of his or her own style.

What Merleau-Ponty means by a "mode of formulation" and a "system of equivalences" is nicely illustrated by an anecdote he borrows from André Malraux of a hotelier in Cassis who was (we are told) intrigued to see Renoir working on a painting of riverside washerwomen while standing before the sea. Merleau-Ponty explains that Renoir's interest in painting the blue of the sea was to study a "fragment of the world" so as to grasp "a general manner of speaking." His concern was to "interpret the liquid substance, to manifest it, to compose with this itself" and so to master what is "typical of the manifestations of water" (*PdM* 87–88; cf. 104). Renoir was, one might say, sampling an aspect of the visual world, the style of water's visual effects, so as to add to his compositional repertoire as a painter. Or, more prosaically, he was practicing a certain way of painting water.

The point Merleau-Ponty makes with this anecdote can be elucidated more fully by relating it to Ernst Gombrich's strikingly similar account of the relationship between artistic expression and style. Despite thinking of the history of Western art in a progressivist manner, guided by the representational ideal of accurately rendering the appearance of the natural world, Gombrich's *Art and Illusion* argues against the idea of an "innocent eye" that would allow art simply to copy nature. Whether or not this holds as a psychological thesis—which is how Gombrich presents it—is irrelevant to his central argument, which shows that painting constitutively relies on available compositional techniques. Taking landscapes, animals, human bodies, and clouds as examples, Gombrich shows how, in practice, learning to paint has always involved mastering the use of pictorial "schemata" embodying the proportions and physiognomy considered essential to accurate rendition.[32] Such schemata, which were collected in compendia for teaching purposes, function as a "visual vocabulary" (Gombrich 1960, 143) from which representational paintings could be built up. The use of schemata in painting does not, of course, rule out personal and specific variations. Indeed, once mastered, they have an enabling function, underdetermining final products and obviously permitting the personal or context-dependent variations of which art history is replete. Nor is Gombrich's claim simply empirical or historical; rather his point is that the representation of objects in painting is necessarily mediated by techniques of rendition. As he says, the "artist, no less than the writer, needs a vocabulary before he can embark on a 'copy' of reality"

(Gombrich 1960, 75). It is this fact that Gombrich relies on to explain the role of history in art, and to give substance to his claims about the need for and nature of progress in naturalistic representation. For such progress plausibly depends on the development of representational techniques, or schemata, which Gombrich sees (in a Popperian manner) as evolving in an experimental process of trial and error.

It may be objected that Gombrich's progressivism and emphasis on the telos of naturalistic representation make it precisely the kind of position Merleau-Ponty opposes. This is true. Nevertheless, what I take to be Gombrich's central claims—that expression in painting is to be understood in terms of the use of schemata and that these are the vehicle of style—can be generalized independently of his representationalist orientation, so that the parallel with Merleau-Ponty's position is preserved. For although there is no longer a standard "vocabulary" of representational elements, even nonrepresentational or "abstract" painting is usually built up of a visual vocabulary of characteristic formal features such that the works of individual artists—for example, Klee, Kandinsky, Pollock, Stella, Twombly—usually exhibit a distinctive "look," or style, enabling them to be told apart (at least by a so-called connoisseur). To generalize Gombrich's view, schemata need only be thought of as exemplifying formal patterns rather than formal patterns that stand for objects—that is, as being *configurational* rather than representational. In this sense it is the composition and variation of formal elements (schemata) in more or less systematic and sophisticated ways that constitute the characteristic features, the style, of representational and nonrepresentational works alike. As Gombrich (1960, 309) puts it, it "is in the microstructure of movement and shapes that the connoisseur will find the inimitable personal accent of an artist." Merleau-Ponty's Renoir anecdote can be interpreted in this perspective. Rather than, say, simply eliciting the visible essence of water, Renoir is seen to be rehearsing or honing techniques, or configurational schemata, that underpin and so characterize his painting. Moreover, just as Merleau-Ponty intimates, it is regularities or characteristic features of configurational schemata that constitute a painter's distinctive look, so that these can be duly considered the vehicle of style.

There are two senses in which the analogy with painting exemplifies the idea of an "allusive logic." The first is well highlighted by the example of early modern painting, which provides a paradigm for the way configured form functions in presenting the world. As already pointed out, eschewing the idea of "sufficient signs" of the visual—which Merleau-Ponty considers its characteristic feature—does not prevent (say) Cézanne's or Matisse's work from being figurative or representational. The crucial feature of such painting is that (at least some) differences in pictorial articulation correspond to discernible features of visual experience. Yet the significance of these examples is to bring out a functional asymmetry, showing that for features of a painting to be "representational" it suffices that they can be read in a picture-to-world direction. To allow this, painted articulations should succeed—at least roughly—in picking out discernible features of the world. The asymmetry arises, however, in that to meet this requirement the painting need not be recognizable as a direct copy of what it represents.[33] On reflection, this is obvious: since there is no principled limit to the fine-grainedness of visual experience, to insist on correspondence read from world to picture would imply an unworkable criterion. What Merleau-Ponty calls modern art's "toleration" of the *inachevé* is nothing but the acceptance, or even celebration, of its inevitable fate. Hence, the first sense in which paintings embody an "allusive logic" is that they can function representationally without aspiring to capture every feature of what they represent—that is, without the ideal of adequation. Rather than "capturing things themselves in its forms," it suffices that a representational painting exemplifies a certain pattern of articulation and invites us to see the world as articulated in the way the painting presents it.[34]

The second sense in which paintings exhibit an "allusive logic" lies in the way they employ form as a presentational means. For schemata, Gombrich's "visual vocabulary," depict characteristic aspects of the world that have been abstracted out for both reapplication and reinterpretation in the context of diverse works. Alongside obvious schemata—such as faces, body forms, etc.—this can be thought of as including less object-related methods for rendering space, light (shadow), texture, and so on. Although they acquire concrete significance in single works, such schemata function as they do precisely because they are schematic, being neither devoid of nor fully determinate in representational significance. They are, so to speak, adumbrations rather than fully developed embodiments of determinate pictorial meaning and in this sense the "allusive" means on which the presentational feats of paintings rely.

4 Painting as a Model of Deliberative Activity

Underlying Merleau-Ponty's discussions of painting is the assumption that the indirect sense borne by linguistic forms resembles the style of paintings in its mode of intelligibility. This is important in part because it establishes the parallel between these discussions and his treatment of Saussure. Beyond that, as we will now see, the act of painting provides a model of embodied intelligence that helps in understanding why linguistic forms exhibit the inchoate rationality that Merleau-Ponty sees as characterizing indirect sense.

It will help, first, to keep in mind that the intelligibility of the style of paintings is partly due to its impersonality. For if artistic style is conceived as a system of equivalences characterizing works themselves, then Merleau-Ponty is right to suggest that people other than the artist will be able to assess it, just as others are competent to assess the uses to which a scientific technique can be put. To some extent the meaning embodied by an individual painting is already a function of the context in which it is considered. For it clearly makes a difference whether one asks what the significance is of doing such and such for the artist herself, in her contemporary context, or for subsequent art. Although in the first sense an artist presumably knows the answer better than anyone else, in the second sense she has no particular privilege, while in the third sense many others will be in a better position to judge. The same applies, perhaps more so, to general features such as stylistic innovations. Such "acquisitions," as Merleau-Ponty terms them, naturally undergo a *recontextualization*, from the individual to the public and historical perspective, such that once a new painting technique is invented its originator has no more privilege in determining its uses or full significance—in the second and third senses—than the inventor of a scientific theory.

Further important implications of the painting analogy can be brought out by considering another example Merleau-Ponty uses, one highlighting the embodied character of painting activity. Thus to explain the way indirect sense is "between words," Merleau-Ponty discusses a slow-motion film of Matisse at work that shows that, rather than moving with mechanical or surgical precision, Matisse's hand deftly rehearsed different possibilities, appearing to "meditate," before applying each single brushstroke (*PdM* 62). Having characteristically warned against overrationalizing this process, Merleau-Ponty comments: "it is true that the hand hesitated, that it meditated; it is therefore true that there was choice, that the chosen trait was chosen so as to satisfy ten conditions … that were unformulated, informulable for anyone but Matisse, since they were defined and imposed only by the intention to make *that painting, which did not yet exist*" (*PdM* 64). The immediate point of this example is simply that the brushstrokes actually made are situated within and stand out against this field of alternative possibilities, even though the relevant alternatives did not exist prior to the act of creative expression.

The example also highlights, however, the fact that in the act of painting embodied intelligence is at work. For painting is an operation that is both *intentional*—as it is directed toward the work being produced—and that involves *deliberate choice*. Talking of "deliberate choice" here does not, of course, mean that Matisse engaged in a process of conscious and explicit reasoning to arrive at his choice of brushstroke. As Merleau-Ponty puts it elsewhere, in discussing Cézanne, the concern is not with "a thought that is already clear, ... 'conception' cannot precede 'execution'" (Merleau-Ponty 1996b, 24). Rather there is what might be called "motoric deliberation," a preconceptual weighing up, (literally) going through the motions so as to get the right feel for the stroke required. The reason this is deliberation, rather than haphazard movements, is that the resultant brushstroke is nonetheless subject to both considerations of appropriateness and, where needed, correction in the context of the nascent work, and hence to some degree of conscious control.

This example suggests that the choice processes governing features of a painting are rationally ambivalent in the sense that they stand between two poles hinted at by the expression "motoric deliberation." On the one hand, the way these features are determined may tend to be discourse-led—that is, guided by the intention to embody certain ideas. An extreme example of this is perhaps minimalist painting, in which certain artistic features arguably acquire their significance from a framework of (theoretical) discourse. On the other hand, the painting process may flow more contingently, as the (dispositional) product of having learned in a certain way—the artist's "natural" way of painting—or simply so as to achieve a particular visual appearance. In the former case reasons precede and determine pictorial features, in the latter they succeed and seek to interpret them. The point of this rather artificial contrast is not to assert that the two kinds of influence can be cleanly pried apart in artistic practice. Quite the contrary, the point is precisely that it is artificial—that reasons and contingent factors are interwoven in the painting process so that such a distinction cannot be made clearly. Hence what I have called rational ambivalence: it will generally be unclear whether particular features of any painting are discourse-led or discourse-leading, whether they result from or provide a basis for reasoning.

The rational ambivalence of the painting process clearly bears on the interpretability or intelligibility of a painting. To be sure, every aspect of a painting—in contrast to the causality of photographs—is in some way the product of deliberation. Correspondingly, everything about a painting is potentially intelligible: not only is each brushstroke part of a pictorial whole, say in rendering objects as part of a motif, but stylistic choices too are often integral to realizing a work's aim.[35] So in attempting (seriously) to interpret the meaning inherent in a painting, the default assumption must be that every feature is deliberate. The painting is treated as a complex of interpretable forms, with the aim being to articulate reasons for particular features' having been produced in the way they were rather than in some other way. Yet, at the same time, the assumed interpretability of a painting is inhibited by its rational ambivalence: as a product of motoric deliberation, a hybrid of dispositional and reasoned activity, it is impossible in principle to ascertain a cutoff between causal and rational factors in the painting process. Indeed, even if one is tempted to think of an artist's testimony as an appropriate criterion for making such a distinction, this will be at best a limited guide, because a painter too will generally face similar difficulties in establishing such a cutoff in his or her own experience. The net result is that although an interpreter of a painting must by default assume that every feature is deliberate, this assumption cannot point toward a system of underlying reasoning. To put it another way, as a product of motoric deliberation, paintings exhibit one of the features—nonretrojectability—identified above as characterizing the inchoate rationality of motivations.

There are several important parallels, underlying Merleau-Ponty's views, that allow the features of painting just outlined to be seen as relevant to language. The basic reason for thinking of language use and painting as analogous bearers of indirect sense is that both are expressive acts by embodied agents. As such, both are the product of deliberate intentional activity, so that there is prima facie reason to think of them as being intelligible. There is often a temptation, particularly with regard to language, to assimilate all deliberative activities to highly reflective, maximally explicit cases such as solving mathematical problems. The significance of Merleau-Ponty's painting analogy, in particular his Matisse example, is therefore to provide an alternative paradigm of deliberate intentional activity that better reflects its character as "incarnate logic" or "logic within contingency" (*S* 110). A further parallel is that in language, as in painting, the contingent and arbitrary blends seamlessly with the rational. For although the formation of lexical features will often be due to explicit reasoning, contingency is involved—as Saussure (1972, 211ff.) emphasizes—through both changes in word parsing due to phonetic drift and the ultimate arbitrariness of the phonetic basis of linguistic signs.

With these parallels in mind, the comparison with painting provides means for understanding why linguistic forms should be thought of as having the inchoately rational character of motivations described in section 2. Comparison with the production of a single painting (as just described) already suggests that linguistic articulations should be thought of as exhibiting the first feature of nonretrojectability. That is, while we might find that a system of signs makes sense in particular ways, we should be wary of thinking we have unlocked its underlying or inherent rationality—as we may be overinterpreting the influence of contingent causal influences. This conclusion is reinforced by the fact that linguistic articulations, just as stylistic or other impersonal techniques, are subject to recontextualization of the kind described at the start of this section. Once coined, linguistic terms are open to reinterpretation and reuse, potentially in many different contexts that will (could) not have been considered by their inventors, diminishing further any hope of linking them with a determinate pattern of reasoning.[36]

It is also possible to appreciate how semideterminacy—the second feature of motivations—arises with linguistic expressions. This is linked with the important difference between individual paintings, as concrete artifacts, and an articulate lexical system, as what I will call a *type artifact*. The lexical system as a whole, and each of its constituents, are not just one-off products, but—in Merleau-Ponty's terms—the "sediment" of "iterated" expressive operations (*PdP* 229, 222). Individual words are thus both conditioned and conditioning, both products of and the material for deliberate intentional acts that (in the vast majority of cases) have been reused or recycled countless times. Such recycling has feedback effects that imbue linguistic articulations with motivational character: "having been employed in different contexts ... the word is gradually charged with a sense that it is not possible to fix absolutely" (*PdP* 445). For, on the one hand, our propensity to find words meaningful or appropriate for intentional use is explained by the fact that they have (usually) been used previously. In contrast to that of individual paintings, the intelligibility of linguistic articulations, as type artifacts, is an evolved intelligibility: words are tried and tested expressive tools.[37] On the other hand, with such iteration the effects of rational ambivalence and recontextualization on the overall product are compounded, so that there can be no sensible aspiration to uncover "the" rationality inherent in the sediment of linguistic articulation. By thus both compounding the supposition of intelligibility and frustrating any hope of unique underlying rationality, the—distributed, iterative, evolutive—mode of production brings about the semideterminacy of linguistic articulation as a whole. The need to think of linguistic forms as an inchoately rational field of (relative) motivations thus results from their being both a product of motoric deliberation, which introduces nonretrojectability, and type artifacts, which generates semideterminacy.

The difference between an individual painting, as a concrete artifact, and linguistic articulations, as type artifacts, can be usefully summed up by saying that although each can be thought of as an arrangement of interpretable form, they differ in their respective modes of interpretability. With an individual painting—as a single object of determinate origin—an art historian's interpretation may aim to rediscover its original meaning, even though this aspiration must contend with the rational ambivalence of painting highlighted here. The important point, nevertheless, is that its interpretability is *retrospective*, based on the supposition of antecedent rationality. Interpretation here aims at recovery. By contrast, with linguistic articulation—a sedimented type artifact of indeterminately many origins—such retrospective interpretation would be pointless given its typological and pervasive motivational character. In this case, to talk of interpretable forms instead means that words are open to constant (re)interpretation in acts of speech, thus exhibiting *in situ* or *situated interpretability*. Here interpretation is application to the given circumstances. The difference between these two modes of interpretability has the important implication—discussed in section 6—that linguistic forms (e.g., words) should not be thought of as having recoverable "original" meanings. Their intelligibility is instead inscribed in more complex and protracted processes of sense generation in which there is no reason to think that first uses are preserved or privileged in any way.

It is perhaps worth highlighting, finally, that the interrelated features discussed here—inchoate rationality, motivational character, and situated interpretability—are not imperfections, but are fundamental to the way words work in the real world. Philosophers have often been tempted to think of words, at least "ideally," as bearing a determinate meaning and as playing a well-defined role in some fully rational system. Against such preconceived ideals it should be emphasized that the evolved intelligibility outlined here and the consequent situated interpretability are required for words to function as they do. They underlie both the aptness of words, the fact that these seem antecedently suited for use in presenting features of the world, and their plasticity, the fact that they can be adapted to new situations and applications.

5 Style as a Preconceptual Generality

The analogy with painting can also shed light on Merleau-Ponty's descriptions of style as a "preconceptual generality" and as "that which renders all meaningful signs [*signification*] possible" (*PdM* 63n, 81). These claims are clearly intended to correspond to the role of sublexical factors in Saussure's conception of signs, in which context talk of differential and preconceptual structures may appear theoretical and somewhat abstract. Painting, however, serves as a visual analogue of Saussure's approach, literally illustrating the interplay of different levels of structure—in particular Merleau-Ponty's "expressive system"—in a different sense modality. Thus in the case of paintings the above descriptions seem easier to follow, since it is obvious that any given object, a chair say, could be rendered in indefinitely many different ways, different visual styles, without falling under a different concept.

To isolate what the preconceptual features of paintings are it will be helpful to consider Panofsky's (1993, esp. 51–67) well-known distinction between three aspects of pictorial meaning: the preiconographic, which catalogs or inventories the objects depicted; the iconographic, which identifies, where appropriate, the conventional symbolism of those objects; and the iconological, which is concerned with interpreted content. A conspicuous feature of this approach is that it applies—albeit intentionally—only to representational, conventional, and highly codified works. Even accepting this limitation, however, in concentrating on represented objects and narrative Panofsky's approach effectively assimilates pictorial meaning to conceptual and propositional content. In so doing it fails to accommodate two essential aspects of painting. First, it disregards the fact that preiconographic items are built up from formal elements that in themselves could not be assigned any determinate meaning. Yet anything in a picture that is conceptually interpretable must consist of some configuration of forms; it must, so to speak, be *configured*. It is this feature, I suggest, that Merleau-Ponty is drawing attention to in describing style as "preconceptual" and as a precondition for "signification." Second, Panofsky's approach fails to reflect the fact that arrangements of form and color are continuously variable, or what Goodman calls the "syntactic density" of pictures.[38] This means, as in the chair example above, that any number of compositional or configurational differences could fail to make a difference to the "content" of a painting. However, what matters here is that we nonetheless *see* differences in formal configuration (i.e., in style).[39] This is why the example of paintings is so attractive for Merleau-Ponty. Because we can see differences in (syntactically dense) configuration, and because configured forms must underlie any conceptual content paintings have, painting serves as a paradigm in which the seeing of stylistic differences is plausibly prior to conceptual awareness.[40]

How might these thoughts apply to language? As already mentioned, Merleau-Ponty's analogy with painting and his notion of style as a "system of equivalences" are intended to converge with the Saussurean conception of signs, as complementary perspectives on the process of linguistic articulation and indirect sense. Hence in both cases the focus is on certain aspects of structure—configurational schemata in painting, and differential signifiers and signifieds in language—the role of which is to be subconceptual. Less obviously perhaps, style choices in painting can be viewed as taking place in a differential realm in Saussure's sense insofar as stylistically relevant choices are made on a "syntactically dense" continuum of possibilities. Moreover, in a slightly extenuated, and somewhat intuitive, sense Saussure's notion of value could also be applied to the use of configurational or stylistic features. For clearly the use of various materials, methods, and forms in a given work can be considered in terms of interchangeability and comparison—as expressed in the maxim that in a successful work nothing could be added or taken away without aesthetic compromise.

Nonetheless, there are limits to the analogy with painting, and in this respect considering Panofsky's approach is particularly instructive. In assimilating pictorial meaning to conceptual and propositional meaning Panofsky might appear to be assimilating paintings to *language*. Accordingly, the features of pictures Panofsky's approach misses—configuredness and denseness—might be thought not to apply to language, and so to show up the analogy's limits. Yet this would be too hasty. First, as just hinted, in the case of language too the formation of linguistic signs from sublexical elements can be considered a process in which configurations are built up that underlie conceptual meanings. The second feature, syntactic density, is less clear-cut. Although the phonetic aspect of speech is itself continuously variable, its organization through use in a sign system results in sublexical elements—phonemes or morphemes—characterized by discontinuities. Hence, the resultant configuration of linguistic signs is not syntactically dense, but syntactically discrete—or, to borrow another of Goodman's terms, "finitely differentiated."[41]

This brings out a difficulty not discussed by Merleau-Ponty. For it is precisely the fact that indefinitely small differences count, that there is syntactic density, which makes style such an important feature of paintings. Indeed the limitless potential for variation of every feature makes it practically inevitable that each painter will exhibit an individual artistic style. The same cannot be said regarding language. As language users we cannot freely vary the form of signifiers; or rather we can to the extent that others are able to neglect such variations and map them onto a familiar and discrete phonological/morphological scheme. Far from being a sociological observation, this identifies a fundamental disanalogy between language and painting: insensitivity to syntactic variations beyond a certain degree is a requirement for the practical and communicative functioning of linguistic signs. Indeed, such insensitivity is presumably a necessary condition for the use of language to organize experience in a manageable way. As a result style, in Merleau-Ponty's sense, is a more strongly constrained and presumably less salient phenomenon with language than it is with pictures. Moreover, because it cannot rely on continuous malleability, the development of individual linguistic style will generally turn on factors other than syntactic variations.[42]

Nonetheless, as long as these qualifications are kept in mind, there is no problem in principle with applying Merleau-Ponty's notion of style to language. Sublexical composition can then be thought of as the primary (syntactic) vehicle of style in much the same way as use of a characteristic palette or configurational schemata defines style in painting. Like painting, each language can then be seen to have its own characteristic "mode of formulation" or "system of equivalences." Hence, where differently configured syntactic schemes *are* found—say between English, French, Russian, and so on—this can clearly be described as a difference in "style" in Merleau-Ponty's sense.

6 Presentational Sense as Indirect Sense

In discussing Heidegger's view of the disclosive function of linguistic signs (chapter 3, section 3), I introduced the notion of presentational sense to refer to an aspect of their meaning that is due to their form and in virtue of which they function to present features of the world in an articulate manner. The aim of this chapter has been to draw on Merleau-Ponty's discussions of indirect sense to fill out that somewhat schematic characterization. Thus Saussure's conception of signs provides a detailed account of the kind of form inherent in linguistic signs, and correspondingly of the *structure* of presentational sense. Further, the analogy with painting provides a model for understanding the presentational *function* of linguistic signs as an allusive "pointing out" of features of the world.

Finally, Merleau-Ponty's discussions of Saussure and painting converge in suggesting that as bearers of presentational sense linguistic forms have an inchoately rational and preconceptual mode of intelligibility. This closing section will focus mainly on highlighting how presentational sense, understood in the way just summarized, can contribute to a phenomenologically plausible conception of language. Toward the end I show how it meets the requirements identified in chapter 2 for prepredicative factors, before identifying a basic inadequacy that Merleau-Ponty's discussions of language share with Heidegger's, leading to the need for an account of what I previously called pragmatic sense.

What defines a conception of language as (minimally) phenomenological, I suggested in the introduction, is the aspiration to accurately describe features of the experience speakers have of language. One straightforward way Merleau-Ponty's conception of language does this is in its ontological commitments. Being set out in terms of behavior, gestures, words, sublexical features, and acts of speech, his conception of language assiduously avoids positing any linguistic entities that are not available phenomenologically in speakers' lived experience (such as propositions or abstract entities). Presumably no one would deny that these are the ontological features of which language consists. The distinctive feature of Merleau-Ponty's approach, however, is that rather than merely acknowledging these features en route to a supposedly more basic or powerful theoretical framework, he seeks to do justice to the fact that these items, in particular words, are literally *involved in* the constitution of linguistic meaning. As seen in the last chapter, this was already a key aim for *PdP*'s discussion of language, even though its account of the way language is a "vehicle" of sense rested on inadequate metaphors. With the appropriation of Saussure's conception of signs, however, this aim could draw on a more literal and detailed accommodation of the fact that sublexical structure is integral to the functioning of linguistic signs. Thus it becomes possible to describe systematically—rather than metaphorically—two further phenomenological features, namely, the perceptible differences between (natural, "national") languages and the intuitively obvious significance of many grammatical features. For example, the "grammatical" ordering of words into verbs, adjectives, and adverbs typically relies on the use of standardized sublexical features that identify how their referents are being presented, say as a temporally occurrent process or event (verbs), a property (adjectives), or a manner of doing such and such (adverbs). Of course, there is no necessary link between such features and their referents—using a noun to refer to a process, for example, does not entail blindness to its temporal character. They do, however, have a disclosive significance—as the means we use to organize and present features of the world in such ways—that a phenomenologically accurate conception of language ought to respect as having a distinct function in meaningful language use rather than being merely "superficial" features with the potential to mislead unsuspecting philosophers.

Merleau-Ponty's conception of language also provides the means to address, in a phenomenologically plausible fashion, two issues initially highlighted in discussing Heidegger's view of signs as formal indications (chapter 3, section 2). The first is that Heidegger, both prior to and in *SZ*, relied on an undefined notion of linguistic or Articulate form. Saussure's views can therefore be seen as adding to Heidegger's position in much the same way as they did to Merleau-Ponty's—namely, in providing a detailed conception of linguistic form. On Saussure's model linguistic articulation is understood to take place at the lexical and sublexical level, rather than centering on higher-level structures such as sentences, texts, or discourses. In this respect it coheres well with Heidegger's view of propositions as "derivative." More importantly, it is phenomenologically plausible to think of linguistic articulation in terms of the distinct elements found at the lexical and sublexical levels, since higher-level structures—sentences, texts, discourses—are clearly composed of these. At the same time Saussure's model also provides an answer to the (Socratic) question of how far down presentational sense goes, since by definition morphemes are the smallest syntactic units that can be considered to articulate a significant difference.

The second issue concerns Heidegger's view that signs are internally linked with determinate grounding experiences and have the essentially atemporal function of simple preservation of original meanings—to which Merleau-Ponty provides an important corrective.[43] To be sure, if skepticism about meaning or the possibility of communication is to be avoided, then at some level (say of texts) or under certain conditions (e.g., constant patterns of use) linguistic signs must be assumed to be able to preserve meaning. However, Merleau-Ponty's view of the way indirect sense functions makes clear why this assumption cannot be made without qualification. Once we consider how the processes of production and transmission of linguistic expressions introduce and compound the effects of rational ambivalence, the naïveté of assuming that individual signs stand in some recoverable internal relation to determinate phenomena is exposed. Heidegger's view rests on an oversimplification that might be described—in terms used in section 4 above—by saying he treats linguistic signs as though they were concrete artifacts rather than type artifacts, and so attributes to them retrospective rather than situated interpretability. Although recurrent phenomena or actions might provide a basis for stable patterns of expression, Merleau-Ponty's picture makes clear that there can be no underlying general assumption of simple preservation. Language use instead generates a field of type artifacts characterized by inchoate rationality and situated interpretability. The relationship between linguistic expressions and phenomena is hence reconfigured: rather than being internally linked with specific phenomena or contexts of use, these characteristics engender the flexibility that makes individual linguistic expressions appropriate to a range of varying situations.

The result is a crucial difference in the way the relationship between language and time is conceived. For Merleau-Ponty words do not have original contexts of sense; their expressive power is instead rooted in the here and now of lived experience. As a consequence Merleau-Ponty, unlike Heidegger, is able to acknowledge the genuinely temporal character of language. To be sure, under Saussure's influence he valorizes the living present of speech acts (*parole*) in claiming that "the power of language ... is entirely in its present."[44] While linguistic articulations remain semideterminate on principle, determinacy of meaning is maximized in particular utterances: "The act of speaking is clear only for whoever speaks or hears effectively" (*PdP* 448). Nonetheless, he conceives of acts of speech as embedded in a temporal horizon. The speaking present itself is not pointlike, but temporally extended: "the [synchronic] system that is realized is never entirely in the act." The linguistic present is thus privileged only as conditioned by a background of sedimentation or acquisition, such that "the contingency of the linguistic past intrudes through to the synchronic system."[45] Similarly, this present opens onto future language according to the "blurred logic of a system of expression that bears the traces of another past and the germs of another future" (*PdP* 52–53).

Thinking of presentational sense on the model of Merleau-Ponty's indirect sense thus brings two important modifications—an articulate conception of linguistic form and the temporalization of linguistic phenomena—to Heidegger's view of the disclosive function of linguistic signs. With these modifications, I want to suggest, the main importance of the notion of presentational sense is to allow the function of linguistic form in disclosing the world to be conceived in a phenomenologically accurate manner.

One aspect of this is that it avoids imposing artificial or idealized order on the intelligibility inherent in linguistic forms. In *SdC* Merleau-Ponty had suggested that symbolic behavior involves "structural correspondence" between signifiers and signifieds based on a "systematic principle" that he left unexplicated (*SdC* 132–133; cf. chapter 4, section 1). This idea makes no appearance in *PdP*, but having adopted Saussure's views, Merleau-Ponty was later able to draw on the idea of a sign system (*langue*) with explicitly and exhaustively identifiable structural components. However, the organization of this sign system is now due to iterated intentional activity, yielding a differential scheme that is type artifactual and characterized by an inchoate mode of intelligibility. Thus, although the structure of the sign system is thought of as inherently meaningful, it no longer embodies an idealized system organized around one or more "systematic principles."

Most importantly, perhaps, the example of modern painting provides a model of schematic, allusive presentation of the world that can be plausibly applied to language. In chapter 3, section 3, I suggested that Heidegger must *in some sense* be right in claiming that linguistic signs "point to" features of the world and so "allow them to be seen." This must be right insofar as in any given context of use it is the articulations inherent in the sign system's overall pattern of difference and distinctness that are used to organize experience of the world by picking out features of the world. One might be tempted to think that in order to do this, the structure of the sign system must "correspond" to or replicate the structure of the world in some precisely specifiable way. However, modern painting of the kind considered by Merleau-Ponty shows how this function can be achieved by a pattern of forms that are not a complete or precise copy of the world and that are schematic in character rather than embodying fully determinate sense. Thus it illustrates how linguistic articulations can fulfill the presentational function—pointing out features of the world, letting them be seen directly—that Heidegger thought of as proper to Articulacy in an allusive or "indirect" way.

The role of linguistic form in disclosing the world is sometimes described by saying that a given language embodies a "way of seeing" the world. Humboldt (1995, 53–54), for example, claims that "a specific worldview lies in every language" and treats the terms "worldview" (*Weltansicht*) and "language view" (*Sprachansicht*) as synonymous. The same thought is implicit in Whorf's (1956, 213) infamous claim that each language "dissect[s] nature" in a different way. Viewing linguistic signs as bearers of two kinds of sense in the way advocated here allows this thought to be acknowledged without exaggerating its importance or seeing it as entailing relativism. Thus the capacity of presentational sense to point out features of the world in the allusive manner just described makes sense of the idea of seeing the world in the perspective of a language with a specific form. However, this idea is counterbalanced by a recognition that this is only one aspect of the function of linguistic signs and so does not suffice to establish a linguistic relativism.

In what kind of situations, then, *does* presentational sense play an important role? How does it enter into our experience as language users? Generally speaking, these will be situations in which the form of linguistic expressions becomes important in conveying specific "meanings," those in which we rely on the "relative motivations" inherent in our language's sign system. But what examples are there of such situations?

The activity that perhaps best illustrates the need to be attentive to the role of linguistic form is translation work. Not only does this demand attention to every word, but it is often differences in the form of specific words that make it difficult to render nuances or associations found in one language in another or underlie the claim that something is "untranslatable." This kind of difficulty is pervasive in Heidegger's philosophical writing. One example, mentioned in the introduction, is his use of the verb *stellen* to highlight connections between the "essence" of modern technology and a specific post-Cartesian mode of thought in a way that cannot be reproduced (say) in English. Another obvious example is Heidegger's exploitation in *SZ* of the form of the word *Dasein*—"being there"—to suggest epigrammatically the situatedness of human disclosure of the world in an existential space. Similarly ingenious and subtle exploitation of the form of words is characteristic of Derrida's writing. A well-known example here would be his coining of the term *différance,* which relies on two senses of the verb "différer"—"to differ" and "to defer"—to link the production of preconceptual differences (*différences*) with an indefinite spatial and temporal deferral of any supposedly meaning-grounding original presences (Derrida 1972, 12, 19). A second example is Derrida's (1978, 36) suggestion that style is a form of weapon, based on the connection between the word *style* and *stylet,* the latter of which can be either a stylus (a writing implement) or a stiletto (a dagger). Rather than being isolated cases, these examples are typical of techniques of language use—often derided as "punning"—of which both authors make constant use. However, my point here is not to suggest (absurdly) that this kind of language use is commonplace, nor even to endorse such use of language as a philosophical method. The point is simply that such uses of language are effectively untranslatable precisely due to their sophisticated exploitation of nuances and associations that are anchored in specific linguistic forms.

Our sensitivity to the implications bound up in linguistic forms is not limited to highly deliberate and extreme cases of the kind just mentioned. A less artificial example is found in Nietzsche, who objects to pity as a basic ethical notion on the grounds (inter alia) that it would increase suffering in the world.[46] Whether or not one agrees with it, this thought is easier to follow when one realizes that the word he uses for pity is *Mitleid*. Because the verb *leiden* means "to suffer," and the prefix *mit* suggests "along with others," the word *Mitleid* intimates that pity is a form of suffering along with others, such that pitying others increases the total amount of suffering in the world. Rather than engineering or highlighting a complex network of associations in the way Heidegger and Derrida do, Nietzsche simply interprets the word in a way that appears obvious to German speakers (and which again complicates translation). One final example: During a tour at the Tate Modern I once heard a curator make a point of explaining that the German term *entartet* implies the art that the Nazis used this term to denigrate no longer belongs to the genus or kind (*Art*) of art. While it is correct that the German term suggests this, the curator was apparently unaware that the corresponding English term *degenerate* has precisely the same implication (deterioration so as no longer to fall under the genus). In this case, what is of interest is not that *other* languages can use sublexical form to anchor specific associations, or that this leads to translation difficulties, but that such features are present in words established in our own language (English) even though we are often insensitive to them.

The last two examples illustrate that an expressive potential can reside in the form of the linguistic signs even if we are unaware of it in routine use. In this respect Merleau-Ponty is right to emphasize the importance of creative language use, which is more likely to involve exploiting subtle nuances or implications of words in producing allusions, metaphors, and perhaps even poetic effect. It is surely right to think that such effects are conditioned by the formal structure of the language we are using and that they rely on an awareness of the overall economy of linguistic forms and the potential for the meaning of established words to be transcended or exceeded. Although Merleau-Ponty focuses primarily on literary language use, even creative use of language is typically more prosaic. Thus the invention of linguistic expressions often involves recognizable combinations or modifications of existing words, and even where this is not the case new terms may be implicitly conditioned by existing forms from which they are to be distinguished. A good example of this is the invention of technical computing terms. Often these initially state outright what (say) a device does, but over time become abbreviated or elided to retain only a more obscure reference to the sense initially expressed (e.g., a modulator-demodulator becomes a *modem*; electronic mail becomes *e-mail*, and then simply *email*). Finally, there are less overtly creative uses of language in which it is equally important either to evoke or to avoid suggesting certain nuances or associations. Thus in technical or legal writing, or perhaps simply in emotionally delicate communication, we might expect to be held accountable for our choice of words and so invest effort (e.g., by making use of dictionaries or thesauri) to understand the differences and similarities between alternative terms.

In each of the above kinds of situation—translation, creative language use, inventing new terms, writing carefully—it is phenomenologically plausible to suggest we are (at least sometimes) attentive to the role of linguistic form in anchoring specific associations or nuances of meaning. In other words, these are situations in which words and their sublexical constituents are experienced as literally the means of which thoughts are composed, as expressive "materials" that embody certain constraints, as Merleau-Ponty unsurprisingly points out, in much the same way as an artist's colors (*PdP* 446). These are the types of situation in relation to which it makes sense to think of linguistic expressions as bearers of presentational sense.

A further advantage of the notion of presentational sense, based on Merleau-Ponty's recognition of the genuine temporality of language, is that it accounts plausibly for the significance of what John Austin (1979, 201) fittingly describes as "trailing clouds of etymology," the fact that "a word never—well, hardly ever—shakes off its etymology and its formation." Etymology adheres to words in the sense that their morphological features anchor certain associations or relations to other terms, in a manner typically elucidated by considering language historically. Although a Merleau-Pontian picture of the evolution of linguistic expressions undermines the idea of accountability to determinate and recoverable sense origins, its reliance on the idea of relative motivation still allows an expression's form to be thought of as intelligibly linked with its meaning. On this picture, etymological or historical facts can shed light on "the meaning" of a word by clarifying how it is presenting its objects (i.e., which contrasts and associations are anchored in the word's structure). It therefore allows that such facts—say in cases of doubt, subtlety, ambiguity, neologism, or metaphor—can tell us something important about the way the world is being disclosed to us. What we learn in such cases, however, is something not about a mythological "origin" of sense, but about the presentational sense of an expression. Further, Merleau-Ponty's view of the pervasive indeterminacy in the way language's past shapes its present implies that etymological facts can be neither an unfailing nor an adequate guide to what an expression means in present use. Thus it echoes Austin's insight that while words do not generally shake off their etymology and formation, it is equally true that the bind is not very tight—that the concern is with clouds rather than chains.

This also allows a better understanding of Heidegger's controversial reliance—especially in later works—on "etymology," which appears to depend on claims about historically original meanings.[47] Some clarification is required here, as it not obvious that Heidegger was specifically interested in the historical originality of meanings. Admittedly, already in *SZ* he describes his project as aiming at "originality" and the ultimate "business of philosophy" as being to "preserve the *force of the most elemental words*" (*SZ* 220). But Heidegger's indifference to *historical* originality is strikingly illustrated by his treatment of the term *phenomenology*: while apparently appealing to the Greek etymology of its two "constituent parts" to explain its meaning, Heidegger shrugs off the "history of the word itself" as insignificant.[48] This is not obviously inconsistent, because Heidegger's primary aspiration (cf. chapter 3, section 2) is to sense-genetic originality, the sense inherent in "grounding experiences" or phenomena, rather than to historical accuracy. Moreover, his apparent fascination with word histories can be explained by his model of signs' functioning, which implies that sense-genetic and historical originality cannot be fully dissociated. For if linguistic signs were internally related to and simply preserved their link with determinate phenomena, historical and sense-genetic originality would be correlative, and historical origins simply the first instantiation of a sense genesis. Accordingly, if etymology was indicative of historically original understanding, as the simple preservation model suggests, it would also be a useful pointer to the corresponding grounding experiences or phenomena. However, by recognizing the genuine temporality of language, Merleau-Ponty's approach provides a more plausible view of Heidegger's method than Heidegger himself. On this view, Heidegger's "etymological" claims do not recover—either sense-genetic or historical—"original" understanding, but instead shed light on the presentational sense of the words he is using, allowing them to be given a powerfully resonant interpretation in the present.[49]

One might be tempted to object that the examples of expressive use cited and relied on here are inappropriately *recherché*—so to speak, a phenomenology of exceptions rather than the rule. This objection misconstrues the point of these examples here. I am not suggesting (as Merleau-Ponty indeed does) that such relatively unusual situations are explanatorily distinguished over more common ones or that all language use relies on presentational sense. My claim, to reiterate, is rather that the notion of presentational sense is appropriate in describing some kinds of language use, those to which Merleau-Ponty's views are specifically suited. Moreover, despite contrasting with more quotidian language use, the situations described here highlight features present, though not saliently relied on, in all use of language. Thus it seems obvious that even when we use language routinely or "inauthentically" in Heidegger's sense—that is, while lacking awareness of its presentational import—the sublexical-lexical articulations to which presentational sense is due are nonetheless pervasive. Part of the point of appealing to such apparently untypical uses of language is therefore that they serve to highlight and so allow us to recognize neglected features of our own prior experience, just as a phenomenological approach to language should.

One constraint I have suggested presentational sense would need to meet is to be able to function prepredicatively. While it is obvious that much of Merleau-Ponty's discussion explicitly centers on prepredicative factors, I want to highlight briefly how his views meet the three conditions identified in chapter 2. First, and most obviously, in concentrating on the function of words and sublexical elements, the notion of indirect sense clearly focuses on what I called the subpropositional features from which sentences/propositions are constructed.

It is also, second, subinferential in both senses required for it to function foundationally. On the one hand, even when presentational sense is not being treated as inferentially relevant, features of linguistic form are still pervasively involved and grasped—so to speak, as mere syntax—in all language use. On the other hand, as illustrated by the above examples, the differential structure of presentational sense can also be drawn on and developed as a basis for justifying specific uses of language. In this respect, the relatively motivated differences in the overall pattern of linguistic articulations—Saussure's *langue*—are the basis of words' interpretability. These differences comprise a kind of "sense" insofar as they are present in all language use and generate the possibility of developed interpretation. Possessing language, as Merleau-Ponty puts it, as a "principle of distinction" (*PdM* 46) thus extends over different levels of competence, from the most basic (merely "syntactic") mastery of sublexical forms to the exploitation of structurally inherent differences in inferentially relevant ways.[50]

Finally, the idea that presentational sense comprises a functionally discontinuous foundation of predicative awareness is reflected in the distinction between sublexical and lexical (sign) levels of linguistic structure.

Merleau-Ponty's claim that the "logic of [language's] construction" cannot be "put in concepts" (*PdM* 52–53) points in this direction. In fact, as explained in section 2 above, his own view is not that the predicative and prepredicative function in genuinely different (mutually irreducible) ways, but that conceptual or predicative properties are reductively explained by the dynamics of sublexical elements. Against this, I argued that this reduction is implausible and that the possibility of constituted, distinct signs being the bearers of—for example, inferential or referential—properties should also be acknowledged. Distinguishing sublexical elements and distinct signs as different levels of organization in this way provides a basis for the claim that prepredicative and predicative factors function differently. However, the idea of functionally discontinuous foundation requires not only that lexical properties presuppose the dynamics of sublexical elements, which I take to be true, but also that these two levels of semiotic organization and function cannot be straightforwardly—that is, in a 1:1 manner—mapped onto each other. Yet it is quite straightforward to see that there can be no such mapping, because the two levels differ in the topology of their constituent units. Not only does a language have many more lexical than sublexical items, but lexical items are themselves composed of sublexical forms, ruling out any straightforward mapping between the two levels.[51]

So far in this section I have been highlighting the advantages of the notion of presentational sense modeled on Merleau-Ponty's views for a phenomenological conception of language. All the same, Merleau-Ponty's conception of language has significant inadequacies. To begin with, it provides an altogether more partial picture than Merleau-Ponty seems to have supposed. One example, already highlighted, is that in concentrating on the role of sublexical elements, Merleau-Ponty neglects the possibility inherent in Saussure's conception of signs that the functioning of constituted signs might require describing in a way not reducible to sublexical dynamics. Similarly, while Merleau-Ponty recognizes that language has a genuinely temporal character, it would be wrong to think that the question of how language's past carries over into its present and future can be addressed simply at the sublexical level. Rather, the historical transmission of linguistic meaning involves a range of processes at different levels such as sentences or discourses and texts, as well as the "linguistic units" focused on by structuralist linguistics.[52]

Two more trenchant problems with Merleau-Ponty's approach result from its systematic bias in favor of creative expression. The first is that it fails to identify any checks on the unfolding of expressive potentials—that is, how words are involved in expansive *and limiting* processes that result, as Ricoeur (1969b, 71) aptly terms it, in a "regulated polysemy." This one-sided emphasis on openness and creative expression, which casts any sense-preserving language use simply as lacking proper expressiveness and hence "secondary," fails to account for the stability of many expressions' meanings and to explicate the constraints effecting such stability. It fails, so to speak, to explain why language is not in expressive free fall.

Merleau-Ponty also fails, second, to account for the role played by a background of established use. The need to do this can easily be appreciated since, given the ultimate arbitrariness of the phonetic aspect of signs, in order to present entities in an articulate manner—that is, to convey differences and similarities—signs must be interpreted in relation to standard patterns of use. To understand the resultant interdependency, however, a more detailed account of what Merleau-Ponty terms "direct" sense is required. In fact, he acknowledges the reliance of indirect on direct sense, and of creative on established expression (cf. chapter 4, section 3). Yet, surprisingly, he seems not to realize that this undermines his asymmetric claim to explanatory primacy on behalf of creative expression and indirect sense. This might be due to his tendency to associate the notion of "direct" sense with the dubious metaphysical implication that intentional objects are presented in an idealized, nonaspectual manner. However, the correct conclusion to draw is instead that what is needed is a conception of the "direct" background for creative expression that does not require such eccentric metaphysical commitments.

It seems to me that these two problems have a common root in a failure to consider more closely the pragmatic aspect of language. For what holds a pattern of linguistic regularities in place, and so keeps expressive free fall in check, are the exigencies of living and acting in certain ways. And since what characterizes living in a certain way is a certain pattern of regularities, it seems natural to link an account of established ("direct") patterns of language use with the forms of life in which they arise. Indeed, it is puzzling that Merleau-Ponty neglects practice in the way he does. For given his emphasis on existential and bodily significance (or lived sense) in *PdP*, it is far from clear that already constituted or established forms of expression should qualify as "secondary." Inauthentic, nonoriginal, constituted speech instead bears meaning in the same ways as—and presumably with more vital urgency than—what Merleau-Ponty calls "original" or "authentic" speech. So one would expect a consistent phenomenological commitment to the lived present to focus on the world of everyday practice. In this light Merleau-Ponty's preference for highbrow literary language use as a paradigm evidences a disregard for lived experience—the world as experienced (*le monde vécu*; *PdP* iii)—and day-to-day embodiment that is incongruent with his own commitment to phenomenology.[53]

The failure to consider in any detail the link between language and action is a deficit Merleau-Ponty shares with Heidegger. Again this is to some extent surprising, given *SZ*'s emphasis on the foundational function of purposive understanding and its corresponding twofold instrumentalist view of (linguistic) signs. Thus while both Merleau-Ponty and Heidegger's conceptions of language are grounded in recognition of the pragmatic dimension of language, neither thought this sufficiently important or involved to deserve detailed discussion. Hence the solution to the two problems just identified with Merleau-Ponty's conception of language converges with my attempt to fill out an account of pragmatic sense within the Heideggerian framework. What is needed is an articulate conception of the way linguistic regularities are connected with practice, without reliance on metaphysically bizarre "direct" encounters with intentional objects. It is this we will find with Wittgenstein.

III Wittgenstein: The Pragmatic Aspect of Language

6 Language and the Structure of Practice

The Heideggerian framework set out in the first part of this book provides a general picture of the role of language and schematically characterizes the disclosive function of linguistic signs in terms of two distinct kinds of feat. One of these was their "distinctive" function as signs of exploiting linguistic form so as to point out features of the lived world in an articulate way. This kind of feat gave rise to the idea of presentational sense, which the last chapter tried to fill out in greater detail by drawing on Merleau-Ponty's notion of indirect sense. Despite complementing Heidegger's view in this way and by highlighting the general importance of embodiment, Merleau-Ponty does not help in understanding either the general connection between language and practice that Heidegger recognizes or the second kind of feat defining the disclosive function of linguistic signs, namely, the variety of roles they play as instruments in established practices. Hence the need remains for a more detailed account of both this connection and what I previously called pragmatic sense, a need this chapter and the next attempt to meet by drawing on the later Wittgenstein's conception of language.

Two features of Wittgenstein's view make it particularly suited to phenomenologically accurate description of the character of rules in linguistic practice, and so to the task of explicating the notion of pragmatic sense. The first is commitment to the idea that the use of certain concepts is internal to corresponding forms of practice, such that the "logic" of those practices is reflected structurally in linguistic articulation. As Wittgenstein puts it: "*Practice* gives words their sense"—"the concept ... is at home in the language-game" (Wittgenstein 1989h, 571; 1989i, 363 [§391]). The point of this, I will be arguing, is that not only is the use of signs in a particular way constitutive of corresponding practices, but also, conversely, that acting in certain ways imposes practical requirements that condition the use of signs. To highlight this literal embedding and shaping of concepts within practice I will refer to Wittgenstein's late conception of language as a "praxeological" one.[1] The second feature important here is the characteristic modesty and flexibility that Wittgenstein's conception of rules takes on as the language-game analogy matures. As will be seen, the late Wittgenstein not only affords rules a circumscribed role in linguistic processes, but reconfigures his conception of rules, including the notion of determinacy bearing on them, so as to reflect their attunement to real-world surroundings.

The task for this chapter will be to examine these two features in some detail so as to show that Wittgenstein's conception of language can explicate the notion of pragmatic sense in a phenomenologically plausible manner. In particular, this discussion will make clear how the notion of rules helps in understanding the various kinds of pattern or form that the world takes on in the perspective of practice. The next chapter will then show how Wittgenstein's views on rule-following allow for the resultant notion of pragmatic sense to be grasped prepredicatively. I begin this chapter by clarifying my approach to Wittgenstein, addressing several potential difficulties in attributing to him a phenomenological conception of language. The second and third sections introduce Wittgenstein's notion of language-games and explore the internal linking of practice and language this analogy suggests. The following three sections look at the evolution of Wittgenstein's conception of linguistic rules in his later thought, particularly the implications of the maturing language-game analogy. Thus the fourth section considers why Wittgenstein gave up on the idea that language is fully governed by a set of rules that might be modeled as a kind of calculus. The following section argues that in later texts Wittgenstein modified his understanding of rules so as to accord better with the form and inherent requirements of real-life practices, while the sixth section highlights several ways in which the role of rules is circumscribed and qualified under the influence of the language-game analogy. The final section sets out how these developments allow the notion of pragmatic sense to be explicated in a phenomenologically plausible manner.

1 Appropriating Wittgenstein

To begin with I want to say something about the approach to Wittgenstein I will be taking here. To read Wittgenstein as the author of a praxeological conception of language is obviously suggestive of a particular textual focus, as the relevant features of his thought could emerge only once he had given up on the idea that language is a calculus and come to rely on the idea of language-games as a model for philosophical understanding of language. Accordingly, my interest in the following will be on Wittgenstein's "late" texts, more specifically those from the *Philosophical Investigations* (*PU*) onward. While these comments might so far seem obvious, it is important for my purposes that these writings contrast not only with the *Tractatus*, but also with texts of the early 1930s, up to and including the *Philosophical Grammar* (*PG*) of 1933. This might seem less obvious, because the positions of *PG* and *PU* are often treated as belonging together. However, without denying that there are many continuities, what makes Wittgenstein's views of particular interest for my purposes is a subtle but important change in the role of rules in his thinking that occurred between these two works.

The change I have in mind can be summed up as a shift in emphasis, such that Wittgenstein's initial commitment to the idea that Rules Constitute Meaning (RCM) became subordinate to the idea that Practice Constitutes Meaning (PCM). Many readers of Wittgenstein see him as guided by the first idea—that is, the conviction that meaningful language (or actions, etc.) is defined by adherence to a pattern of rules.[2] On such a view, the importance of the language-game analogy is that it provides a model of what it is for rules to be embedded in a practice, how practices constitute a "normative" rather than a merely causal order. However, I want to claim that Wittgenstein came to appreciate that rules do not in themselves make language use meaningful, but are simply symptomatic of humans acting (repetitively) in meaningful ways. To put it another way, he recognized that RCM gets things the wrong way round, that particular actions are meaningful not in virtue of being repeated, but are repeated because they are meaningful. As a result of this shift in emphasis, Wittgenstein modified and circumscribed the role he saw rules playing, and this is what makes his position particularly suited to explicating the pragmatic sense of linguistic signs in a phenomenologically sensitive manner.

The following sections will set out these claims in more detail, but for the moment I want to highlight what is at stake, why this shift in emphasis results in a distinctive reading of Wittgenstein. Commitment to RCM suggests—as the early and mid-period Wittgenstein indeed did—thinking of language as a system of rules and of linguistic competence solely in terms of rule-following, and so thinking of language as pervasively intellectual in character and as embodying a rational system. However, it seems to me that doing this reveals a residual tendency to systematize and idealize language in a way that Wittgenstein later came to see as a mistake. By clarifying some of the modifications to Wittgenstein's position between the two texts, the following sections therefore aim to oppose an intellectualist reading that overassimilates the later position of *PU* to that of *PG*.

At this point I should perhaps reiterate my priorities in discussing Wittgenstein. Part of the aim in the following, as just outlined, is to advance various exegetic claims about the development of Wittgenstein's thinking. However, while I will be aiming to remain true to Wittgenstein's own development, these exegetic claims nonetheless remain of secondary importance. My main motivation for following this trajectory here is to present a phenomenologically plausible conception of the relation between language and practice, one which I take Wittgenstein to have developed by realizing over time various implications of the language-game analogy. Should it, however, turn out that this view is not Wittgenstein's own, then my response would simply be that it should have been, that working out the implications of the language-game analogy ought to have led him to the view presented here.

The approach to Wittgenstein I am proposing might be thought to face two important challenges on general methodological grounds. First, it might be thought that to talk as though Wittgenstein advocated various conceptions of language at different times is to seriously misunderstand his own views. Given his infamous suggestion that philosophy is not about putting forward "theses" (*PU* §128), it seems necessary to ask whether any determinate conception of language can in fact be attributed to Wittgenstein. A simple response to this might be that in making use of his writings we are free to depart from Wittgenstein's own metaphilosophical views. Although true, this response fails to address the potential difficulty that no position, whatever its content, could be said to represent Wittgenstein's own views. For my purposes here it is therefore worth getting clearer about Wittgenstein's commitments and the sense in which a conception of language *can* be attributed to him.

There is much debate about how best to interpret Wittgenstein's "no-thesis thesis." Here I want to distinguish and comment briefly on three approaches, each of which could function as part of a "therapeutic" philosophy aiming to liberate one from some, or even all, philosophical views. Perhaps the most straightforward, first, is to read Wittgenstein as an "ordinary language philosopher." On this approach, philosophical problems are the result of "misunderstandings" fostered by linguistic forms, to be eliminated by returning words from a "metaphysical" to their "everyday" use.[3] This, Wittgenstein thinks, involves simply pointing out the obvious—that which is already open to view—so that any putative "theses" would be accepted by everyone as commonplaces.[4] No "theses" are involved because no theory is advanced; the method is simply to remind us of what we already know in making "ordinary" or "everyday" use of language. On this reading, Wittgenstein's position would face formidable challenges as a way of doing philosophy. It is not, for example, obvious that "everyday" use of language can be delimited; nor, if it could, that this should be considered intrinsically more meaningful than other, "noneveryday," uses of language; nor, even if that were allowed, why everyday language should be considered to eliminate philosophical problems, rather than simply failing to address them.[5] However, the plausibility or otherwise of this way of doing philosophy is of little importance for my present purpose. What matters here is rather that it relies on a conception of language—that applying to, and perhaps manifested in, everyday use—as the basis for debunking "metaphysical" abuses of language. Thus if Wittgenstein is read in this first way, there should be no problem in seeing him as committed to a specific view of language.

A second approach to the no-thesis thesis is to read Wittgenstein as setting out a self-undermining conception of linguistic meaning. On this reading Wittgenstein advances no "theses" because any claims he appears to be making prove meaningless in light of his own claims about language and meaning. This approach has been taken by so-called resolute readings of Wittgenstein's work, which see this breakdown as integral to respecting his stated metaphilosophical aims.[6] If read this way, Wittgenstein's position would presumably be of no use for my purposes here. For the implication would be either that no view of language (or anything else) can be attributed to Wittgenstein, or that his view of language is incoherent and hence of little interest. I will ignore this approach here, however, for two reasons. One is that it seems more suited to the *Tractatus* than to Wittgenstein's later work. The other is that it is far from clear that it yields an intelligible reading of Wittgenstein's position. How, for example, can his claims about meaning self-undermine without being meaningful? Why are his metaphilosophical claims not undermined in the same way as his prima facie substantive claims?[7]

A third, more sophisticated, approach to the no-thesis thesis is to treat Wittgenstein's philosophical method, in particular his style of exposition, as entertaining or discussing various views without positing them or asserting their truth. An interesting example of this approach is provided by David Stern. Adopting Cavell's view of *PU* as a confessional dialogue between the "voice of temptation" and the "voice of correctness," Stern (1996, 444) claims that neither represents Wittgenstein's own views and that these are "two opposing voices, opposing trains of argument, which form part of a larger dialogue in which they ultimately cancel each other out." In a later text, he talks more generally of a voice that puts forward philosophical theories being opposed by a voice that destabilizes those theories, either by satirizing or correcting them. The result is again to be "an unresolved struggle between these voices ... that gives the book its hold over us" (Stern 2004, 50; cf. 53–54). In both variants the claims made in each of the two voices make sense in themselves, but the text acquires a nonthetic character overall through the dialogical arrangement of dissenting voices.

All the same, it is difficult to reconcile the aspiration to ultimate neutrality supposed by this approach with Wittgenstein's clear and pervasive preference for certain views. In Stern's first variant these are expounded, after all, by the "voice of correctness." Indeed, the aspiration to neutrality here appears puzzling, because the whole point of (intellectual) confession is presumably to rid oneself of certain views, the temptations, and so to gain self-knowledge.[8] However, in the second variant too there is a voice presenting theoretical views that appear to be Wittgenstein's own, a voice he uses to express claims that show up the inadequacy of both his own earlier views and the views of others.[9] So although the use of multiple voices might be a means he uses to withhold full assent, in neither case does it seem accurate to suggest that Wittgenstein simply retains neutrality with regard to the philosophical claims he discusses. Hence, insofar as it is possible to discern in Wittgenstein's discussions a coherent set of views about language (rather than ad hoc or inconsistent claims), these can be attributed to his preferred voice and thought of as views to which Wittgenstein was in some sense committed.

A more convincing approach to the no-thesis thesis should strike a balance between acknowledging Wittgenstein's preference for a specific conception of language and recognizing his reluctance to advance this conception as a philosophical theory. The key to doing this—as will emerge more fully in the following—is to keep in mind that Wittgenstein's stated method is what might be described as "illuminating comparison," —that is, offering comparisons or analogies to eliminate misunderstandings, say about the nature of language. Because the aim of this is purely negative, and because this procedure is analogical in character, Wittgenstein's method can be said to be positing nothing. Yet this method clearly involves a conception of language as a means of comparison, and at least some degree of commitment to this conception in that it is considered capable of correcting misunderstandings.

It is of course still significant that Wittgenstein sought to avoid positive theses. This aim should be kept in mind in interpreting *PU*, and it is presumably one factor underlying the work's unorthodox and cryptic literary form—which is surely intended to make it difficult to discern positive views and so to attribute to Wittgenstein a systematic theory.[10] However, the difficulty presented here by Wittgenstein's no-thesis thesis is a practical rather than a principled one: although complicating exegesis, it does not preclude treating *PU* as having a determinate and determinable conception of language—as I will be doing in the following.

The approach to Wittgenstein I am suggesting here also faces a second challenge on general methodological grounds. Even if it is accepted, as just argued, that Wittgenstein is committed to a particular conception of language, it might be questioned whether his approach can be thought of as a "phenomenological" one. In this respect it is relevant to note that there have been various attempts to interpret Wittgenstein as a "phenomenological" philosopher.[11] This claim can be motivated in several ways. An obvious starting point is Wittgenstein's apparently phenomenological emphasis on description rather than explanation (*PU* §109; cf. §§124, 496). In addition, his focus on language as a "spatial and temporal phenomenon" and insistence on the need to understand what is already manifest in linguistic phenomena—what "already lies open before our eyes" (*PU* §126; cf. §89)—might give the impression that Wittgenstein was himself committed to the aim of accurately describing phenomena. An apparently plausible basis for developing interpretations of this kind is then provided by the fact that following his return to Cambridge in 1929 Wittgenstein was for a time interested in the development of what he himself called a "phenomenology."[12] But such interpretations have also been developed on the basis of perceived parallels between Wittgenstein's thinking and the phenomenological movement—typically relying on various of Husserl's claims about phenomenological method.[13]

Despite these attempts, it seems to me that there are good grounds for denying that Wittgenstein himself intended to be a "phenomenological" philosopher in any significant sense. To begin with, his notebooks from around 1929 make clear that his interest was in describing immediate sense data (i.e., phenomenalism) and that he soon rejected his own "phenomenological" project as incoherent.[14] Further, in the absence of explicit references, or any other evidence of influence, there is no reason to suppose that perceived parallels between Wittgenstein's methods and, say, Husserl's go beyond the kind of similarities one would expect to find between any two philosophers. More acutely, Wittgenstein explicitly and unequivocally claims in *PU* §383 to be "analyzing not a phenomenon (e.g., thinking), but a concept (e.g., that of thinking), and so the application of a word." Although the tenability of this distinction might be challenged—isn't the investigation of language simultaneously an investigation of phenomena, as Austin (1979, 182) would later claim?—Wittgenstein's self-understanding is clear: the object of his descriptive study is language, not phenomena as such.

For my purposes, however, it is not necessary to establish that Wittgenstein understood himself as a "phenomenologist." All that is required here is that his conception of language is compatible with a phenomenological approach.[15] That is, to the extent that it proves recognizably suited to describing actual linguistic processes, it can be treated *as though* it were guided by the idea of accountability to phenomena. While this remains to be shown in the following, it seems reasonable to expect this requirement to be met, as Wittgenstein's method of illuminating comparison generates an indirect commitment to the aim of phenomenological accountability. After all, if the comparisons he proposes are to yield clarification and eliminate misunderstandings, the underlying conception of language must cohere recognizably with actual linguistic phenomena. Finally, it seems reasonable to suggest that Wittgenstein's claim not to be analyzing phenomena as such cannot apply in the special case of language, since, whatever else it might exclude, analyzing the "application of a word" presumably just is to study the phenomenon of language.

2 Language-Games

The intrinsic link between the use of language and practice that defines Wittgenstein's praxeological conception of language is best understood by considering the notion of language-games. Although constantly relied on, and perhaps the most distinctive feature of his thinking from 1935 onward, Wittgenstein's characterizations of language-games are terse and somewhat vague. In the earliest of these, in the *Blue Book* from 1933 to 1934, language-games are introduced as "ways of using signs simpler than those in which we use the signs of our highly complicated everyday language"; they are "primitive" or "simple forms of language," from which "we can build up the complicated forms ... by gradually adding new forms."[16]

In *PU* the term *language-game* is introduced, in §7, against the background of an example of a "system of communication," or "complete primitive language," used to coordinate work on a building site (§§3, 2). Repeating a claim made in both the *Blue* and *Brown Books*, Wittgenstein goes on to suggest that it is through such simple "games" that children first acquire language.[17] The term *language-game* is then introduced as applying to "those games by means of which children learn their mother tongue," including simple rote-learning drills and "primitive languages." Further explication comes in §23 with a sizable list of examples of language-games such as giving orders, describing the appearance of objects, inventing stories, telling jokes, translation, and so on. Finally, in both these passages Wittgenstein highlights that language-games concern "the whole process of using words," "the whole: of language and of the activities with which it is interwoven" (§7). Indeed, the expression "language-game" is intended to convey precisely that "the speaking of language is part of an activity, or of a form of life" (§23).

To understand what is important about the idea of language-games, it will be instructive to proceed by considering three apparent problems. First, it is strange that Wittgenstein links language-games with children's acquisition of language. Not only does this look like an (uncharacteristically) empirical claim—for which Wittgenstein offers no evidence—but it is far from obviously true that the examples of §23 are in any way specific to, and in some cases even part of, children's learning of language. Understandably perhaps, Baker and Hacker (1980, 52) claim that by the time of *PU*, Wittgenstein had rejected the idea of an "analytic-genetic" connection between language-games and language acquisition. However, the textual evidence is far from clear, with both §5 and §7 talking as though such a connection exists.[18] Nonetheless, what does seem beyond question is that becoming a competent language user involves acquiring the ability to participate in the various kinds of practice Wittgenstein enumerates in §23, examples that do not in any way reflect the specificity of (children's) learning situations. It therefore seems that Baker and Hacker's intuition is correct and that, despite the way Wittgenstein presents it, the idea of language-games does not require any specific link with language-acquisition scenarios. The point Wittgenstein is making, somewhat misleadingly, in talking of language acquisition is thus best taken as the claim that the constitution of linguistic competence—what this consists in—can be described in terms of the ability to participate in the various kinds of language-game.

Second, Wittgenstein's suggestion that language-games are "complete"—albeit primitive—languages is often thought problematic.[19] Surely, the objection goes, nothing that qualifies as (human) language could be so impoverished. So isn't Wittgenstein simply wrong (e.g., §6) to suggest that a system of communication of the kind described in §2 could even conceivably be a "complete" language? An indication of the weakness of this objection is that is so obvious: if it were a problem for his conception of language-games, one must wonder, how could Wittgenstein have failed to notice? Often the objection also has a question-begging air in relying on unstated or undefended assumptions about what language requires for completeness—simply brushing aside Wittgenstein's opposing thoughts on this issue.[20] But the main reason, I suggest, Wittgenstein was unconcerned about the conceivability of real human communities living in such impoverished ways is its irrelevance to his intended use of language-games. On this Wittgenstein is quite explicit: "Our clear and simple language-games are ... *objects for comparison* which, through similarity and dissimilarity, are to cast light on the conditions of our language."[21] Simplified language-games are thus a methodological device, typological or schematic sketches of situations intended to focus attention on certain features of linguistic phenomena. Their ability to fulfill this function does not depend on the conceivability of human life comprising only a given language-game. Rather, as Wittgenstein explains in the *Brown Book* (Wittgenstein 1960, 81), to think of language-games as "complete systems of human communication ..., it very often is useful to imagine such a simple language to be the entire system of communication of a tribe in a primitive state of society." It is not necessary, but *very often useful*. Thus to think of language-games as "complete" languages is simply a heuristic that is helpful in conceiving them as isolable linguistic subsystems.

Third, at first glance it might be doubted that Wittgenstein tells us anything informative about language-games. Having initially been somewhat casually linked with language learning and primitive languages, Wittgenstein himself implies that the "countless different kinds of use of everything we call 'signs,' 'words,' 'sentences'" are underlain by just as many types of language-game (*PU* §23). Hence, beyond the obvious connotation of having something to do with using linguistic signs, the notion of a language-game looks open-ended, unbounded, or undefined. Before either embracing this as part of Wittgenstein's theory-free approach, or complaining about the apparent indeterminacy of the language-game notion, it is relevant to bear in mind his views about the nature of concepts and concept acquisition. For instance, it should be remembered that Wittgenstein does not think that vagueness in a term's application renders it useless.[22] Moreover, the discussion of "games" in terms of family resemblance is intended to apply precisely to language-games. It is these that paradigmatically have nothing essential, "not one thing at all in common on account of which we use the same word for all," and instead comprise a cluster of elements linked by case-to-case similarities (*PU* §65). Wittgenstein's view of concept acquisition corresponds to this, such that concepts are to be explained or defined by giving examples, suggesting how these can be extended by analogy, and expecting this to allow the term to be used in certain ways (*PU* §§69, 71, 75, 208). Which is clearly what Wittgenstein does in explaining the notion of language-games: he gives examples, hints at its analogical extension, and expects us just to get the hang of it.

Nonetheless, if it is to mean anything at all, there is a need to be clear about the point of speaking of "language-games." A direct answer, found in the *Brown Book* (Wittgenstein 1960, 108), is that the function of a word "can easily be seen if we look at the role this word really plays in our usage of language, but it is obscured when instead of looking at the *whole language-game*, we only look at the contexts, the phrases of language in which the word is used." In *PU* the same thought recurs in both Wittgenstein's early use of language-games—in arguing that the uniformity (*Gleichförmigkeit*) of words conceals the diversity of their functions and that sentences do not have a single underlying form[23]—and their methodological role in eliminating misunderstandings (as Wittgenstein supposed) due to linguistic forms. At the very least, therefore, to talk of "language-games" is Wittgenstein's antidote to the philosophical inadequacy of considering language in a reductive formalist manner.

Against this background it seems reasonable to interpret the notion of a language-game as having two principal features. The first is to emphasize the embedding of language, qua the use of spoken or written signs, within the broader context of human activity. The point of this emphasis is not that linguistic acts are "speech acts" (e.g., promising), acts that would plausibly be impossible without the use of language or that take place "within" language; nor does it allow them contingently to accompany "extralinguistic" actions, running in parallel like a film's soundtrack. Rather, as Wittgenstein states, the point in talking of "language-games" is that the use of signs is interwoven with, or part of, activity more generally. This is precisely what the term *praxeological* is intended to express: that the logic inherent in patterns of language use is determined by the "logic" of practices, or simply that the structure of language is intrinsically linked with that of practice, such that "everything that describes a language-game belongs to logic."[24] As long as this intrinsic link between sign use and practice is borne in mind, it seems to me that Wittgenstein's earliest characterization of language-games—as "simple forms" of language use, from which "we can build up the complicated forms ... by gradually adding new forms" (Wittgenstein 1960, 17)—is also the best. The intrinsic link with practice means that simple forms "of language" are simultaneously simple forms of activity or practice, which is why Wittgenstein can, and indeed does, treat language-games as the kind of components that make up language.

The second principal feature of language-games is therefore to isolate simplified forms of language-using activity. However, while Wittgenstein thought that language could—at least for the purposes of comparison and eliminating misunderstanding—be decomposed into or modeled in terms of pared-down typological situations, the general link between sign use and practice is treated as intrinsic and irreducible, as a feature indispensable for language-games to function as "objects for comparison."

3 Practice Constitutes Meaning

I now want to look more closely at what the claim that the use of certain concepts is internal to certain forms of practice amounts to. To do this, it will be helpful to consider the difference between the language-games of §2 and §8 in *PU*. In §2 Wittgenstein describes a primitive language that "consists of the words *cube*, *column*, *slab*, and *beam*." This language is to allow "communication" between two builders, such that when builder A calls one of the terms, builder B hands A the corresponding object. In §8 this language is extended to include "a series of words used ... like numbers," the terms *this* and *over there*, and color patterns. This enables, as Wittgenstein points out, an extended range of commands to be deployed in the task of fetching building blocks.

The language-game of §8 obviously differs from §2 in introducing new "kinds of words" (*PU* §17). But rather than characterizing these in grammatical or logical terms, which he conspicuously avoids, Wittgenstein is more concerned with the underlying issue of what it is to differ in these ways. His general point is reasonably clear: by emphasizing the use (*Gebrauch*) made of expressions, their role in human practice or language-games, and the variety of functions served by linguistic "instruments," he aims to highlight the inadequacy of the conception of language attributed to Augustine in §1 that models the function of all linguistic terms on the paradigm of naming or referring to objects. The point of §8 thus seems to be to introduce terms that are difficult to assimilate to the paradigms of ostensive definition and naming.

But how exactly does this count against the "Augustinian" view? One might simply respond that in §8 too the basic function of linguistic tools is to refer, that the key difference between the concepts Wittgenstein discusses lies in their referential properties, and that precisely these explain their varying roles in language-games. In other words, why shouldn't one insist on a uniform, reference-based explanation, and downplay Wittgenstein's pragmatic perspective on language as prima facie appearance?

Wittgenstein's answer to this challenge must be taken to include three claims. First, that referential (naming) relations are not primitive, or something about which nothing further of "semantic" relevance can be said. For while intimating that it will always be possible to say that a word "stands for" (*bezeichnet*) something or other, Wittgenstein claims that generally speaking this is uninformative (*PU* §§10, 13; cf. §14). This is reminiscent of Heidegger's view that relationships between signs are "easily formalized" in a way that "levels off" their phenomenal content (*SZ* 78, 88). Both also agree that the antidote to formalism is an awareness of the background of practical activities in which language use is embedded. However, Wittgenstein differs from Heidegger in being more explicit about what gets lost in suggesting that what it is for something to be a referent can be explicated by describing corresponding language-games.[25]

Second, Wittgenstein must be taken to be denying that extensions have explanatory primacy. This is to be expected, if describing language-games shows us what it is to be a referent, as this suggests that the role of expressions in language-games is extension-determining rather than extension-determined. Together these two claims imply that to describe the functioning of expressions merely in extensional terms is a superficial, philosophically unsatisfactory approach. The inadequacy Wittgenstein highlights lies in that, were it not for the different roles they play in our actions, we would have no grasp of the different ways in which, for example, concepts of number, color, form, etc., refer to features of the world. The upshot is a clear asymmetry: someone who understands how the language-game works will understand what reference is in its context, whereas someone who understands which referents are being picked out need not understand what is going on.

Third, *PU* in fact challenges the very idea of what might be called *simple referring*. Wittgenstein not only relates the intelligibility of ostensive definitions to the context of language-games, and discredits the idea of primitive simples that might be referred to in an ideal language, but also suggests that merely to point to something in an inarticulate way is not to understand it (cf. *PU* §§28–37, 46–64, 261). The implication is that any reference to entities or features of the world—*articulate referring*—picks out a referent as a such and such, and that the language-game environment determines what this such-and-such-ness is.[26]

Wittgenstein's own proposal can be understood only by clarifying what he means by the function or the role of words. It might otherwise be felt that his talk of "roles" in a language-game, or "forms of life," is just as uninformative as general talk of "reference." To get clearer about the important difference between the language-games of §2 and §8, and so what it is for words to function in different ways, it will be instructive to consider a variation on the builders' language-game that I will call §2*. Imagine the builders of §2 always saying "red brick" instead of "cube," "white marble" instead of "column," "gray concrete" instead of "slab," and "brown wood" instead of "beam." Each word of §2 is thus replaced by two, one (we can imagine) being used as a color word, the other to identify kinds of material. What difference does this make? Does the use, or the role, of the composite expressions in §2* differ from that of the simple expressions in §2?

This question can be approached in two ways. On the one hand, to an onlooker attending to the conditions of utterance and the effects of various expressions their roles would look the same. To that person the difference between the two language-games is merely notational.[27] On the other hand, if the expressions in the revised language-game are genuinely being used to refer to colors and materials, their roles differ in that the building blocks are picked out in different ways in the two language-games: What in §2 is identified as a cube simpliciter, is picked out in the new variation as a color-material composite—as a red something and as a brick something, and so on.[28] So although the commands uttered in both cases are in some sense coextensive, the referring in §2* is the result of a twofold determination. An indication of this difference in roles is that in §2* one would expect speakers straightforwardly to understand hitherto unheard combinations such as "red marble" and "brown brick." That is, one would expect §2* to have a potential for extension or connection with further practices that §2 lacks. Yet even if not obviously manifested in such ways, the two language-games nonetheless differ in their internal structure.[29]

Though §8 brings an extension of, not simply a variation on §2, it is the second of the approaches above that sheds light on the relevant difference. As in §2*, the positive point of the §8 extension involves not simply new notation (i.e., different words that function in the same way), but new "kinds of words" (*PU* §17) (i.e., words that function in different ways). And what it is for words to function in different ways—as the comparison of §2 and §2* shows—is to be involved in practices of differing internal structure. An effective way of highlighting this difference is to recall Wittgenstein's claim that language and its concepts are instruments (*PU* §569). Instruments can be characterized in terms of both their internal and outward functioning—in other words, both how they work and what they are for.[30] The language-games §2 and §2* differ in their internal but not in their external functioning, just as two clocks might work in different ways while doing the same job outwardly. Introducing new kinds of word in §8 is thus tantamount to the introduction of both new (linguistic) instruments and new forms of practice—such as counting and identifying objects spatially or by color, even where there is little difference in their outward function.

This kind of difference in the way language-games work brings into focus what is meant by saying that the use of certain concepts is intrinsically linked with certain forms of practice. This link—which is the basis of describing Wittgenstein's conception of language as praxeological—has two aspects. On the one hand is the fact that deploying different concepts (not mere notations) amounts to acting differently: "What one calls a change in concepts is of course not only a change in talking, but also one in doing" (*BPP I* §910). A clarification is needed here: I have so far suggested that two approaches can be taken in comparing §2 and §2*, and that the role of expressions remains unchanged or is changed accordingly. These apparently conflicting claims can be reconciled by explicitly distinguishing the "role" of language-games from that of expressions within them and saying—in terms used by Wittgenstein—that a language-game overall has a "point," whereas words have "techniques" of use.[31] Thus a language-game's point is its outward function (what the game is for), whereas the techniques governing an expression's use tell us about its internal functioning (how it works), so that comparison of §2 and §2* illustrates different techniques of word use with the same point. The first aspect of the internal link can therefore be restated by saying that differences in the internal functioning of a language-game, in the way words are used as instruments, constitute differences in the practice.

On the other hand, the comparison of §2 and §2* also illustrates quite clearly the second aspect—that is, how engagement in certain forms of practice imposes constraints on the concepts to be used: "When language-games change, concepts change, and with the concepts the meanings of words" (*ÜG* §65). The basic idea here is that if you want to communicate on a building site, you will need names for the different kinds of object; similarly, if you want to buy apples, you'll need some kind of quantity words and type words for different kinds of fruit, and so on. To put it more precisely, there is an implicit practical requirement that the language used be sufficiently unambiguous for the practice in question to work. The builders' task, for example, requires distinct instructions for bringing each type of building block. In other words, getting the job done implies the need for an appropriate degree of distinction. This kind of requirement does not, as comparing §2 and §2* shows, imply that the same end could not be met with different concepts—although to do so is *eo ipso* to engage in different practices. But it does imply that the concepts used are pragmatically adequate—that is, that they meet the requirements implicit in the overall task of the language-game. With §2 the notion of practical requirements looks like a very clear-cut structural constraint: all that this simple task requires are four distinct commands. This might make the more differentiated language of §2* (sixteen possible expressions) appear to have redundancy as against the optimized (four possible expressions) language of §2. However, the transparency of this optimum is due to the simplified scenario, and in general the question of what is pragmatically adequate ("what works") and efficient is hardly likely to be tractable in the full economy of language.[32] Nonetheless, the idea of practical requirements can be generalized in two points. First, if it is to proceed smoothly, any practice will tend to impose the need for a specific degree of determination. Second, to avoid needless inefficiency, we would expect there to be specific terms to refer to the objects, processes, agents, and so on, that are characteristically involved in that practice.

To conclude this section I want to mention a couple of implications of the praxeological link between language and practice. First, in the vague sense of according some kind of primacy to practice, Wittgenstein's view can be described as a form of pragmatism. Wittgenstein himself polemically disavowed pragmatism, taking this to be the view that to be true is to have the right consequences or to be useful (*PG* §133; cf. *BPP I* §266). The use of this label is warranted, nevertheless, by both Wittgenstein's focus on human practice as constituting linguistic meaning and his frequent likening of words and sentences to tools or instruments.[33] However, as described here, the praxeological conception is a quite specific form of pragmatism, according to which human practices literally shape word use, so that language-games comprise the locus of conceptual articulation.

Second, the idea of a structural link between practice and language can also be followed from the microlevel of individual language-games through to the macrolevel, such that (as previously mentioned) linguistic competence as a whole can be thought of as incorporating the ability to participate in a corresponding set of language-games, as the sum total of articulatory abilities required by these individual practices. In this perspective the "form of life" that manifests itself in a language can, again literally, be thought of as composed of the forms of individual language-games—as the "complicated form" of language built up from the "simple forms" of language-games.[34] This composition of language-games into the whole of a language should not, however, be pictured as the juxtaposition of atomic or hermetic units. Rather language-games are, so to speak, in praxeological arrangement—that is, combined such that they structurally and functionally interlock in roughly the way individual subroutines (functions) within a computer program do.[35]

An important feature of this view of language—relied on the next chapter—is the stratification of language-games: once language is viewed as a functionally interdependent aggregate of constituent language-games, there will be different layers of practice, some of which build on or exploit, and so presuppose, others. Wittgenstein himself often relies on appeals to such stratification: the ability to speak to oneself requires the ability to speak (*PU* §344; *LSPP* §855), to carry out arithmetic "in one's head" presupposes arithmetic skills (*PU II*, 563), pretending to be in pain presupposes possession of the concept of pain (*LSPP* §§861–876), language-games of doubt presuppose those governed by certainty, and so on.[36] As Wittgenstein explains at one point, "it is characteristic of our language that it arises on the basis of solid forms of life, regular doing. ... We have a concept of which forms of life are primitive and which could only arise from such [primitive forms]" (Wittgenstein 1976, 403–404).

4 The Incoherence of Full Determinacy

Wittgenstein's emphasis on the praxeological structure of language naturally connects with his interest in linguistic rules, because whatever structural or functional features characterize language-games must be expressible in terms of regularities that competent speakers grasp. Accordingly, Wittgenstein's notion of rules provides the principal means for conceiving both techniques of word use and the typological structure of language-games—the means, so to speak, for describing the "shape" of what is grasped in linguistic practice.

Another important aspect of the notion of rules is that it connects with a traditional view of intellectual activity that sees, as Kant (1983a, 184 [B171]) succinctly puts it, the intellect as the "faculty of rules" and judgment as the ability to subsume under rules. On this traditional view, what makes something—actions, thoughts, concepts, and so on—meaningful is precisely the fact that it belongs to a rule-governed pattern of use. This thought—the principle that Rules Constitute Meaning (RCM)—underlies Wittgenstein's frequent allusions to chess and is explicit in his likening of words to chess figures, notably in his claim to be talking of language "as of the pieces in the game of chess, by stating rules of play."[37] It is therefore tempting, indeed I think quite common, to interpret Wittgenstein as relying on RCM in a straightforward and strict manner, holding that rules are both necessary and sufficient for meaningful use of words. On this view, rules lay down the legitimate use of words just as rules define the "possible moves" of chess pieces, thus distinguishing meaningful from nonsensical use. Although this view clearly appealed to Wittgenstein for a long time, the next three sections of this chapter will argue that his later views are characterized by a more relaxed attitude to rules, shaped by his abandonment of the calculus model of language and the maturing of the language-game analogy. The task will therefore be to clarify why this change in attitude came about, what it involved, and how it impacts on what the notion of rules can tell us about the structure of linguistic practices. The resultant view of the form and scope of language-game rules will provide a clearer picture of the general functional topology of pragmatic sense as explicated by such rules.

A good place to start in considering Wittgenstein's view of rules is the *Philosophical Grammar* (*PG*), which is characterized by both straightforward commitment to RCM and the attempt to combine the idea of language as a calculus with that of language as a game. The dominant, indeed pervasive, picture is that of language as a calculus: "understanding of language … is of the same kind as understanding, mastering a calculus"; understanding a sign is to make "a step in a calculus (a calculation, as it were)"; "The proposition has its content as part of a calculus. Meaning is the role that a word plays in the calculus."[38] The analogy between language and games is also found, but plays a subordinate role, in Wittgenstein's exposition. And although the term occurs, the idea of "language-games" is not systematically developed.[39] The hybrid of rule system and activity that Wittgenstein had in mind here is epitomized in the following comment: "Language is for us a calculus; it is characterized by *linguistic actions* [*Sprachhandlungen*]" (*PG* §140). Given the fully rule-governed nature of a calculus, it is no surprise that *PG* foregrounds what Wittgenstein calls the "grammatical" rules that are analogously to govern language: "We are interested in language as a process according to explicit rules"; "What interests *us* about the sign, the meaning that is for us definitive, is that which is laid down in the grammar of the sign."[40] Nor is it any surprise that chess—a game played out in an a priori possibility space—is the example of a game that best suits the paratactic picture of language as calculus-game that Wittgenstein here envisaged.

Already in *PG* Wittgenstein's reliance on rules is qualified in two ways to allow for differences between the ideal picture of a calculus and real linguistic phenomena. First, the vagueness, or "haziness of the normal use of our language's concept words," is acknowledged. Wittgenstein highlights that this renders such words neither "unusable" nor inadequate "to their purpose," which would be like saying "the warmth this oven provides is no good because you don't know where it begins and where it finishes."[41] Second, *PG* acknowledges the phenomenon of inexplicitness, that one can both learn a language without the use of explicitly stated rules and make meaningful use of a term without being explicitly aware of its definition (*PG* §§26, 13).

At the time of *PG* Wittgenstein seems to have thought of inexplicitness as unproblematic. His concern, as in the *Tractatus*, was how meaningful use of language is constituted by semantic rules, not whether speakers have explicit knowledge of these—that is, with the constitution and not the epistemology of meaning (thus assuming that the former can be treated independently of the latter). Nevertheless, there is some tension. Wittgenstein describes the sense of a sentence as both the role it plays in a calculus and what one says when asked about it (*PG* §84), as if these were in preestablished harmony. Accordingly, some passages (§§13, 84, 28) seem to suggest that speakers are able, when asked, to explicate the rules they are following. Assuming this ability by default reduces inexplicitness to a superficial phenomenon and preserves the idea that language is in principle rationally transparent. Elsewhere (§26), however, Wittgenstein seems to accept that speakers might have difficulties in explaining rules.[42] In this case inexplicitness is a significant phenomenon, but when combined with the assumption that language is a calculus, it implies that speakers have "implicit" or "tacit" knowledge of its governing rules: our language use is thought of as manifesting rules that we must be assumed in some way to know, even though we might never have heard them explicitly stated and ourselves be unable to state them. Hence the question of *PU* §75: "What does it mean to know what a game is? What does it mean to know it and not be able to say it?"

It is worth considering how in *PG* Wittgenstein was able to reconcile these phenomena with his then dominant picture of language as a calculus. The answer lies in a methodological stance anticipating that of *PU*: already in *PG* Wittgenstein sees his philosophy as aiming to eliminate misunderstandings by clarifying our use of language (*PG* §§72, 32). To do this, he relies on a method of comparison: "We consider language *from the viewpoint* of a game according to fixed rules. We compare it with, measure it according to, such a game" (*PG* §36; cf. §26). Thus Wittgenstein does not assert the factual reality of the calculus model of language, but offers it as an idealized picture for comparison, with the aim of clearing up certain misunderstandings. Once this concession is made, it is simply business as usual for the RCM principle. The phenomena of vagueness and inexplicitness of speakers' knowledge can be accommodated, since neither directly challenges either the necessity or the sufficiency of rules to meaning constitution. Although suggesting discrepancies between the ideal picture of rule-governance and phenomenal facts, these phenomena can be thought of as concessions or anomalies, ways in which natural languages merely approximate to the calculus ideal.[43]

We can now start to bring into focus the relationship between *PG* and the later *PU*. It should be noted first that although often looking like conventional presentation of a systematic theory, *PG*'s clear statements about the nature of meaning are rendered ambiguous by the comparative method just described and must be taken not as straightforwardly advancing "theses," but as offering a supposedly illuminating comparison. In other words, although not embellished with the rhetoric of the no-thesis thesis, a nonthetic approach is already found in *PG* and is not new to *PU*. All the same, as pointed out in section 1, this nonthetic approach relies on an underlying conception of language as a means for comparison, which, although its empirical reality is not asserted, is obviously assumed to approximate well enough to the reality of language to dispel misunderstandings. It is here that a significant difference between *PG* and *PU* emerges, with the calculus model being eclipsed by the language-game analogy. *PG*'s reliance on the calculus model is therefore significant in assuming both that this offers an illuminating comparison (i.e., that it is a good model for dispelling misunderstandings) and that it is straightforwardly compatible with the games analogy and the latter's implications. The question is therefore: What changed? Why did Wittgenstein give up the calculus model as an analogy of language?

The answer is that he came to the conclusion that the calculus model is incoherent. This conclusion emerges in *PU*'s first discussion of rules

(§§81–88) and their relation to the nature of concepts (§§65–88). The main target there is Frege's idea that concepts should be fully determinate, an idea that also animated Wittgenstein's own earlier model of language as a calculus (cf. *PU* §71; Frege 1994a, 31). The idea in question is that for an expression to have determinate meaning there should be no possible doubt as to its conditions of applicability, so that rules for its use will cover all conceivable circumstances. This idea is challenged in §84 of *PU*: "But what, then, does a game look like that is everywhere bounded by rules? Whose rules allow no doubt to creep in, block up all its holes?—Can we not think of a rule governing the application of the rule? And a doubt which *that* rule removes—and so on?" It is here, in §§85–87, that Wittgenstein introduces the difficulty of an infinite regress of rules: seemingly any statement of a rule can be misunderstood, requiring disambiguation by a further rule (statement), which can in turn be misunderstood, and so on. Blocking the possibility of such a regress in principle would seem to require rules that admit of no possible misunderstanding, or as Wittgenstein puts it, "doubt." Yet, as Wittgenstein concedes, there can be no such regress-blocking rules.

The question of how this regress-of-rules argument is to be interpreted is particularly vexed. One interpretation is that Wittgenstein accepts the conclusion of this argument, so that rules are afflicted by principled indeterminacy, resulting in a general skepticism about meaning.[44] I want to suggest instead that Wittgenstein himself intends the regress-of-rules argument as a standard reductio ad absurdum, such that the absurd conclusion leads to a rejection of the premises that generate it. Thus viewed, the regress-of-rules argument reveals that the demand for a system of univocal, fully determinate rules—a calculus—underlying language is unstable because it renders inconceivable the ideal of determinacy that would be required to satisfy it. Its point is that, because it generates a regress and hence incoherence, the demand for full determination by rules is to be rejected.[45]

This interpretation of the argument is directly supported by both the way it is introduced (in the challenge of §84, as above) and concluded in §88. In §69 Wittgenstein had conceded that his view *might*, if a suitable definition of exactitude could be given, render concepts "inexact." However, returning to this question, §88 concludes that there is no provision for "*one* idea of exactitude; we do not know what we are supposed to imagine this to be." That is the implication of the regress-of-rules argument: we simply do not know what the demand for full determinacy, or exactitude, could amount to. This recognition is what separates *PU* from *PG*. For the regress argument implies that the calculus model is built around an unintelligible ideal and an incoherent conception of linguistic rules. Moreover, these deficiencies strike at the heart of *PG*'s vision and cannot be protected by the nonthetic method. Since it is based on an incoherent conception of linguistic rules, the calculus model cannot function as an illuminating comparison—rather its intimation of ideal determinacy is precisely what sows the seeds of confusion.[46] Hence the difference between *PU* and *PG*: the calculus model is dropped—cannot coexist with the language-game model—simply because it is, or at least Wittgenstein believed it to be, incoherent.[47]

If the regress-of-rules argument is, as I have suggested, intended as a reductio, the incoherence of ideal determinacy ought to be taken not as denying the phenomenon of linguistic rules, but as requiring them to be conceived in some other way. The following two sections attempt to show that this was Wittgenstein's response, that having dropped the calculus model he developed the language-game analogy further, leading to a revised overall picture of rules and their role in language. Tracking the way Wittgenstein reconfigures his conception of rules to avoid recourse to an ideal of determinacy (section 5) and the limitations he sees in the role of rules in language-games (section 6) will thus serve two purposes. While the main aim remains that of clarifying Wittgenstein's view of how language-games are structured, it will also provide further, indirect support for the interpretation of the regress-of-rules argument just proposed.

5 Rules Reconfigured

I want to characterize the way Wittgenstein reconfigured his conception of rules following *PG* in terms of two main features, which I will describe as the *empirical attunement* of rules and a *pragmatization of determinacy*. Wittgenstein's general acceptance of the empirical attunement of rules is signaled in *PU*'s conspicuously mundane characterizations of rules: they are calibrated to "normal conditions" and explanations of them are somewhat ad hoc, serving to eliminate particular, rather than all conceivable uncertainties (*PU* §87; cf. §142). It also emerges in Wittgenstein's terse response to the regress-of-rules argument in §85: whether a given sign, explanation, definition, and so on, is sufficiently clear or prone to misunderstandings, he says here, is not settled by an underlying calculus, but is an empirical matter. Given Wittgenstein's frequent emphasis on the difference between grammatical and empirical claims, this grounding of his conception of rules in empirical facts might appear puzzling. So what is the point of this claim?

Superficially it might seem to be either an acquiescent matter-of-fact response to the regress-of-rules danger, or to suggest that it is just a contingent fact that we and the signpost manage to do the right thing most of the time. But there is a more subtle point to the claim that signs are "in order" when they fulfill their purpose "in normal conditions" (*PU* §87). For if there were no general agreement in judgments about, say, how to interpret signposts, then signposts would not be fit for their purpose, in which case either we would have a different solution to the problem of indicating directions, or that kind of practice would not be possible in human life. Thus it is presupposed with regard to anything correctly described as a rule characterizing human action that there is some such normal agreement in judgments; if there were not, there would simply be no rule.[48] Accepting causal facts as the "framework" underlying the operation of language (*PU* §240) is therefore not simply an intellectual quietism, nor indeed an indirect solution to a supposed rule skepticism.[49] It is instead part of Wittgenstein's response to the incoherence of his earlier conception of rules. Rather than seeing the notion of a rule as being endangered by the mere possibility of conceivable alternative interpretations, Wittgenstein now emphasizes that the notion of a rule is only ever in play when convergence in judgments and regularities in use are presupposed. Thus the reference to "normal conditions" indicates that the requirement for a background of regularity, which the later Wittgenstein often highlights as necessary in talking of rules, is being met.[50]

Wittgenstein's recognition of the empirical attunement of rules is more pronounced still in his post-1945 thinking, where linguistic rules are accepted as exhibiting features characteristic of empirical rules. This can be well illustrated with several comments from his *Last Writings on the Philosophy of Psychology* (*LSPP*). In describing the use of words, "we must find something characteristic in these individual cases, a *kind* of regularity"; bearing in mind that "we don't learn the use of words with the help of rules," Wittgenstein explains that there is a "ROUGH lawfulness" in the use people actually make of words.[51]

The way to interpret these comments, I suggest, is that Wittgenstein now saw linguistic rules as being similar in shape to empirical rules, as having the same kind of structure, form, or shape as the latter. This is not to say he now equated linguistic rules with empirical or observed regularities, which would deny the normative character he often emphasized as distinguishing rules from mere causation, nor that he gave up his distinction between "grammatical" and "empirical" propositions. The point of the claim is perhaps best seen by tapping into scientific common sense: as any trained scientist or engineer (such as Wittgenstein) will know, experimental data is messier than the theoretical laws that model it. The results of measurements are "imperfect," more or less statistically scattered or laden with experimental margins of "error," and it is only through processes of extrapolation, smoothing, and approximation that useful, and perhaps simple, representations of rules are attained. Conversely, because they cannot express the messiness or inexactitude of experimental results that support them, theoretically formulated laws invariably look better—simpler, cleaner, more precise—than they are. In this respect they "idealize" experimental data in a way that might tempt us to misinterpret them as exact laws. Hence, in saying that linguistic rules have the same "shape" as empirical rules the thought is that real linguistic phenomena stand in the same kind of messy or imprecise relation to any representations of linguistic rules—they are, as Wittgenstein suggests, "roughly lawlike."

Another way of approaching this thought is to see real language use as building up patterns that are analogous to statistically normal distributions in being spread around an average value. Thus in a so-called normal (or Gaussian) distribution, the probability of occurrence of some variable (event) is represented for a continuous range of its values by a bell-shaped curve, which can be more or less sharply focused according to the amount of scatter or spread of such occurrences around its mean value. Wittgenstein's thought, I am claiming, is that linguistic rules are to be conceived analogously, such that the "normal" conditions governing possible uses of terms similarly exhibit a characteristic degree of tolerance or spread. As in the statistical case, these normal conditions can be more or less narrowly focused: occasionally linguistic rules will be sharply defined, but in general "it is not fixed in advance that there is such a thing as 'a *general* description of the use of a word.' And if there is such a thing—then it is not fixed how determinate such a description has to be."[52] In this vein Wittgenstein even declares that the "greatest difficulty" in his "investigations is to find a way of representing vagueness."[53] As in the scientific case, I suggest, the difficulty is that any statement of a linguistic rule (e.g., in a "grammatical" proposition) will be unable to capture the "messiness" of linguistic phenomena and so remain open to misinterpretation as standing for an exact law. Nevertheless, Wittgenstein's frequent talk of "characteristic" features of language-games and his description of words as having a "physiognomy" can be seen as a response to this difficulty.[54] For in the physiognomic analogy characteristic features are not well-defined, constant shapes, but distinctive forms that emerge from the field of continuously varying momentary (facial) expressions.

Recognizing their empirical shape has an important implication for the functional character of linguistic rules. Wittgenstein has suggested that linguistic rules, like statistical rules, are generally characterized by some tolerance or vagueness (*Unschärfe*). This suggests that, rather than being a yes/no matter, conformity with rules is gradual in character and should be thought of in terms of how close given instances are to the statistical or empirical average. While this graduated picture does not undermine the distinction between clear cases of conformity and nonconformity to rules, it detaches the idea of linguistic rules from that of a sharply defined functional space that might be characterized in terms of discrete valencies, in particular bivalence. Yet to protest that such empirically attuned rules are "not rules," or not usable as rules, would be to make precisely the mistake *PU* inveighs against, namely, to lapse into thinking that (all) rules must be ideally determinate or precise.[55]

The empirical attunement of rules is also reflected in *PU*'s picture of the language-world relation. An apparent complication here is that for Wittgenstein language-games clearly have some kind of operational structure or logic that cannot be reduced to causal facts or regularities. This view, sometimes referred to as the "autonomy" of grammar or language, is (unsurprisingly) advanced most clearly and most forcefully in *PG*: "Grammar is accountable to no reality. Grammatical rules first determine meaning (constitute it) and are hence responsible to no meaning and [are] to this extent arbitrary."[56] This "autonomy" vis-à-vis reality encompasses four claims: The structure of language, as "grammar," cannot be understood in terms of (a) purposes that language has, (b) the effects of language use, either intended or actual, including on agents, or (c) true representation of the world; rules of language are hence (d) arbitrary and conventional.[57] These claims are an important part of the overall picture of language in *PG* and its strong commitment to a scheme-content dualism: as a calculus, language is there considered an a priori medium of meaning constitution, with language-world "connections" being set up by "ostensive explanation" (*PG* §138).

There are admittedly vestiges of these views in *PU*, with comments suggesting that language-games should not be understood in relation to effects (*Wirkung*) and a passing allusion to the conventional character of language.[58] And in §497 Wittgenstein suggestively writes: "The rules of grammar can be called 'arbitrary,' if this is supposed to say that the purpose of grammar is only that of language."[59] However, this is manifestly *not* to claim that language (hence grammar also) does not have purpose(s) in terms of which it might be described, and indeed the requirement that sentences should have some function is used to challenge the idea that grammar is "arbitrary" in §520. It seems to me that this signals a significant, and judicious, weakening in *PU* of claims to the "autonomy" of language. For whereas *PG*'s scheme-content dualism makes it look like a mere coincidence that language has anything to do with the real world, the whole idea of the notion of language-games is to emphasize that language is intrinsically bound up with the rest of the world. Consequently, although the structure of language-games cannot be described straightforwardly in causal terms, neither can language be thought of as "autonomous." It must instead be thought of as *attuned to the world*, being largely shaped by the purposes that language-games serve and—on pain of language-games' breaking down (*PU* §142)—calibrated to "normal" causal conditions.[60]

The second main feature of Wittgenstein's reconfigured conception of rules results from their being conceived without any recourse to an ideal of determinacy. As mentioned above, in *PU* §85 Wittgenstein states that whether a rule gives rise to doubt or is open to interpretation is an empirical claim. The immediate point of this comment is clearly that although in practice we usually understand signs in a certain way, it is possible to think of alternative interpretations of them (cf. *PU* §201). However, as it comes in the context of a critique of the ideal of full determinacy (cf. *PU* §84), the key question is what this comment tells us about the relationship between language and determinacy. In this respect Wittgenstein's claim is that there are not rules to cover every eventuality, but that there is a degree of determination that usually suffices for practical purposes.

This claim can be interpreted in two ways. The first interpretation makes it an expression of resignation in the face of intellectual anxiety: "sufficiently determinate for practical purposes" here sounds like "well, we get by, but isn't that surprising, given that the rules we follow are in principle indeterminate." This suggests a perpetual difficulty of principled indeterminacy, such that language use involves constantly deciding among an inexhaustible range of options. The vertiginous character of this interpretation results, however, from acknowledging the desideratum of greater determinacy than that actually required. Yet this seems to miss the point of Wittgenstein's critique of the idea of full determinacy. For acknowledging this desideratum is a slippery slope, hinting at an iterative tendency that could be halted only by some ideal terminus. Hence this first interpretation faces two problems. First, it seems to be holding actual language-games accountable to an impossible, indeed unintelligible, standard of determinacy. Second, because it was the point of Wittgenstein's regress-of-rules argument to discredit the idea of full determinacy, it is difficult to see how this first interpretation can claim to capture Wittgenstein's own intentions.

Against this, Wittgenstein's praxeological conception of language suggests a second, less timid interpretation of the claim that a certain degree of determination usually suffices for practical purposes. Recall that a similar claim was encountered above (section 2) with the thought that acting in particular ways imposes practical requirements on language use. The importance of such requirements, I now want to suggest, is that determinacy is a practice-immanent and practice-relative notion, such that what suffices for practical purposes (i.e., is pragmatically adequate) becomes the *criterion* of determinacy. To say that the rules we follow ("normally") suffice for practical purposes is then not to express resignation, but to affirm that they meet the *standard* of determinacy required to act in the relevant way. It is in this sense that Wittgenstein effects a *pragmatization of determinacy*, the second main feature of his reconfiguration of rules.

It should be noted that Wittgenstein does not, strictly speaking, present things in these terms. Nonetheless, there is evidence for attributing to him this pragmatization of the notion of determinacy. For example, *PU* §69 claims that strict definition of concepts is relative to particular purposes, while §88 strongly hints that exactitude should be understood as practice-relative.[61] Wittgenstein also insists that possible doubt does not entail actual doubt (§84), or even that one "could have doubted" (§213; cf. Wittgenstein 1976, 399), and that doubts—particularly skeptical doubts—stand in need of motivation (*ÜG* §392). These claims can be understood in light of Wittgenstein's view that taking some things to be certain—that is, in a particular way, at a particular degree of determinacy—is constitutive of any language-game. That one "could not" have had "conceivable" doubts is a praxeological impossibility in the sense that acting in such and such a way constitutively excludes certain questions.[62]

On this second interpretation, rather than accepting and stoically facing up to principled indeterminacy, Wittgenstein provides an alternative view of what constitutes semantic determinacy. It remains the case, of course, that where rules are found, a greater degree of determinacy can be thought up, a "doubt" requiring a further rule to settle it. On this second view, however, someone who proposes the "doubt" is not identifying an ambiguity operant in the established practice, but is entertaining the prospect of a different, extended, form of practice in which a new choice, or a new degree of freedom, does play some role.[63] If this new form of practice has some practical point or importance, one would expect it to lead to an improvement to, or refinement of, the original language-game. But once a language-game fulfills its purpose (e.g., as the use of signposts does), then the new doubt simply introduces redundant "determinacy," the kind of decision that plays no role in actual practice. Hence, Wittgenstein concedes that one could imagine people thinking with a far higher degree of determinacy than "we" do, but highlights that more precise concepts might not have the same practical value (*PU II* 510, *LSPP* §267). This is the basis of his attitude toward philosophical "doubts": if these play no role in and lead to no improvement in practice, the determinacy they aspire to is redundant—a spinning of wheels not connected to the relevant mechanism.

6 Rules Constrained

In addition to the reconfiguration of his conception of rules, a second major aspect of Wittgenstein's response to the breakdown of the calculus model is to see rules as having a more circumscribed role in linguistic phenomena. His recognition of various constraints—I will discuss four—shows how Wittgenstein's commitment to the principle that Rules Constitute Meaning (RCM) became subordinate to the idea that practical considerations are what make language use meaningful.

Two qualifications to the role of rules—vagueness and inexplicitness—had already been acknowledged in *PG*, despite looking anomalous in the optic of the calculus model (see section 4). Freed of this model's constraints, however, these phenomena are cast in a different light in *PU* and become integral parts of Wittgenstein's revised view of rules. Thus the notion of vagueness plays a prominent role in *PU*.[64] For example, in discussing games Wittgenstein allows "that the extension of the concept is *not* closed off by a limit" (*PU* §68) and suggests that, though the use of terms can be fixed for certain purposes, this is not normally the case. With his rejection of ideal determinacy, Wittgenstein's emphasis that "inexactness" does not make a term "unusable" becomes particularly poignant (*PU* §88). In *PG*, despite acknowledging the practical acceptability of vague terms (§76), Wittgenstein had not allowed this insight to affect his reliance on the calculus model of language, with its constant intimations of the desirability of full determination. There vagueness was an inconvenience, to be reconciled with the calculus model via a noncommittal method of comparison, rather than a positive and integral aspect of *PG*'s conception of language. The situation is fundamentally different in *PU*. For once the ideal of univocal "exactness" is discredited by the regress-of-rules argument, "vagueness" can, and indeed must, become an integral part of Wittgenstein's positive conception of language, a fact reflected both in his discussions of determinacy and in his recognition of the empirical shape of linguistic rules.

Similar considerations apply to the theme of inexplicitness. In *PU*, as in *PG*, Wittgenstein allows both that language can be learned without the use of explicit rules, by observing and getting the hang of it, and that speakers need not have an explicit awareness of the rules they are following in competent use of language.[65] The implication is that there are two kinds of criterion for adherence to rules: the ability to produce a correct pattern of use and the ability to explain, or form a conception of, that pattern of use—or, as Wittgenstein himself puts it in *PG* (§42), providing "samples of use" and "stating the rule." In this perspective acknowledging inexplicitness means accepting the possibility of a mismatch between a speaker's abilities to produce examples of use and to define reflectively the rules governing language use. In *PG* the phenomenon of inexplicitness was again anomalous, causing difficulties from which the calculus model needed protection through the nonthetic comparison method. Since in using a calculus each step must be covered by a statable rule, the kind of rational opacity suggested by such a mismatch conflicts with the demand for rational scrutability implicit in the calculus model. On this model there is no provision for the abilities to produce and to conceive of correct moves to come apart, so that inexplicitness can only look like a failure of competence. In addition, as mentioned above, when combined with the assumption that rules underlie all meaningful use of language, the phenomenon of inexplicitness generates the need for an account of "tacit knowledge"—of what it is to know a rule that one cannot state.

The situation is again fundamentally different in *PU*. For once the calculus model is rejected, there is no need to assume that the abilities to produce examples of use and to explain rules generally are necessarily correlative. Accordingly, *PU* allows the possibility of using language in a way we cannot explain, such that we might find our own use of language reflectively opaque.[66] In such cases Wittgenstein's advice is to "let yourself be *taught* the meaning by the use" (*PU II* 550). This kind of mismatch between the abilities to produce and to conceive of correct use is presupposed by the inability to survey (*übersichtlich darstellen*) our use of words that §122 describes as a "main source" of our lack of understanding. Indeed, if these abilities were in perfect equilibrium, such lack of understanding—the Augustinian quandary (cf. *PU* §89) that Wittgenstein sees as characterizing philosophy—could presumably never arise. Moreover, in the *PU* perspective there is no need for confusing terms such as *tacit* or *implicit* knowledge of rules—terms that seem inappropriate in describing rules manifested publicly in linguistic behavior. Instead, so-called tacit knowledge can be equated with the ability to produce a certain pattern of use, and so-called explicit knowledge with the ability to form some conception of those patterns.

As contrasted with *PG*, however, *PU* also introduces two new constraints on the scope of RCM. The first of these, a recognition of *particularity*, is suggested by Wittgenstein's consideration of making up or changing the rules of a game "as we go along" (*PU* §83). He introduces this by imagining people playing with a ball in a way that seems to be interrupted and modified randomly, and suggests that someone might claim the "people are playing a ball game the whole time and so complying with *determinate rules at each throw*" (italics added). Admittedly, the point of this comment is not immediately clear. Taken in isolation, it seems to affirm that rules are present, always defining the activity, even though they are constantly changing. In this way it might be interpreted as a stubborn defense of RCM. Nonetheless, the context of this comment makes clear that its point is rather to highlight the absurdity of such a stubborn defense and the notion of rules it relies on. Wittgenstein not only considers a background of regularity, of normal conditions, a necessary component of rules, but unequivocally excludes application of the notion of a rule to one-off cases in *PU* §199. This is also the message of the immediate surroundings of §83: §81 initiates Wittgenstein's critique of his previous view of language as a calculus "according to determinate rules." §82 then mentions three criteria for the presence of rules, while hinting that none might be fulfilled. Finally, §84 recalls Wittgenstein's view (already expressed in §68) that the use of words is not everywhere bounded by rules, before adducing the regress-of-rules argument against full determination by rules. In this context the idea of each throw being according to determinate rules (§83) is clearly a desperate attempt to find omnipresent rules underlying meaning.

All the same, in talking of "rules" that are made up as we go along, Wittgenstein allows a certain ambiguity to persist. The reason for this, I suggest, is that this possibility picks out a difference he wanted to respect between correctness conditions in a particular context and more broadly established patterns of regularity. For despite introducing a verbal difficulty in interpreting §83, "rules made up as we go along"—that is, particular in-context correctness conditions—would allow things to be done appropriately or rationally without their being covered by an overarching system of rules, an allowance that would accord with the dominant line of argument in §§81–88. Thus I am suggesting that in *PU* Wittgenstein has a bipartite notion of correctness conditions—covering both general patterns (rules) and particular in-context correctness ("rules made up as we go along")[67]—and that conforming to rules, in Wittgenstein's sense of general patterns, is not necessary for meaningful language use (or actions).

The final constraint on the scope of RCM can be called *nonregulation*: the fact, which Wittgenstein supports with the regress-of-rules argument, that the "application of a word" is not "everywhere bounded by rules" (*PU* §84). Considering the objection that this would leave the use of an expression "unregulated," Wittgenstein responds that "there is also no rule, for example, as to how high, or how hard, the ball may be thrown in tennis, yet tennis is a game and it too has rules" (*PU* §68). As presented, this is the modest claim that an activity can still be said to be determined by rules even where rules do not fully determine the possible moves. However, I think it is important that this comment brings out a crucial aspect of the games analogy, insofar as games are precisely a kind of activity in which not everything is determined by the rules. Some evidence that Wittgenstein shares this view is found in his *Remarks on the Foundations of Mathematics* (*BGM*), where he entertains the idea of a "game" in which it turns out that whoever makes the first move is predetermined by the rules to win. Wittgenstein denies that such a situation—that is, one fully determined by rules—would be called a "game" (*BGM* III §77; VII §§13, 27). What Wittgenstein is rightly acknowledging here is that it is essential to games that there is some *Spielraum*, some openness or room for maneuver, in which the game unfolds.[68]

PU's treatment of the four features just discussed can be seen as developing the implications of the language-game analogy once liberated from its forced coexistence with the calculus model. An important aspect of this liberation is that with the transition from *PG* to *PU* the games analogy is broadened beyond chess to games more generally (cf. *PU* §65). As a result, each of the four features becomes an integral part of Wittgenstein's conception of language, rather than being something anomalous that needs to be reconciled with the calculus model. The transition toward this later position amounts to a shift in emphasis, such that practice rather than rules is what matters in constituting meaning. For while Wittgenstein's reconfiguration of linguistic rules (as empirically attuned and pragmatically determinate) might be reconciled with the RCM principle as some kind of approximation, and while *PG* treated vagueness and inexplicitness as acceptable anomalies, his recognition in *PU* of particularity and nonregulation directly limit and so challenge that principle. Thus Wittgenstein's later position involves concessions that are possible only once the idea that Rules Constitute Meaning (RCM) becomes subordinate to the idea that Practice Constitutes Meaning (PCM) in the way suggested by the language-game analogy.

Although it requires going beyond what Wittgenstein himself says, I want finally to highlight two further implications of the language-game analogy, in particular the thought that playing a game involves more than mere adherence to its constitutive rules. The first is that language-games should be thought of as varying in the degree to which they are rule governed. Some activities, such as mathematics, involve well-defined rules and highly articulate structure; others—Wittgenstein cites aesthetics and ethics as examples (*PU* §77)—do not. This does not imply that loosely regulated activities are deficient. Since, on the language-game conception, rules in language are linked with regularities in practice, we should expect just as much or as little regularity of structure in patterns of language use as there is homogeneity in the kinds of activity in which language is used.[69] At least two factors will influence the degree of regulation. The first is that a uniformity constraint is constitutive of some kinds of activity. Thus mathematics does not differ from, say, aesthetics simply in that mathematicians define terms precisely whereas aestheticians do not. Rather, mathematics is an intrinsically standardized activity, one in which results and agents must agree if what they are doing is to count as mathematics, whereas aesthetics does not constitutively require such uniformity (and each kind of activity might otherwise lose its point).[70] The second factor is how closely linked particular concepts are with the successful execution of corresponding kinds of action. Typically in highly specialized and/or technical activities the use of specific vocabulary will be essential to successful coordination of the relevant activity. To put it in terms used in section 3, there will generally be a practical requirement for a sufficient degree and appropriate kinds of linguistic differentiation. Thus how much room for maneuver there is in the use of words will generally be a function of the type of activity in question. For these reasons, I suggest, language-games are best thought of as *contexts of regulation* in a dual sense: both the extent to which rules are found and what those rules are—how much regulation and which—are constitutive features of the respective practice.[71]

The second further implication is that the language-game analogy suggests, in contrast to the calculus model of language, a broader notion of competence. Of course, it is true that both bad players and good players, football players say, must adhere to the rules of the game in order to count as playing it at all. But what constitutes competence—what good players have, and poor players lack—is knowing what to do over and above the rules. (And, of course, being physically capable of doing it!) Thus a player's competence involves a general understanding of the game (e.g., knowing what it is to be playing well, to be lucky or unlucky), strategic awareness (how to react in various situations, which tactics might consolidate a strong position or help in a bad one), and dexterity or skill, which typically requires making good decisions and simply doing the right thing on the spur of the moment. This broader notion of competence is another factor—alongside the incoherence of the calculus model—that makes the union of language-games and the calculus model unstable. For, quite clearly, not only do these aspects of competence in playing a game go beyond the rules that define the game, but they cannot be characterized by hard-and-fast "rules."[72] Equally, though not answerable to rules, these aspects of competence are not a matter of chance. Alternatives to a given strategy or move can be entertained and found better or worse than what actually happened (fortunately for sports commentators and pundits). These aspects of competence differ from "rules" in being ad hoc or contextual, yet they remain subject to considerations of appropriateness—that is, graduated and nonexclusive evaluation, rather than a sharp distinction between correct and incorrect. Together they can be described as the "phronetic aspect" of linguistic competence: the disposition, acquired by experience, to do the right thing, all things considered, in particular circumstances.[73]

7 Pragmatic Sense

So far this chapter has focused on setting out the late Wittgenstein's praxeological conception of language. Having examined the inherent connection between language and practice suggested by the language-game analogy, I have argued that between *PG* and *PU* Wittgenstein's views shifted under the influence of the language-game analogy, and that his guiding idea became that Practice Constitutes Meaning rather than Rules Constitute Meaning. This resulted in a more "relaxed" view of linguistic rules than that suggested by his earlier model of language as a calculus, consisting of a revised (reconfigured) notion of linguistic rules and a more circumscribed (constrained) role being attributed to such rules in his conception of language. The motivation for discussing Wittgenstein here was to expand on the general connection between language and practice recognized by Heidegger, and in particular to explicate the notion of pragmatic sense that was introduced in chapter 3 as the sense signs have due to the respective roles they play as instruments in established practices. To conclude this chapter, I therefore want to highlight what this reading of Wittgenstein teaches us about pragmatic sense and how it can be seen as forming part of a phenomenological conception of language guided by the idea of phenomenological accountability.

The core feature that obviously makes Wittgenstein's conception of language suitable for explicating the notion of pragmatic sense is the "praxeological" connection he recognizes between language and practice. This feature is central both to his view of language-games as the context of conceptual articulation, as we have seen, and to the idea of forms of life as functionally interconnected aggregates of language-games. Given this connection, the notion of language-game rules provides a means for describing the constitution or structure of both language use and the broader practices in which this occurs, and so for defining pragmatic sense.

This internal link between language and practice is also one of two features that make Wittgenstein's views particularly suited to phenomenologically accurate description of language. For in contrast to formal, particularly calculus-based, conceptions of language, language-games are clearly recognizable as schematizations of actual linguistic practice. As such they picture language as processual in character, rather than as some kind of abstract object or structure, so that Wittgenstein can plausibly, trivially even, claim to be talking about language as a "spatial and temporal phenomenon," rather than as "some nonspatial and nontemporal nonentity."[74] This general orientation, as the preceding chapters have emphasized, is shared with both Heidegger's and Merleau-Ponty's conceptions of language. However, in terms of phenomenological accountability, an obvious advantage of Wittgenstein's approach is its focus on everyday language use. Thus he directs attention to the most common uses of language, in contrast to Merleau-Ponty, who focuses on the presumably less usual cases of creative expression and tries to pass off the day-to-day life of language as a "secondary" phenomenon. Last but not least, by adopting this focus, Wittgenstein is able to do justice to the fundamental importance of practice (i.e., human action) in shaping the articulatory feats of language—a basic functional connection that Heidegger also recognizes but says very little about.

The second feature that is central to the ability of Wittgenstein's later views to accurately describe linguistic phenomena is what I have been calling his more relaxed view of linguistic rules. The reason for this is that the picture of language Wittgenstein arrives at—on which vagueness, inexplicitness, particularity, and nonregulation become integral features—has important consequences regarding what rules can tell us about the structure or "shape" of linguistic practices and hence pragmatic sense.

One such consequence, as the features of nonregulation and particularity imply, is that rules can be expected to provide only an incomplete picture of language-games. It is true (tautologically) that irregular features are not characteristic aspects of language-games. Such features—corresponding to what I have called the phronetic aspect of linguistic competence—are hence "imponderable" in the sense that they elude a general, rule-based description of language-games.[75] While this imponderability means that Wittgenstein's view of language inevitably focuses on rules, openness to the phronetic aspect of linguistic competence should be thought of as their constant other, as part of the ground of human activity from which the figure of rules emerges. In other words, Wittgenstein's view of rules has two sides. On the one hand, it recognizes the important implication of the language-game analogy that there is more to life—and more, in practice, to language—than rules. On the other hand, because characteristic aspects of language-games are precisely those that occur regularly, the constitutive form, structure, or shape of linguistic practices must be describable in terms of rules. Despite having a more limited role, rules thus remain central on the language-game analogy as those means by which established—and so presumably particularly important or useful—uses of words can be described.

Another important consequence is that Wittgenstein's position has the versatility to allow for practices to be governed by rules in many different ways. For example, the general idea that linguistic rules have an empirical shape—analogous to a statistical distribution—allows for the rules defining practices to have greater/lesser degrees of sharpness or tolerance. Further, as pointed out above, Wittgenstein's position allows for practices (as contexts of regulation) to differ in the degree to which they are rule-governed—that is, how pervasively they are characterized by rules, and how strictly regularities are enforced. Thus at one extreme there will be practices, such as mathematics, that are strictly regulated, being characterized by rules that are well defined, rigorously adhered to, and pervasive; at the other extreme there will be practices, such as interior design or fashion, that are less regulated, being characterized by rules that are less well defined, less standardized, and not pervasive. In the first kind of case, describing established rules might tell us all there is to know about the practice in question, providing a precise and detailed portrait of the practice, so to speak; whereas in the latter kind of case it would perhaps tell us very little, yielding something like a caricature or a rough outline at best.

These consequences can be summed up by saying that, on Wittgenstein's nonidealized picture, rules correspond to general features of language-games or word use *to the extent that* such features are actually found in the relevant practice. As expressed in Wittgenstein's talk of the "physiognomy" of meaning, this reconfigured and constrained notion of rules serves to describe characteristic features of language-games, and hence language. This, as I said, is the second feature that makes Wittgenstein's view suited to the accurate description of linguistic phenomena. Whereas the calculus model had fixed on a certain kind of practice as paradigmatic, and hence intimated an overintellectualized and ultimately incoherent ideal, his later conception of rules has the virtue of being as versatile and structured/unstructured as the forms of activity that make up actual human lives. In this way, Wittgenstein's relaxed view of rules reflects the fact that actual language use can, without deficiency, be more or less sharply delimited and manifest greater or lesser degrees of regulation. Rather than relying on idealized or artificial standards of precision, this conception of rules and their role therefore exhibits the flexibility required to describe the diverse types of language-game that make up actual linguistic phenomena.

So it is by both centering on the right kind of thing (i.e., language-games) to invite comparison with linguistic practices, and doing so in the right ("relaxed") way, that Wittgenstein's praxeological conception of language proves particularly attuned to accurate description of real linguistic phenomena. Nonetheless, as explained at the start of this chapter, in claiming that Wittgenstein's conception of language can form part of a phenomenological approach to language, I am not suggesting that he himself intended such an approach. What establishes the compatibility of his late conception of language with a phenomenological approach is rather that it meets the key requirement of accurately describing linguistic phenomena, the requirement I have labeled phenomenological accountability. However, further evidence of this compatibility is that Wittgenstein's conception of language also exhibits several features identified in chapter 2 as generally characterizing a phenomenological conception of language. For example, the notion of language-games not only, as just noted, conceives of language as a process rather than some kind of stasis, but is specifically intended to block the tendency to think of language as a system of forms. As with Heidegger, this "antiformalism" is linked with a nonreductive approach to semantics, evidenced in Wittgenstein's critique of simple referring and his dictum "everything that describes a language-game belongs to logic" (*ÜG* §56). In addition, while Wittgenstein had earlier conceived of language, so to speak, as a world-independent ("autonomous") medium of meaning constitution, his later thinking was led by the maturing language-game analogy to reflect the embedding of language in human practice and its empirical surroundings (the empirical attunement of rules), and so to a view of language as language-in-the-world.

7 Coping with Language

The last chapter looked in some detail at the later Wittgenstein's praxeological conception of language, characterized by the language-game analogy, and set out how this provides a basis for understanding what I have been calling pragmatic sense. This chapter will complete my development of the Heideggerian framework, as set out in part I of the book, into a phenomenological conception of language by undertaking two main tasks.

The first is to consider what kind of grasp speakers have of pragmatic sense. As a distinctive feature of the Heideggerian framework is to allow that language might be grasped in a prepredicative manner, pragmatic sense should identify structural features in a way that is consistent with such uses. However, it might seem that meeting this requirement poses a problem for Wittgenstein's conception of language. In contrast to Merleau-Ponty's views, which are explicitly developed in terms of prepredicative and preconceptual factors, Wittgenstein talks about the connection between concepts and practice. This constant use of the term *concepts* might be taken to imply that Wittgenstein's interest is in higher-level, more intellectual feats than those targeted by talk of "prepredicative" factors.

While there is no denying Wittgenstein's terminology, I want to argue that his underlying position is sufficiently close to my approach here to make assimilating it unproblematic. In the last chapter I opposed an intellectualist reading of Wittgenstein's conception of rules by showing how his position shifted between *PG* and *PU*. Although some kinds of language use do approximate well to the kind of highly theoretical and intellectual activity that characterizes the use of a mathematical calculus, it is the adoption of a more relaxed view of rules that allows Wittgenstein's position to be plausibly applied to a wider range of linguistic practices. This chapter will show that Wittgenstein's view of rule-following undergoes a similar shift, becoming more versatile and more general by moving away from a narrow focus on highly rational and reflectively transparent practices such as mathematics. As a result, Wittgenstein's later conception of rule-following avoids assimilating all understanding of language to its most theoretical and intellectual modes of use, as his earlier calculus model had done. Together with his view of the limits of justification, this gives rise to the idea of a linguistic form of knowing-how that can sustain the idea of language being grasped prepredicatively.

This first main task is undertaken in the first two sections of the chapter. The first of these looks at the conditions Wittgenstein's account of rule-following is supposed to meet and how his appeal to the notion of customs or institutions does this. This will make clear how the requirements for rule-following can be met by practices that lack reflective awareness of the kind that was central to Wittgenstein's earlier model of language as a calculus. Against this background, the second section draws on Wittgenstein's view of the limits of justification as a model and shows how his late position—terminological differences notwithstanding—meets the conditions identified in chapter 2 to function prepredicatively.

The second main task for this chapter will be to complete the story of how the Heideggerian framework has been filled out by this and the preceding chapters. Thus the third section of the chapter sets out how Wittgenstein's views fit together with and complement those of Merleau-Ponty and Heidegger, while the final section summarizes both the overall conception of language developed in the preceding chapters and the resultant view of the disclosive function of language.

1 Rule-Following Practices

To understand how Wittgenstein's views might be able to meet the requirements on the notion of pragmatic sense, the place to begin is with Wittgenstein's own view of the grasp speakers have of language-game structure, as found in his discussions of rule-following. Although the concept of rules was an enduring feature of Wittgenstein's views on language from the *Tractatus* on—where "rules of logical syntax" were to underlie meaningful language and to be implicit in normal language (*Umgangssprache*; 3.334, 5.5563)—Wittgenstein initially, under the influence of the calculus model, apparently saw no need to explain how such rules are linked with the speakers' abilities in which they are supposedly manifested. This changed with the emergence of the language-game analogy, and already in the *Blue Book* Wittgenstein is troubled by the question of what following a rule consists of in practice, in particular by the (Kantian) difficulty of distinguishing behavior involving rules from that merely conforming to them.[1] However, given the developments described in the last chapter, it is to *PU* that one should look for a more mature view of speakers' grasp of rules.

There are two discussions of rules in *PU*. The first (§§81–88)—on which the last chapter focused—follows Wittgenstein's engagement with the issue of the general nature of language in §65 and can be seen as focusing on the relationship between rules and concepts. The second discussion (§§138–242) addresses what rules are and what grasp speakers have of them. Wittgenstein launches this discussion by highlighting the difference between the meaning of a word, which one can supposedly grasp "in a flash," and its temporally extended use, prompting the question of how the two are related. The general difficulty, as Wittgenstein sees it, is that if a word is to be thought of as meaningful, it seems there should be something about it—its meaning—that determines how it is used and to which speakers' understanding corresponds. Whatever this something is, it might be thought, will explain what rules consist in. Discerning with any precision both the aims of this discussion and the conception of rules it relies on is complicated by *PU*'s attempt to avoid expounding positive "theses," in particular by its aphoristic, discontinuous, and conversational mode of exposition. However, the placing of this discussion provides an important clue to its general aim. Because it follows Wittgenstein's comments on the nature of philosophical problems and philosophical method of §§89–133, it is here, if anywhere, that one would expect the idea to apply that philosophical problems result from misunderstandings that are to be resolved by reminding us of the everyday uses of words. This suggests that whenever Wittgenstein endorses specific views here, he intends this to be a reminder of an antecedently familiar notion of rules. As §235 puts it, his aim is to remind his readers of the "physiognomy" of "what in everyday life we call 'following a rule.'" This comment is of some significance, since—given his view of the role of everyday language use—it underlines Wittgenstein's underlying commitment to a conception of linguistic rules that he takes to be free from misunderstanding.[2]

That conclusion is reinforced by the fact that the general contours at least of Wittgenstein's views, both negative and positive, can be quite easily discerned. On the negative side it is clear that in the course of his discussion Wittgenstein canvasses and rejects several proposals as to how determinate meaning is constituted: the platonist image of a set of antecedently existing rails that guide us, occurrent mental states such as intuitions, or ideal "mechanisms" somehow inherent in sign systems. This is to be expected, since the early parts of *PU* had already relied on Wittgenstein's preferred view of meaning as use in language-games, while nonetheless rejecting the idea of a "general concept of meaning" (§§5, 43). So it would be surprising and clearly problematic if his discussion of rules were subsequently to reveal a philosophically "superlative" type of meaning-constituting fact capable of founding a general concept of meaning (cf. §192).

Yet at the same time, although his discussion is tortuous and oblique in form, positive aspects of Wittgenstein's view of rules are also intimated throughout: understanding is likened to abilities and techniques (§150), such as reading (§§156ff.), predicated on the basis of certain externally observable circumstances (§154); the basis of rules is acting without justification (§§211, 217). Against this background Wittgenstein seems to think that unclarities about the notion of rules are resolved simply by referring to a "constant use, a custom" (§198), "*customs* (uses, institutions)," "a technique" (§199), or "a practice" (§202).

The difficulty is how this proposal is to be understood. Wittgenstein's invocation of these notions—which he treats as broadly synonymous—is at best conspicuously elliptic and enigmatic.[3] So what does this talk of customs and so on amount to? And how does it help in answering the question of what rule-following involves? The remainder of this section will attempt to answer these questions by clarifying the conditions Wittgenstein thought an account of rule-following needs to meet and how the appeal to customs satisfies those conditions. Doing this will provide the background required to argue—in the next section—that a Wittgensteinian conception of pragmatic sense is compatible with the thought that language can be grasped in a prepredicative manner.

To begin with, Wittgenstein thought that two general conditions would bear on a conception of (linguistic) rules. First, within the framework of the language-game conception, "rules" and "rule-following" are correlative concepts, since determining what a rule is amounts to determining how an agent acts in adhering to it. In Wittgenstein's words: "Following a rule is a human activity," which can be described "only by describing in a different way what we do" (*BGM* VI §29, VII §51). Second, descriptions of rule-following activity are subject to Wittgenstein's general requirement that an "'inner process' requires external criteria" (*PU* §580). To some extent the significance of this thought as a methodological artifice is straightforward. From the *Blue Book* onward Wittgenstein often suggested replacing internal images with external ones, in the conviction that any explanatory function an internal representation might fulfill must be equally comprehensible in terms of the use of an external representation, such that the description of mental abilities coincides with a description of the practice(s) in which they are manifested.[4] However, one further point is worth emphasizing. It is perhaps tempting to describe Wittgenstein's insistence on external criteria as an insistence on the "public observability" of criteria, or on "third-personal" as opposed to "first-personal" criteria.[5] Such glosses can be misleading, and it is of some importance to underline that the point of Wittgenstein's exteriorization method is simply to enforce explication of the supposed meaning-conferring role of inner representations. A consequence of this is that rules and criteria for rule-following are essentially *impersonal*: for the exteriorization move and the perspective it implies effectively insist on the objectivity of criteria by eliminating any reference to a (first- *or* third-) personal perspective.[6]

The account Wittgenstein develops of rule-following is governed by two more specific conditions. The first—the *regularity condition*—is that the basis for this to occur is a (recognizable) regularity, since it is only against the background of a repeated pattern of behavior that the question of whether a rule is being followed or not can arise (cf. *PU* §§207, 237). Accordingly, Wittgenstein explicitly excludes describing one-off actions as rules: "Is what we call 'following a rule' something that only *one* person could do only *once* in his life? ... It is not possible for one person to have followed a rule on only a single occasion."[7] However, although necessary, the requirement that there be a background of regularity does not suffice in Wittgenstein's view to speak of rule-following. Rather, although "acting according to a rule presupposes recognizing a uniformity [*Gleichmäßigkeit*]," Wittgenstein thought of regularity of action, or conformity to a rule, as merely a "precursor to acting according to a rule" (*BGM* VI §§44, 43). What this lacks can be brought into focus by considering two examples from *Remarks on the Foundations of Mathematics*. First:

> there could be a caveman who produces regular [*regelmäßige*] sequences of signs for himself. He entertains himself, for example, by drawing --.--.--.--. or -.-..-...-....- on the wall of the cave. But he does not follow the general expression of a rule. And we do not say that he acts regularly [*sic*: *regelmäßig*] because we are able to form such an expression. (*BGM* VI §41)

This indicates that the feature distinguishing acting according to a rule from actions describable in terms of a regularity has something to with the agent's (not a mere observer's) deployment of a "general expression of the rule." I will therefore label the second condition Wittgenstein is trying to meet the *agent's grasp condition*. Clearly, this condition should be taken to refer to features of the agent's behavior that are externally observable, rather than suggesting reliance on a private, first-personal sphere. To see what else this condition involves, consider the contrast Wittgenstein suggests with activities in which a general expression is used such as teaching/learning:

If one of two chimpanzees scratched the figure |--| in the clay soil and another then the series |--||--| etc., then the first would not have provided a rule and the second not have followed it, no matter what happened in the souls of the two. However, if one observed, for example, the phenomenon of a kind of lesson, of demonstration and imitation of successful and unsuccessful attempts, of reward and punishment and suchlike; if the one so taught in the end lined up previously unseen figures as in the first example, then we would indeed say that one chimpanzee is writing down rules, the other following them. (*BGM* VI §42)

There are two differences between this case—in which Wittgenstein allows that rules are being followed—and the former that might seem decisive. First, the case of the chimpanzees differs from that of the caveman in that it involves social interaction. This might seem to suggest that the agent's grasp condition specifies a need to agree with others in one's use of a term, and to support the claim that Wittgenstein's appeal to customs, institutions, and so on, was supposed to convey that rule-following is an essentially social practice. The second difference is that in the chimpanzees' case there is external evidence for the agent's deployment of a general expression of the rule. This would suggest that the agent's grasp condition specifies the need for a certain level of cognitive ability (rather than social skills). The contrast between the two situations then emerges as a paradigm case of Wittgenstein's use of (imaginary) language-games to elucidate what is involved in two different kinds of mental state. What matters with this second difference is the structure of sign-use practices as such, with the difference between rule-conformity and rule-following lying in the level of complexity exhibited in the two kinds of behavior.[8] In this respect the second example involves additional complexity, with the manifestation of corrective behavior serving as evidence that the agent takes something involved in the practice to be a "general expression" of the rule.

The first difference bears on an aspect of Wittgenstein's views that has been widely discussed and remains controversial, namely, the question of whether rule-following is possible only in a social context—as his talk of customs and institutions perhaps implies—or whether it is also possible for an isolated individual ("Robinson Crusoe").[9] In the following, however, I want to focus on the second kind of difference, that relating to outward evidence which indicates the structure and complexity of the cognitive feat involved in rule-following. This is not to deny that language, or even rule-following, has an important social aspect, nor that a complete understanding of linguistic phenomena requires consideration of its social dimension. I simply want to bracket the question of whether rule-following is essentially social or individual, as nothing I will be saying in the following depends on how that question is answered.

It might perhaps be objected that this question cannot be bracketed, because the whole point of a "communitarian" approach is that talking about rule-following is indistinguishable from talking about the social context. This would indeed be the case if the two differences above were internally connected, such that the social context plays a constitutive role in the cognitive feat that rule-following requires. There are some grounds for doubting that Wittgenstein thought this is the case. For example, in one passage he comments: "When, then, do I say that I see the rule—or a rule—in this sequence? ... [Is it] not also simply when I can continue it? No, I explain generally *to myself* or to another [person] how it is to be continued" (*BGM* VI §27; italics added). This passage suggests the social aspect is not essential: I explain to myself *or* to someone else. Moreover, this indifference is to be expected, given the impersonal nature of the "external" criteria Wittgenstein insists on. For this suggests that what counts as explication or justification is governed by the same considerations or criteria in both cases—that is, whether one explains something to oneself or to someone else.

Independently of exegetic considerations, however, it is also difficult to see the two factors as indistinguishable. The cognitive ability highlighted in the above example is the ability to use a general expression of the rule as a standard for correction so as to uphold a specific pattern of sign use. It is important to note here that if an individual does not (generally speaking) have this ability, then a wider community will be unable to help. For calibrating one's own language use to that of the community is an operation of the same complexity, indeed the same kind, as upholding a regularity at an individual level: one involves comparison with community use, the other with one's own use.[10] In other words, the cognitive ability in question here cannot be socially grounded, because any influence a language community might be thought to exert on an individual will presuppose it. In saying this my aim, to reiterate, is not to take a stance on whether rule-following is essentially individual or social, but simply to clarify that the cognitive abilities highlighted by the above example must be in play on both views. Hence, even if one thinks full-fledged rule-following is possible only in a social context, these abilities can be discussed independently of whatever function is attributed to the linguistic community (e.g., making normativity possible, stabilizing judgments, or selecting one regularity among those possible).

The problem with the second example above, as it stands, is that it is too demanding. While clearly sufficing to attribute such abilities, straightforward cases of overt use of a rule expression and conspicuous corrective behavior—as in the chimpanzee example—cannot be considered necessary for rule-following. The advantage of the example is rather to provide a clear case of the kind of cognitive ability rule-following involves: the ability to make comparisons or to offer justification, which in turn implies the ability to discern an aspect of sameness between two things and to make use of something—examples or a general expression—as a model for or standard of that sameness (cf. *PU* §§72–73). Nonetheless, more subtle behavioral indications might equally well suffice as evidence of an agent's responsiveness to the correctness of what she is doing, and so of her ability to engage in normative behavior.[11] This suggests that the condition met by Wittgenstein's chimpanzee example should be formulated in a suitably accommodating form. Corresponding to the idea that the rule must somehow be involved in the process in question, the agent's grasp condition should be seen as the general requirement that, whether or not anything is overtly deployed as a standard, rule-following requires that an agent be both aware of the regularity in question and attempting to maintain it. The difficulty is then to identify, in general terms, what more must be added to a regularity in behavior for an agent's actions to provide evidence that the agent's grasp condition is being met.

Two features of *PU*'s treatment of rules respond to this difficulty. The first is to allow that a wide variety of criteria may be relevant. This is the message of the reflections Wittgenstein interposes between introducing (§143) and returning to (§185) the language-game of forming numerical series. To begin with, the discussion of the concept of understanding (§§146–155) provides a general reminder that criteria for understanding lie in a sign's pattern of use or application (§146), and that to the extent that understanding is accompanied by criterially relevant characteristic processes, these are to lie in certain external circumstances rather than private mental events (§§152–155). The subsequent discussion, however, makes clear that rule-following encompasses complex and disparate phenomena characterized by family resemblance. Wittgenstein does not say this in as many words—that is, advance it in a form that might be taken as a thesis. However, in discussing the example of reading (§§156–178)—a model of "being guided" by linguistic signs—he combines the demand for observable contextual criteria with the insight that "reading" stands not for *one* phenomenon, but a "family of cases" to which "different criteria" are applied "in different circumstances."[12] The pertinence of these comments to the subsequent (from §185) further discussion of rules is established by the transition (§§179–184), in which the orientation toward (external) circumstances as well as the complexity of possible criteria is related back to the topic of rules (cf. §179). The overall implication is that, just as there is no central defining feature of (language) games, there is no simple set of general conditions for meeting the agent's grasp condition, and hence for rule-following.[13] Criteria for the concepts of ability and understanding are, Wittgenstein warns, "much more complicated than it might seem at first glance" (*PU* §182). And precisely in everyday linguistic practice—Wittgenstein's focus—one would expect to find more subtle and variegated manifestations of such abilities. So the difficulty of pinning down Wittgenstein's conception of rules is due not only to his no-thesis thesis and stylistic difficulties, but also because in one sense *PU* does not offer a general answer. Instead of attempting to develop a systematic theory of rules, it remains open to diverse criteria while offering fragmentary indications of the complex "physiognomy" of the concept of rules "in everyday life" (*PU* §235).

The second feature responding to the difficulty of how the agent's grasp condition can be met generally is Wittgenstein's appeal to customs. The significance of this second feature can be understood in light of the first. For although Wittgenstein does not attempt to enumerate all conditions sufficing to meet the agent's grasp condition—presumably thinking that no such enumeration is possible—this appeal does identify the *kinds* of phenomena in which such a condition must be being met. For customs, institutions, and (established) practices are phenomena in which there is transmission, or teaching, of intentionally upheld regularities—even if it is not obvious how to catalog the ways this is done.[14] In other words, Wittgenstein's talk of customs, institutions, and practices is a placeholder for a general answer, offered in lieu of a catalog of specific conditions sufficient to attribute to agents a grasp of the rule. Its apparent vagueness notwithstanding, or rather precisely in virtue of this, such talk provides a way of dealing with the problem of the diversity and complexity of criteria for rule-following. So in this second sense *PU does* provide a general answer to the question of what rule-following comprises. It is generic and vague, but such is the price of its being a general and complete answer.

This, then, is how Wittgenstein's appeal to customs can be understood as responding to the conditions he sees as bearing on a conception of rule-following. As an answer to the question of what the latter involves it might appear elliptic and enigmatic, but this is because Wittgenstein is providing a generic answer to allow for the possibility of the agent's grasp condition being met in many different ways. Against this background, I now want to highlight briefly how this view of rule-following forms part of the later Wittgenstein's general shift away from an overly intellectual picture of language.

One indication of this shift is the mere fact that Wittgenstein singles out "customs" and "institutions" as general characterizations of rule-following practices. These generic terms allow, as just pointed out, that normative practices are more varied in type than suggested by his earlier focus on specific, strictly regulated activities such as the use of a calculus or chess playing. Beyond this, however, Wittgenstein's use of these labels seems intended to signal that rule-following typically lacks—as customs surely do—the highly reflective and distinctively intellectual character suggested by his earlier focus.

This can be appreciated more clearly by considering how the generic proposal I have attributed to Wittgenstein differs from mere rule-conformity. After all, it might be wondered whether the appeal to customs (etc.), understood in the way just outlined, is really able to meet the need identified in §198 to distinguish rules from a mere "causal connection." For, as I have interpreted it, this appeal might look simply like an *insistence* that the agent's grasp condition be met in some way that says nothing substantial about how this is to be done. As a result, invoking the notion of customs might seem frustratingly empty, and the concern justified that Wittgenstein has failed to identify what distinguishes (linguistic) rule-following from the mere rule-conformity exhibited by physical objects.

This concern can be eliminated by recalling the phenomenon of inexplicitness and Wittgenstein's distinction between providing "samples of use" and "stating the rule" as criteria of rule-following (*PG* §42). The former—which I will subsequently refer to as *rule exemplification*—is the ability to use a term in appropriate circumstances, and so to take on corresponding roles in language-games. The latter—*rule stating* hereafter—is the (reflective) ability to define or describe the use of a term (e.g., by specifying conditions in which it would be appropriate or otherwise to use a term). Given this distinction, it might initially seem that—unless one can also state the rule being followed—rule exemplification is equivalent to the "merely causal" instantiation of rules exhibited by inanimate objects. However, it is not difficult to appreciate that these two cases do differ. To begin with, an agent's exemplifying or providing samples of word use is a spontaneous—and so potentially withholdable—feat, rather than a response that is causally or "mechanically" triggered. Further, rule exemplification can also serve as the basis for corrective behavior. As Wittgenstein explains: "I show him, he copies me; and I influence him by expressions of agreement, disapproval, expectation, encouragement. I let him do as he likes, or hold him back etc." (*PU* §208). Insofar as it can form the basis of such corrective behavior, the rule-exemplification ability can suffice to fulfill the agent's grasp condition—that is, to exhibit sensitivity to both the regularity in question and its maintenance. In other words, the rule-exemplification ability suffices to sustain a spontaneous and normative character that distinguishes "customs" (etc.) from merely "causal connections."

Another aspect of Wittgenstein's late position relevant here is that rule exemplification becomes more basic—both practically and logically—than rule stating. This priority, which underlies the phenomenon of inexplicitness, can be better understood with reference to the two specific conditions—the regularity condition and the agent's grasp condition—that Wittgenstein is looking to meet in his account of rule-following. These two conditions do not simply correspond to the two criteria for rule-following: the agent's grasp condition is met by the ability to state the rule only in the more straightforward cases, and can also be met by the exemplification ability. Nonetheless, they allow the logical priority of rule exemplification over rule stating, and hence the possibility of the phenomenon of inexplicitness, to be understood as follows: Rule exemplification suffices in principle to meet both the regularity condition and the agent's grasp condition, which entails that the rule-stating ability is not necessary. Rule exemplification is also necessary, because the regularity condition entails that a rule follower be able to produce instances of the rule: "The application [*Anwendung*]" of a rule "remains a criterion of understanding" (*PU* §146)—which entails that rule stating alone is not sufficient. Thus whereas rule exemplification is both necessary and (when appropriately used) sufficient for rule-following, rule stating is neither necessary nor sufficient.

It should be kept in mind that, although acknowledging sensitivity to correctness without obvious use of a generally stated rule allows for subtlety and variety in the sufficiency conditions for rule-following, this should not be equated with "implicit" awareness of such conditions. Wittgenstein's willingness to attribute rule awareness invariably turns on manifest, or at least manifestable, behavior and in this respect his view of criteria centers on explicit behavior alone. Implicit features, if anything, are simply those that would be explicit in appropriate circumstances—as *PU* §126 puts it, "what might be hidden does not interest us." Accordingly, the distinction between rule exemplification and rule stating should not be thought of as the distinction between implicit and explicit awareness of the rule, but as two different kinds of manifestable ("explicit") awareness of it. Rule exemplification consists in the use of a linguistic expression in appropriate circumstances; rule stating involves a reflective use of language to characterize in some way the (exemplified) use of linguistic expressions.

The preceding considerations allow us to see how Wittgenstein's late view of rule-following shifts away from its earlier intellectualist focus. As we saw in the last chapter, at the time of *PG* he had modeled his conception of language on a mathematical calculus. The highly rational character and reflective transparency that define mathematical activities imply that the rule-exemplification and rule-stating abilities are correlative—such that lacking either would amount to a lack of competence. By contrast, we have just seen that on his later view rule exemplification can suffice for rule-following and has a practical and logical priority over rule stating. Consequently, although it will always be possible in principle to state whatever rules are exemplified in language use, the ability to do this is not required generally on Wittgenstein's later view. Indeed, far from thinking of this kind of case as unusual, Wittgenstein treats it as normal and as central to the conception of linguistic knowledge in *PU*, according to which the use of words is learned and explained primarily through examples, and often cannot be otherwise explained.[15] This later view of linguistic competence is less intellectualist in the sense of allowing that it is quite normal for language to function despite a lack of reflective transparency on the part of speakers. Thus Wittgenstein's later view of rule-following picks out diverse modes of behavior that typically fall short of the high levels of rational transparency characterizing the use of calculi, without—as we have also seen—collapsing into the merely mechanical instantiation of "causal" rules. It is this spectrum of intelligent, albeit not paradigmatically rational, behaviors that Wittgenstein captures with the label "customs."

2 Prepredicative Language-Games

For Wittgenstein's praxeological conception of language to succeed in explicating the notion of pragmatic sense, it must allow for the rule structure inherent in language-games to be grasped in the prepredicative manner required by the Heideggerian framework. The last section progressed toward seeing how this is possible by making clear that Wittgenstein's criteria for rule-following can be met in a variety of ways, even in the absence of rule-stating abilities. Building on that discussion, this section will show how the three requirements for prepredicative factors identified in chapter 2 can be met by a language-game-based conception of language. To some extent the following discussion will depart from Wittgenstein's own stated views—as indeed it must, given that he nowhere explicitly discusses "prepredicative" meaning. Nonetheless, my proposals will be closely modeled on Wittgenstein's stated views, so that they should remain recognizably Wittgensteinian in spirit.

It is relatively straightforward to see the late Wittgenstein as focusing on *subpropositional* factors—that is, factors that functionally and structurally underlie propositional content. To begin with, there are several indications that, like Heidegger, he does not see propositional meaning as having explanatory primacy: Much of *PU*'s discussion focuses directly on words rather than sentences; §§19–20 and §136 also hint that pragmatic rather than syntactic criteria can individuate sentences, so that propositions cannot be taken to have a general form—which Wittgenstein thought would entail commitment to the calculus model of language.[16] Wittgenstein's focus on subpropositional factors becomes clearer by considering his view of the processes that determine word use. First, the rule-following abilities he concentrates on are in no way specifically tied to the use of expressions in sentential contexts or the idea of propositional content. Indeed, while emphasizing that "following a rule is at the BASIS of our language-game" (*BGM* VI §28), Wittgenstein does not think of rule-following as a specifically linguistic competence. Rather, it can also be applied to nonlinguistic behaviors, as illustrated by his examples using cavemen and chimpanzees in the preceding section. In addition, second, it is implicit in the language-game model that the meaning of terms is determined directly by practical requirements—the need to have a word for this object, that process, or whatever—and not mediated by participation in propositional units (cf. chapter 6, section 3).

Somewhat more involved is understanding how pragmatic sense can be *subinferential* in the dual sense previously identified. This requires that pragmatic sense—that is, the rules corresponding to practice-inherent structure—can be grasped without awareness of inferential properties, while at the same time functioning as the basis of inferential properties. Given the traditional link between rule-following and predication as characteristics of the understanding or intellect, the difficulty is how praxeological rules can play a foundational (rather than a directly constitutive) role in relation to predicative awareness. Although Wittgenstein does not discuss this difficulty specifically, looking at his view of the limits of justification will provide a model for how the subinferential function can be accounted for in a Wittgensteinian manner.

Wittgenstein expresses the view that justification is finite in a number of ways, including the well-known metaphor of the bending spade in §217 and several comments suggesting resignative passivity: "What has to be accepted, the given ... are *forms of life*," "the everyday language-game is *to be accepted*," "Our mistake is to search for an explanation where we should view the facts as 'primal phenomena.' That is, where we should say: *this language-game is played*."[17] The underlying thought—in itself not an original one—is that if justification did not somehow come to an end, an infinite regress would result, implying that there is no such thing as justification (cf. *PU* §485). The distinctive feature of Wittgenstein's position is that he sees justification as coming to an end not in evident truths, in some special way of seeing (intuiting), or even in "unjustified assumptions," but in an "unjustified way of acting."[18] The existence of such basic practices, as Wittgenstein puts it, simply has to be accepted as something "beyond justified and unjustified; i.e., so to speak, as something animal" (*ÜG* §359; cf. §559).

In what sense are language-games, or forms of life, to be "accepted" as "given"? Certainly Wittgenstein does not think that the language-games constituting a language cannot change or evolve (*PU* §23). Nor does he think that no reasons can be given for acting in existing ways. Rather, in claiming that language use is founded in ways of acting that are themselves unjustified his thought is that at some stage justification for speaking or acting in certain ways loses its force (*ÜG* §307). The relevant sense of *acceptance* is therefore best understood in the perspective of "conceptual investigations" of the kind Wittgenstein was interested in conducting—that is, in describing the structural relationships between concepts.[19] For here there is a sense in which established language-games do have to be "accepted" as "given," namely, as the object of description, and insofar as concepts are viewed as standing in rationally motivated relationships, this mode of inquiry points toward something that must be considered unjustified or conceptually "basic." Moreover because, on Wittgenstein's approach, concepts are articulated in language-games, they must be grounded in forms of action (rather than, say, basic logical forms) that are unjustified.

Rather than either defending or exploring the implications of Wittgenstein's claims for a general theory of rationality, I want here simply to highlight two aspects of his view.[20] The first is the stratification of language-games assumed by these claims. It is implicit in the idea of language-games that cannot be justified—those "beyond justified and unjustified"—that justification does not, on principle, play any role in such games. In other words, the overall praxeological architecture of language-games will bottom out into a basic stratum of language-games that underlie all reason-giving, primitive forms of linguistic practice that are "something animal." Wittgenstein thus relies on a distinction between what I will call "prejustificatory" and "justificatory" language-games—that is, between language-games that do not (cannot) involve justification and language-games that do (or at least might) involve justification. Further, employing this distinction, his claim is that justificatory language-games presuppose, or are based on, an underlying level of prejustificatory language-games.[21]

The second aspect I want to highlight is that the ability to participate in prejustificatory language-games identifies a basic grasp of language. In understanding what this involves, it is important to keep in mind that Wittgenstein is concerned with "justification" in the specific sense of explicating the conceptual or sense structure inherent in practices. In this respect "justification" is analogous to other practices such as description, explanation, or teaching that involve rule stating so as to elucidate the structure of some underlying practice (that being described, explained, or learned). With this in mind, prejustificatory practices can be understood in terms of the relationship between rule stating and rule exemplification discussed in the last section. Prejustificatory language-games can thus be thought of as language-games in which it is possible to participate on the basis of rule-exemplification abilities alone. Further, while the phenomenon of inexplicitness allows the (logical) priority of rule exemplification over rule stating to be discerned, there is nonetheless a link between these two abilities. For the rule stating involved in justificatory language-games is constitutively referred to an underlying pattern of use (rule exemplifications), with the latter founding the former in the sense that, if there were no such regular pattern, then there would no rule to state. Hence, Wittgenstein's idea that justification is finite can be interpreted as the claim that justificatory language-games, which involve rule stating, presuppose prejustificatory language-games, for participation in which rule-exemplification abilities suffice.[22]

To illustrate this stratification of language-games consider two kinds of coffee drinkers. One group, the carefree coffee consumers, are able to use the word *coffee* in a way that is adequate in a range of everyday situations: they use it to refer to various products they buy in the supermarket or cafés, they understand the social role of drinking coffee in their society, and so on. But they do not, for example, know how the product they buy is produced, or which factors distinguish good from bad coffee, and they have no interest in talking about coffee itself. A second group, the cultivated coffee consumers, motivated by considerations of health or social status, are passionate about such matters and often discourse on them, raising coffee talk to a whole new level of sophistication. Nonetheless, in many situations the use made by both groups of the term *coffee* will be indistinguishable: cultivated drinkers will use the term in the same carefree manner when buying or ordering coffee, or arranging to meet friends for their preferred drink. In this sense the language use of the cultivated coffee drinkers can be thought of as made up of two distinct levels: a set of simple language-games also mastered by the carefree consumers in which coffee-related explanation and justification play no role, and a set of further language-games in which they do and which sets them apart as coffee connoisseurs. On the Wittgensteinian model I have been outlining, the practices of carefree coffee consumption are prejustificatory, while those of the cultivated coffee drinkers build on these and also involve justificatory language-games.

This view of finite justification provides a model for how prepredicative language use can be understood in a Wittgensteinian manner. Consider first the basic grasp of language involved in prejustificatory language-games. The rule-exemplification ability consists of a grasp of which practical circumstances—which situations in which language-games—given words can be appropriately used in, an ability that clearly corresponds to Heidegger's view of purposive awareness (knowing what ... is for). However, it does not require an ability to describe an expression's use (rule stating) or to generalize over language-games. In other words, it does not require predicative awareness in the sense Heidegger distinguishes—that is, an awareness of purpose-independent properties underlying the inferential behavior of expressions. By contrast, the switch to predicative awareness will be closely linked with the ability to participate in justificatory language-games. For while the two will not coincide completely—for example, not all acts of predicative awareness are instances of justification—both will typically involve a rule-stating operation in Wittgenstein's sense.[23] Whatever the precise relation between the two, Wittgenstein's general approach requires that the switch to predicative awareness—which Heidegger attributes to the use of Statements—must be characterizable in terms of corresponding types of language-games. In other words, using Wittgenstein's view of justification as a model, we can think of language as being made up of prepredicative language-games that rely on context-relative rule-exemplification abilities alone, and predicative language-games that are based on these and in which an awareness of context-independent inferential properties are developed.

This founding of predicative in prepredicative language-games can be understood—as in the case of prejustificatory and justificatory language-games—as a general point about the functional stratification of language-games. This has the implication that prepredicative language-games can, in principle, exist empirically without ever being developed into more sophisticated practices involving predicative awareness. More generally, however, this foundational relation will pervade the praxeological structure of language as a whole, such that predicative (e.g., justificatory) practices can always be understood as constitutively referred to some underlying form of prepredicative language-game. This dependence can be pictured such that any form of practice involving justification is based on corresponding, but simpler, justification-free forms of practice—for example, routine performances of the rule in question (as in the above coffee cultures example). In this way the foundational relationship Heidegger sees holding generally between purposive and predicative awareness of language can in principle be spelled out in terms of Wittgensteinian language-games.

With the distinction between prepredicative and predicative language-games in place, the two conditions for language to have a subinferential function can be met. First, Wittgenstein's views on justification imply that not all language-games, and hence not all linguistic competence, can be (let alone need be) understood in terms of justificatory or predicative activity. Accordingly, as the phenomenon of inexplicitness attests, the ability simply to exemplify appropriate use of expressions—even without being able to elucidate their use—would suffice to participate in many (though of course not all) established language-games. Second, it becomes clear that Wittgenstein has the conceptual means to allow that practice-inherent rules can be grasped both prepredicatively and predicatively. Thus a basic level of rule-following can be sustained by (prepredicative) rule-exemplification abilities, while a reflective grasp of rules can be understood in terms of (predicative) rule-stating abilities.

Finally, to appreciate how pragmatic sense founds predicative awareness in a functionally discontinuous manner, it is useful to think of the prepredicative ability to exemplify appropriate use of linguistic expressions as a linguistic form of knowing-how. As a characterization of intelligent or skilled behavior, Ryle's classic notion of "knowing-how" was intended to contrast with both the mere regularity of habits and the propositional competence of "knowing-that." Perhaps unsurprisingly, therefore, most of Ryle's examples of knowing-how are nonlinguistic activities, such that "doing is an overt muscular affair" and there is a clear sense in which "[e]fficient practice precedes the theory of it" (Ryle 1949, 32, 30). However, as noted above, Wittgenstein treats nonlinguistic and linguistic behaviors as continuous in applying the ideas of rules and rule-following to both. It is also implicit in the (Heideggerian/Wittgensteinian) view developed here that the distinction between knowing-how and knowing-that—corresponding to that between prepredicative and predicative, or purposive and propositional awareness—falls within language use. It therefore seems appropriate to extend the notion of knowing-how to language, recognizing the fact that our prereflective engagement with the practical world often encompasses the use of language.

Without wanting to be committed to the further details of Ryle's discussion, the notion of pragmatic sense suggested here can be seen to exhibit two features that for Ryle distinguished knowing-how from knowing-that.[24] First, whereas knowing-that is something for which reasons can be required, reasons are not involved in knowing-how (Ryle 1949, 28). As has been seen, this is paralleled both in Wittgenstein's view (inexplicitness) that picking up the use of expressions does not require familiarity with explicit expressions of rules, and in his idea of prejustificatory language-games. Second, whereas knowing-that is bivalent (one either does or does not know *p*), knowing-how comes by degree, as something one does more or less well (Ryle 1949, 59). The praxeological view of linguistic competence presented here exhibits such graduation in two ways. To begin with, insofar as rules have an empirical (statistically distributed) shape, there can be varying degrees of conformity or nonconformity with normal conditions (which themselves can be more or less sharply delimited). In addition, some aspects of the broader conception of competence suggested by the language-game analogy are governed by considerations of appropriateness, admitting different kinds and degrees of response, rather than the polarity of "right" and "wrong."[25] These two features indicate that purposive awareness—including that of language—has a functional topology that is distinct from but is the basis of propositional content, thus giving substance to the claim that predicative awareness has a functionally discontinuous foundation in pragmatic sense.

Why does this matter? Is such linguistic knowing-how just an abstract possibility for a Heideggerian/Wittgensteinian position, or do we have experience of such language use? I want to suggest that in fact much language use is like this, and that this is what makes it important for a phenomenological conception of language. However, we might fail to notice this because our default experience of language parallels that which Heidegger suggests we have of tools such as hammers. Thus usually, when all is going well, our use of language feels effortless and smooth. We might say it has the feel of "smooth coping," as long as this is not taken to suggest that we lack a structured (circumspective) grasp of the world around us.[26] Conversely, it is only when something goes wrong or something unusual happens to provide a contrast that we become aware of the limitations of our default grasp of language.

Wittgenstein himself focuses on one example of this happening. In doing philosophy, he suggests, we often find ourselves in the same situation as Augustine did in attempting to describe what time is—we think we know what something is until we are asked to define it (*PU* §89). But "breakdowns" of this kind do not only occur in philosophy. For anyone teaching their own language to foreign learners it is a common experience to be caught out by questions such as "What does ... mean?" or "Why it is right or wrong to say ... ?" in specific circumstances. Such breakdowns are also a common experience for translators. The day-to-day reality of such work is to be confronted with the inadequacies of your own understanding of language—both "native" and "second" language—that leave you constantly reliant on monolingual and bilingual dictionaries, thesauri, or the Internet to plug the gaps.

These various examples are, of course, relatively unusual linguistic situations—as indeed they must be to highlight the contrast with, and limitations of, more usual abilities. All of them suggest that the need to explain our use of words is itself relatively unusual, that language use is typically prereflective, such that we simply use words rather than talking about our use of them. The Wittgensteinian view proposed here allows such situations to be described in a phenomenologically plausible way, tracking both the discontinuity and the limitations of our abilities suggested by such experiences. First, our prereflective linguistic ability can be understood as the ability to produce contextually appropriate language use, a form of knowing-how that can be described in terms of prepredicative language-games and rule-exemplification abilities. Both Augustine and the language teacher have, and retain throughout, this kind of linguistic know-how. Second, Wittgenstein's view allows us to understand what they—and we—lack by default: Augustine and the language teacher lack a reflective awareness of their own language-use patterns, the ability to state the rules they are following, while the translator often (also) lacks a sufficiently complete grasp of the relevant contexts of use. At the same time, third, it straightforwardly accommodates the fact that, when asked, we are generally able to take up a reflective stance on our own language use. This is not because reflective awareness is "tacit" or lurking in subpersonal systems, but because mature language users understand the kind of language-games involved (e.g., answering the question "why?," making one's views clearer to others) and are able to formulate spontaneously ad hoc explanations or theories about their own language-use patterns.

To highlight the distinctiveness and relevance of the Wittgensteinian position described here, it will be instructive to contrast it briefly with Robert Brandom's position in *Making it Explicit.* Describing himself as an "anti-intellectualist about norms" and an "antiformalist about logic" (1994, 135), in a manner reminiscent of Heidegger, Brandom acknowledges the need to found semantics in a "pragmatics"—that is, to cash out semantic and logical vocabulary in terms of practices, or what we do. The "critical criterion of adequacy" he suggests for a foundational pragmatics is that "the core linguistic practices it specifies be *sufficient* to confer propositional and other conceptual contents on the expressions, performances, and deontic statuses that play appropriate roles in those practices" (Brandom 1994, 159). Brandom makes a generic proposal as to the kind of practice fulfilling this task for which two ideas are central. The first, adopted from Wilfrid Sellars, is to consider language as a "game of giving and asking for reasons." The second, adopted from David Lewis, is to model the way participants keep track of this game as a form of "scorekeeping." In broad outline Brandom's approach is as follows: The fundamental kind of move made in such games is the act of asserting some proposition. In making an assertion one becomes committed to some content, both that expressed and that inferentially linked with this, and undertakes a responsibility, if challenged, to show that one is rationally entitled to make the relevant commitment. The basis of scorekeeping activities are what Brandom calls "deontic statuses": commitment to propositions, the exclusion of propositions on grounds of consistency, along with the entitlement to make the claims one is making. "Talking and thinking," Brandom (1994, 183) concisely assures us, "is keeping score in this sort of game."

The perspective developed here allows several criticisms of Brandom's position. First, it seems to me that in presenting "scorekeeping" as some kind of pervasive universal language-game it fails the test of phenomenological adequacy. On the one hand, it seems implausible to suggest that we are usually aware of keeping tabs on other people's propositional commitments (scorekeeping). On the other hand, this does not seem to be the kind of thing we could do without being aware of it: surely "keeping score" requires conscious and explicit awareness of what the score is. In addition, the scorekeeping model's assimilation of all language use to (tacit?) "giving and asking for reasons" seems to disregard the fact that not all linguistic practices involve reason giving.[27]

In fact, Brandom himself makes some allowance for the unphenomenological character of his foundational pragmatics. At one point he describes the notion of commitment on which his idea of deontic statuses and hence scorekeeping depends as "an artificial, scorekeeping device," adding that language-games also have a "material aspect" (Brandom 1994, 183). To understand this material aspect adequately would require describing phenomenologically plausible language-games, the characteristics and structure of activities or practices in which "scorekeeping" actually takes place. Accordingly, deontic statuses and the sense of accountability involved—what they are commitments to do, what it is to be held rationally accountable, and what counts as meeting or failing to meet justificatory commitments—are commitments that can be understood in terms of language-games.

From a phenomenological perspective this fact invites two objections. First, from this perspective the very intelligibility of the notion of scorekeeping rests on this possibility of being cashed out in phenomenal or "material" terms. Indeed, the whole point of phenomenological antiformalism (say Heidegger's) is that, rather than being thought of as basic, propositional content or commitments must be understood in terms of their embedding in underlying phenomena. For Brandom, however, propositional content and commitments remain fundamental—as clearly reflected in both the criterion of adequacy he identifies and his emphasis on the supposed "*pragmatic priority of the propositional*" (Brandom 1994, 79). As a result, despite recognizing the important desideratum of relating formal semantics to linguistic practice, it seems inaccurate to describe Brandom's position as "antiformalist" or "anti-intellectualist." Rather than taking seriously the "material aspect" of such phenomena, the scorekeeping model is a rational reconstruction that projects onto linguistic phenomena precisely the features required for these to cohere with semantics.[28] In this sense Brandom offers not an antidote to formalism, but simply a further ramification or extension of a formalist approach. The second objection is that, even if its intelligibility is not thought to be threatened, the fact that "scorekeeping" could be understood in terms of its material aspect nonetheless makes it an unnecessary artifice. Why not instead think of the foundation of semantics directly in praxeological terms—as the foundation of justificatory in prejustificatory language-games—in the way set out here?

3 The Heideggerian Framework Completed

We are now in a position to see how these Wittgensteinian views converge with and complement the views of Merleau-Ponty and Heidegger respectively, so as to complete the overall picture of language set out in the preceding chapters.

In addition to their shared compatibility with the general commitments defining the Heideggerian framework, an important point of convergence between Wittgenstein's and Merleau-Ponty's views is the attempt to conceive of language without recourse to an ideal of full, nontemporal determinacy. As seen in chapter 4, Merleau-Ponty's focus on finite embodied agency led him to reject any such ideal of final determinacy and to see all meaning, including linguistic meaning, as characterized by constitutively open horizons formed in ongoing processes. Wittgenstein arrived at a corresponding conclusion, becoming convinced that the commitment to full determinacy implicit in his earlier calculus model of language leads to incoherence. Merleau-Ponty's response was to emphasize the creative openness of language, while suggesting that what passes for univocity in language use is governed by the pragmatic criterion of successful communication (see chapter 4, section 4). A similar connection between practical efficacy and determinacy is found in Wittgenstein's pragmatization of determinacy, according to which the criteria for conceptual determinacy are a function of the practical requirements inherent in a language-game. Common to both authors' thinking about determinacy is a rejection of any ideal end state and the attempt to reconceive it in the context of human practices.

As one might expect, given the way I have presented them within the Heideggerian framework, there is not just convergence, but also complementarity between Wittgenstein's praxeological conception of language and Merleau-Ponty's view of indirect linguistic expression. To begin with, the view of pragmatic sense presented here specifically addresses the deficiencies of Merleau-Ponty's position highlighted at the end of chapter 5. As then anticipated, explicating the way linguistic regularities are linked with practice not only gives a convincing picture of the background of established "direct" language use, but clearly brings out the kind of pragmatic, practical, or vital constraints that necessarily check the unfolding of the expressive potential of language. That is, it identifies a counterbalance to expressive openness that yields (in Ricoeur's terms) a polysemy that is regulated—in the dual sense of being both constrained and describable by rules.

In this respect, an important advantage of Wittgenstein's praxeological approach is to provide a more sophisticated conception of rule-governed established language use than Merleau-Ponty's somewhat peripheral characterizations of "direct" sense. Thus Wittgenstein's critique of the ideal of determinacy and his subsequently reconfigured notion of rules first make clear how direct sense can be understood without eccentric metaphysical assumptions about nonperspectival presentations of intentional objects of the kind that Merleau-Ponty apparently thought "direct" sense implies. Wittgenstein's position also, second, provides a more detailed and versatile conception of linguistic rules than Merleau-Ponty offers. Having freed himself from the simple and rigid paradigm of rules on which the calculus model of language was based, Wittgenstein's later, more relaxed view of rules is much more flexible, and can be plausibly applied to the many varied forms of linguistic practice found in real life. Closely linked with this, third, is the fact that it allows a more balanced view of the relation between creative and established uses of language. Whereas Merleau-Ponty sees established patterns of "direct" language use as distinguished primarily by their lack of creativity and almost mechanical character, Wittgenstein's position allows for different degrees of more/less strictly regulated forms of practice, and recognizes that some practically meaningful language use may elude a rule-based characterization (the phronetic aspect of linguistic competence, as I called it). This suggests that established language use will, in practice, be characterized by varying degrees of regulation, with conspicuously creative and apparently humdrum cases marking different points on a continuous scale rather than standing in simple opposition.[29]

A second kind of complementarity between Wittgenstein's and Merleau-Ponty's views is that they each concentrate on a different aspect of the functioning of linguistic signs. From one point of view, as just emphasized, Wittgenstein's focus on the pragmatic aspect of language is an important counterbalance to the account of creative and expressive aspects of language use on which Merleau-Ponty's view concentrates. But at the same time Wittgenstein's somewhat dismissive approach to the misleading "images" bound up in linguistic forms is not altogether satisfactory.[30] For this seems to accept that linguistic form does have some representational—or, as I would prefer to say, presentational—significance, while treating the sole basis of linguistic meaning as praxeological. Yet if linguistic forms can be misleading, they can presumably also assume a positive role in mediating understanding, leading to the question of how misleading and illuminating uses of linguistic form are distinguished from one another. In this respect the notion of presentational sense developed above, on the basis of Merleau-Ponty's views, can be seen as complementing Wittgenstein's foundational pragmatism by setting out how fine (sublexical) differences can be interpreted as meaningful in the context of an established system of signs—and so be put to work in revealing features of the world.[31]

Another kind of complementarity concerns the level of linguistic articulation on which Merleau-Ponty's and Wittgenstein's respective conceptions of language concentrate. Whereas Merleau-Ponty's notion of indirect sense is based on the differential operation of sublexical elements, the language-game approach operates at the level of words. In Saussurean terms, Wittgenstein thus deals with the level of constituted signs, of distinct rather than differential features. An important consequence of this complementary focus is that it makes clear how the notion of reference can be accommodated within the Heideggerian framework. For a notable feature of the language-game conception is that it provides an "account of" reference that effectively undermines the idea of reference as a basic semantic notion. Thus, as seen in the last chapter (section 3), Wittgenstein's critique of pure referential relations, of simple referring or simple referents, leads to the idea of articulate referring, according to which what it is to be a referent is a function of the way an entity is referred to in the language-game context. Furthermore, what it is to be a referent must be understood as corresponding to the kind of rules found in language-games, and hence—without ideal determinacy—as subject to the pragmatization of determinacy discussed in section 5 of the preceding chapter. The upshot of Wittgenstein's approach is therefore that how, and how sharply, something is picked out as "a referent" is governed by the language-game in question.[32]

In emphasizing the foundational role of practice and its nonreductive approach to semantic questions, Wittgenstein's approach is also clearly similar in spirit to Heidegger's. Both conceive of language in a purposive perspective, such that Wittgensteinian language-games can be seen to comprise the context of Heidegger's instrumental relations, at least to the extent that the latter involve the use of linguistic signs. Similarly, both treat purposive activity as the basic context of the articulation or formation of linguistic sense, with the intrinsic link Wittgenstein identifies between the use of signs and forms of practice corresponding to Heidegger's view that the significance of words results from circumspective Setting-out.

Given these similarities, what Wittgenstein's praxeological conception of language adds to the Heideggerian framework are richer, phenomenologically sensitive means for envisaging and describing contexts of purposive awareness through their varying degrees of organization. That is, whereas Heidegger's talk of what ready-to-hand Equipment "is for" (its *Wozu*) or Dasein-relative aims (the *Worumwillen*) treats purposive notions as descriptively basic, Wittgenstein's notion of practice-inherent rules provides the means for structurally characterizing the various forms of purposive activity. Such rules can be seen as describing both the structure of practices (i.e., of purposive contexts) themselves and the role words play in language-games. In the latter respect, Wittgenstein's conception of praxeological rules also provides the means to understand the specific level of instrumentality built into Heidegger's conception of linguistic signs. For, to the extent that this follows a consistent pattern, what a sign does in a particular language-game will be describable in terms of such rules. Hence, despite the multitude of such roles, what it is for signs to be instruments in the sense of having specific functions can generally be conceived of in terms of praxeological rules.

Finally, one notable example of the way Wittgenstein's views complement those of Heidegger is by elucidating the idea the latter describes as a "modification" of the "hermeneutic *as*" into the "apophantic *as*," the move from readiness-to-hand to presence-at-hand. While Heidegger provides some idea of what kind of move this "modification" involves—that is, from purposive to objective, properties-based individuation of entities—the processes or phenomena in which this transition is effected remain obscure. In particular, Heidegger tells us that this transition is achieved by the use of "Statements." However, since, by his own account, the use of a declarative sentence does not suffice as a criterion for predicative awareness, some detail on how these are to be used so as to qualify as Statements seems necessary.[33] Wittgenstein's views help here by suggesting, at least in outline, how this "*as* modification" can be cashed out in phenomenal terms. At a general methodological level he would insist on the need to explicate the difference between prepredicative and predicative abilities in terms of corresponding types of language-game. A clear example of how this difference, and with it Heidegger's *as* modification, can be understood is provided by the distinction between prepredicative and predicative language-games discussed in the preceding section of this chapter. Thus on the prepredicative side (hermeneutic *as*) are simple, ungrounded language-games in which doing things linguistically requires only rule exemplification. On the predicative side (apophantic *as*) are more complex, founded language-games in which doing things linguistically involves rule stating, justification, explanation, and so on.

4 The Disclosive Function of Language

It is now possible to summarize the overall view of language that results from using the positions of Merleau-Ponty and Wittgenstein to fill out the Heideggerian framework. Doing this will at the same time sum up the answer provided by the position developed here to Taylor's question as to wherein the power of language to make things manifest lies, or what it is that language does in disclosing the world.

Taking Heidegger's analysis of disclosure in *SZ* as a starting point, the first major feature of my view was to situate the role of language in an overall view of how the world is disclosed that recognizes both its continuity with nonlinguistic behaviors and its distinctness. This view was taken to exemplify a pattern of general commitments that should characterize a phenomenological conception of language guided by the goal of accurately describing speakers' experience. I suggested that such a conception should conceive of language not as having an inside-outside topology, but as embedded in or distributed over other aspects of the world, as well as adopting an antireductionist and antiformalist attitude toward linguistic phenomena and treating language as a process rather than some static structure. These general features were summed up in the claim that language should be understood as language-in-the-world. This might be initially be taken as merely emphasizing that language is used in the midst of objects and shaped by (human) interaction with the physical surroundings, rather than being conceivable as an autonomous or abstract entity that comes into contact with the world only accidentally. However, the intervening chapters have shown how Wittgenstein and Merleau-Ponty furnish a more complete picture of what language is embedded in. Whereas the embedding in human practices alluded to in Heidegger's talk of ready-to-hand Equipment can be understood more fully using Wittgenstein's language-game analogy, Merleau-Ponty draws attention to two further aspects of the embedding of language: first the fact that language use is, as "lived sense," a phenomenon of embodied agency, second that language is shaped by the "expressive system" of lexical and sublexical differentiation. The resultant overall picture thus sees languages as embedded in human practice, in embodiment, and in their own semiotic horizon.

A consequence of this, reflected in Heidegger's antireductionism and antiformalism, is that language should not be conceived of one-sidedly in terms of "representational content." To be sure, one of the inputs determining the sense bound up in language is the way "the world" is—that is, the constitution or state of whatever our words are taken to refer to. Further, because human practices do not take place in a vacuum, the ability to act plausibly entails correctly understanding some fragment or aspects of our surroundings. Even where successful action does not require the ability to formulate that understanding linguistically, it is likely to involve commitments that can generally be interpreted as embodying an awareness of various facts. Nonetheless, the role of practical requirements, or the perspective of purpose, in shaping linguistic articulation renders unclear the extent to which practices can be thought to embody accurate representation of the surrounding world. After all, many practices are sustained either without needing to get the world relevantly right, or without our understanding correctly or completely what it is their practical perspective is getting right.[34] In the traditional language of representation, the "ideas" they embody are "confused" rather than "clear and distinct." So although it would be philosophically convenient if linguistic articulation could be thought of as straightforwardly encoding features of "the world," the idea is phenomenologically unconvincing. A phenomenologically plausible approach must instead view the "impurities" due to its embedding in the world not as inessential accretions, but as features essential to an understanding of what language is and does. Accordingly, both the process and the product of linguistic articulation should be seen as the result of a complex structuration involving both practice and the sedimented semiotic horizon, as well as other ("represented") aspects of the environing world.

This blending of different structural inputs in linguistic articulation is reflected in the second major feature of the position developed here. This is the proposal—at what I called the more specific level of the Heideggerian framework—that linguistic signs are instruments in two senses. To recall, Heidegger describes the disclosive function of linguistic signs both generically, as enabling agents to noninferentially grasp features of the world; and more particularly, as the means for performing certain tasks just like other tools. I suggested that these two modes of instrumentality are each linked with a kind of sense, presentational and pragmatic sense respectively. The last four chapters have since set out how these two kinds of sense can be understood based on the views of Merleau-Ponty and Wittgenstein. The presentational sense of expressions, I have claimed, is due to the sublexical-lexical structure of an iteratively evolved sign system. As a result of this structure, words are characterized by the inchoate rationality and situated interpretability that allow them to function—in the generic instrumental sense—as a means for making distinctions and thus pointing out features of the world. By contrast, the pragmatic sense of expressions can be understood in terms of the structure of language-using practices. Provided they are conceived in the suitably versatile ("relaxed") way that I have attributed to Wittgenstein, rather than being narrowly modeled on a mathematical calculus, the rules or regularities exhibited in word usage will provide a general means for describing both the structure of such practices and the various tasks performed by word instruments. Both these kinds of sense, as previously anticipated, suggest that linguistic signs function as instruments in an intrinsic rather than the extrinsic Lockean sense (see chapter 3, section 4). That is, language functions in both cases as an instrument essentially involved in meaning constitution: on Merleau-Ponty's view words are the very materials of which thoughts are constructed, whereas for Wittgenstein their use in certain ways is constitutive of the corresponding language-games.

This dual instrumentality implies that linguistic signs should be thought of as units that connect presentational and pragmatic sense. Despite this connection, the distinction between the two kinds of sense is motivated in two ways by the requirements of phenomenological accuracy. First, each corresponds to a basic phenomenological feature: the idea of presentational sense reflects the fact that sublexical structure is integral to the way linguistic signs function; that of pragmatic sense does justice to the way linguistic meaning is shaped by human practices (language-games). They thus correspond individually to the obvious facts that the entities of which languages consist are words and that words are used to get things done, and collectively to the fact that words are literally *involved in* the constitution of linguistic meaning. Second, each comes into play in different types of situation. In many practical contexts, the kind of expressive potential a word might have in virtue of its form will be irrelevant to its functioning. Conversely, the way a word's lexical form encodes associations and "imagery," rather than adhering to established use, might be a decisive factor in expressive contexts where finding the "right word" is important. In the former case one relies on standard patterns of use (pragmatic sense), in the latter on how the word presents features of the world (presentational sense).[35]

These comments suggest that the relationship between pragmatic and presentational sense should be thought of as one of symbiotic complementarity. Symbiosis involves distinct organisms living together in a close union, often one that is mutually beneficial. Similarly, while we can consider the functions of presentational and pragmatic sense in isolation, it is a constitutive and distinctive feature of natural languages—in contrast to artificial languages—that these two kinds of sense coinhabit linguistic signs. In connecting the two, words can in general be thought of as bringing both a presentational perspective and a background of use to each situation in which they are used. In much the same way as the two aspects of a duck-rabbit figure, each comes to the fore in different situations in light of different needs.[36] It is by combining these two distinct aspects of sense articulation that the dual instrumentality of linguistic signs provides a basis for explaining the range of phenomena described at the start of this book's introduction. On the one hand, the notion of pragmatic sense provides a means for describing the range of abilities in terms of which practical linguistic competence can be gauged, those abilities one must possess in sufficient measure in order to integrate inconspicuously into a "language community" and to count as "mastering" a natural language. On the other hand, the notion of presentational sense identifies the cumulative process and the differential factors that make linguistic expressions (re)interpretable for use in both existing and novel ways, and on which one relies in careful use or in developing new expressions.

A third major feature of the position developed here is the distinctive claim that there are uses of language that are prepredicative in character. This takes up Heidegger's view that all disclosure of the world is embedded or founded in prepredicative Equipmental or purposive awareness, a view that I argued extends to the use of language. The intervening chapters have attempted to give this thought greater substance and plausibility by showing how presentational and pragmatic sense are not only themselves phenomenologically plausible, but also each capable of being grasped prepredicatively (as well as providing a foundation for predicative abilities). In the case of presentational sense this involves grasping the differential force of a word as a means of expression in the context of a sign system; in that of pragmatic sense it amounts to a linguistic form of knowing-how, corresponding to the ability to produce a pattern of use in the context of a language-game. Neither of these abilities, I have argued, entails awareness of the expression's predicative or inferential import.

Together with the dual instrumentality of linguistic signs, this gives rise to the idea of a basic linguistic sensitivity in which both kinds of sense are simultaneously grasped prepredicatively.[37] Such basic linguistic sensitivity consists in the ability to distinguish between different word forms and to discern the right circumstances in which to apply them, and suffices for participation in the kind of prejustificatory language-games on which Wittgenstein's view of language is ultimately based. Employing Wittgenstein's picture of language as made up of a complex of language-games that can be analyzed into a stratified functional hierarchy, prepredicative language use can be understood both as itself sufficing to characterize competence in many language-games, and as identifying a foundational stratum of language-games on which predicative abilities are praxeologically based. Understood in this way, the prepredicative feats of language can be seen to constitute—to recall Seel's (2002, 50) apt term—a basic "domain-opening understanding of things."

This characterization helps to highlight both the specific importance and the limitations of the view I have been developing, as well as pointing to the task for the remainder of this book. Its specific importance lies in providing a more detailed understanding of the basic or "low-level" feats of language, the sense of first bringing to awareness that is intimated by terms such as "disclosure" or "revelation"—both of which are used to render Heidegger's *Erschlossenheit.* By contrast, some limitations of the view developed here are due simply to the limited scope of the issues considered. For example, I have said very little about the social aspect of language, so that it would be usefully complemented by consideration of the ways linguistic communication with others or the role of a linguistic community might have a world-disclosing function. Similarly, there are many ways to study and classify the functions of linguistic signs—say as speech acts, or the functions of speech events identified by Jakobsen (1990, 69–79)—that go beyond the presentational and pragmatic functions focused on here. Part of the interest of the present study, however, is that for these issues (among others) to arise language must already be performing a world-disclosing function, because the phenomena in question occur in the general context identified by the Heideggerian framework, using constituted signs that couple presentational with pragmatic sense.

Another kind of limitation lies in that prepredicative feats are clearly not the whole story with regard to the disclosive or cognitive feats of language use. Predicative or other "higher-level" cognitive feats, more complex and reflective language-games, remain possible, and may be felt to reflect more fully the power of language to reveal the world to agents. However, in that case, it might be wondered, why not simply focus on those higher-level feats rather than mere "domain-opening" forms of understanding? Part of the answer here would be that that is precisely the kind of approach that Heidegger, Merleau-Ponty, and Wittgenstein were all opposing. The motivation for their respective forms of antireductionism and antiformalism was in each case to prevent language being viewed in an overly intellectualized way as a complete, rationally functioning system of signs. Of course, this cannot suffice as an answer, because it simply restates their views, which have already been expounded here in detail—besides which, perhaps all three were mistaken in their motivations. Hence we are left with questions that need to be addressed directly: What is the philosophical importance of the phenomenological conception of language developed here? What role should it—particularly its emphasis on prepredicative factors—play in an overall philosophical understanding of language? Addressing these questions will be the task of the following chapters.

IV Some Philosophical Implications

8 The World Disclosed

Until now I have focused on building up a phenomenological conception of language by exploiting the views of Heidegger, Merleau-Ponty, and Wittgenstein. In this final part of the book I consider some of the broader philosophical implications of the resultant view, in particular how it relates to other theoretical approaches to language and what implications it has for linguistically mediated knowledge. With the latter in mind, this chapter's task will be to clarify what is meant by talking of disclosure of the world. It might seem that there is a tension within the Heideggerian framework that renders this idiom ambiguous or even incoherent. On the one hand, the Heideggerian term *disclosure* suggests that language genuinely reveals or uncovers what is around us just as it is (the real, objective world, so to speak). On the other hand, on Heidegger's phenomenological conception *world* refers to the world as experienced and—as the nexus of "Significance," instrumental relations, involvements, and so on—as permeated with human projects and purposes. This might be taken to suggest that the term *world* in fact refers to an anthropocentric, pragmatic interpretation that refracts and potentially distorts our grasp of our surroundings. In other words, reading all the relevant terms in a Heideggerian manner might appear to suggest a problematic tension between genuine revelation and mere interpretation of what is around us.

The kind of issue just referred to is usually discussed in terms of an opposition between realism and some nonrealist alternative (e.g., idealism, antirealism). A realist view is one that claims we can come to know the objects around us just as they are in themselves; a nonrealist alternative denies this on the basis of some reason or features that make it impossible to grasp the world around us in a fully objective or undistorted manner. Often there is a default assumption that realism is to be preferred to whatever the alternative is taken to be. This chapter might therefore also be seen as addressing the question of whether the phenomenological conception of language developed in the preceding chapters has realist or nonrealist implications.

Once the question is posed in this way, there are prima facie grounds for supposing that this conception is opposed to realism. This can be illustrated by considering the first of three theses Putnam takes to define "metaphysical realism," the assumption that "the world consists of some fixed totality of mind-independent objects" (Putnam 1981, 49). One challenge to this assumption comes with my rejection of ideal or final determinacy and my insistence on the constitutive openness of language's functioning. This directly opposes the metaphysical realist's appeal to a "fixed totality" of objects, because the whole point of this appeal is to identify a standard of determinacy and an end point for inquiry. A second challenge to the metaphysical realist's first thesis results from my emphasis on the "involvement" of words in the formation of propositional content, as this seems to undermine the mind-independence of objects referred to with language. The concern here is not whether objects can generally be thought of as existing independently of thought or language in a causal sense. Although many entities in the human lifeworld would not exist without language use, many other entities (e.g., naturally occurring ones) would exist just as they do without language. Rather the concern is with the sense attaching to the word *object* in knowledge "of objects"—that is, whether this means objects-as-represented or objects-as-they-are-in-themselves, or in Kant's terms "appearances" or "things-in-themselves." Emphasizing the role of presentational and pragmatic sense might seem to imply that using language to disclose the world leaves us on the wrong side of a philosophically problematic gap, with something akin to Kantian appearances, or objects-for-us, rather than objects-as-they-are-in-themselves.

In the following I try to show that a framework of Heideggerian commitments of the kind I have been assuming eliminates such concerns. The first section looks at Heidegger's discussion of the implications of his emphasis on Dasein's being-in-the-world for the realism/idealism issue, arguing that there are two main lessons to be drawn from this discussion. The first is that Heidegger sees traditional questions about realism and idealism as having been based on mistaken ontological presuppositions and hence as superseded. The second is that the expression "being-in-the-world" marks genuine and direct contact with our surroundings, such that there is no principled gap between the way the world appears to us in the perspective of our projects and the way the world is "in itself." The two subsequent sections apply these lessons to a phenomenological conception of language. The chapter's second section thus sets out how feats of linguistic disclosure can be understood as a way of being in genuine and direct contact with our surroundings. The final section aims to prevent the present phenomenological view from being confused with either realism or nonrealism by distinguishing it from familiar positions, highlighting its Heideggerian virtue of moving beyond this supposed alternative.

1 Heidegger on the "Reality Problem"

One striking feature of Putnam's approach is that metaphysical realism's basic postulate is characterized in an explicitly dualist manner as "mind-independent objects." This naturally evokes the familiar traditional picture—found, for example, in Descartes, British Empiricism, and Kant—of a fundamental subject-object opposition and the corresponding inside-outside imagery. Given this basic picture, the poignancy of the realism/idealism issue is easily appreciated, since knowledge is then bound to depend on the relationship between internal representations and external objects, and the two extremes of confinement within representation and cognitively reaching out to external reality.

One of Heidegger's principal aims, however, was to overcome this basic picture of humans' epistemological situation. Explicit discussion of how his position bears on traditional debates about realism and idealism, or as Heidegger terms it the "Reality Problem" (*Realitätsproblem*), comes in §43 of *SZ*. His response to this problem is based on the fundamental objection that both realist and idealist strands of modern philosophy lack requisite ontological understanding in failing to explicate what it is for the things they presuppose to exist. Realism suffers from a "lack of ontological understanding," Heidegger charges, since "adequate ontological analysis of reality" depends on the extent to which "that *from which* independence is to exist [*bestehen*], *what* is to be transcended, is *itself* clarified with respect to its *being*."[1] Similarly, while recognizing that "reality is possible only in understanding of being," idealism still stands in need of "ontological analysis of consciousness itself" to understand what it is to be real (*SZ* 207). While his formulation betrays Heidegger's conviction that of the two idealism is the philosophically superior position, his objections are intended to apply quite generally to setting up philosophical problems on the basis of an inside-outside (or subject-object) opposition.[2] For, as his preceding analytic of Dasein makes clear, Heidegger believes that attempting to spell out the ontological assumptions on which such a distinction is predicated would reveal the untenability of this basic picture and its "demolition of the original phenomenon of being-in-the-world."[3] The significance of this is not simply that both realism and idealism are faulted options, but that to set up debate in terms of these two alternatives betrays a fundamental misunderstanding of the human epistemological situation.

Although by this point in the work his general approach to the problem comes as little surprise to the reader, Heidegger's direct response to the Reality Problem has two sides, one negative, the other positive. On the negative side he cursorily and polemically dismisses the Cartesian questions of whether "consciousness-transcendent" entities exist and whether this can be proved as making no sense (being "ohne Sinn"), and claims that the real "scandal of philosophy"—borrowing Kant's phrase—is that such proofs are demanded (cf. *SZ* 202, 205). The basis of this claim, set out in the preceding sections of *SZ*, is Heidegger's conviction that the existence of "external" objects is implicit in the disclosedness that Dasein is, such that it comprises a condition of possibility for the question of their existence even to be raised (*SZ* 202, 205). Although it may be disputed that to question the external world's existence makes no sense, Heidegger's underlying point is that making sense of the supposed problem would require explicit defense of the ontological assumptions—about what it is to be a subject or object—on which these questions are based. (And once we are clear about those assumptions, Heidegger plausibly believes, we would recognize them as failing to reflect the kind of beings we are.)

The positive side of Heidegger's treatment of the Reality Problem is to reinterpret what this problem is in an unorthodox way. Crucially, Heidegger takes the term *reality* (*Realität*), as traditionally used, to refer to "nothing other" than what he has otherwise called "*Vorhandenheit*"—that is, the being of "Things" or presence-at-hand.[4] With this stipulation in place, the Reality Problem is no longer that of whether objects exist (whether Things are present-at-hand), but instead becomes that of how present-at-hand Things feature in our understanding of Being. This second aspect of Heidegger's response is somewhat more subtle than the first, and reflects a recognition that the question of how objective knowledge (as we might call it) comes about remains even when the existence of the external world is taken to be beyond doubt. And his answer to *this* question, already provided by Heidegger's analytic of Dasein, is that "objective reality" is founded in readiness-to-hand, and hence in the purposive perspective of Dasein's concerns. Thus, as Heidegger understands it, the Reality Problem is about the relationship between entities accessed or individuated in different ways (the perspectives of purpose and properties respectively).

To appreciate the significance of this interpretation of the problem, it is important to keep in mind that Heidegger does not question our ability to know "real" Things. His view is simply that this ability needs to be situated in a more adequate overall picture of humans' relation to their surroundings—that knowledge is a "founded mode" of Dasein's being-in-the-world (*SZ* §13). More specifically, this ability involves—as set out in his discussion of Statements (see chapter 1, section 4)—a switch from a purposive grasp of the world to predicative awareness in which the "instrumental relations of Significance" are "cut off" (*SZ* 158). Thus Heidegger allows that properties-based awareness of entities can free itself or abstract from the underlying contexts of involvement and purposive conditioning (hence "objective knowledge" is possible). Indeed, this is for Heidegger what the "independence" of "reality" consists in—rather than an unexplained notion of "mind"-independence on which he claims realists rely. Yet at the same time his talk of "foundation" reflects a conviction that such independence must be acquired the hard way, by "cutting off" the antecedent links with human purposiveness. That is, Heidegger thinks that context-invariant features (properties) can emerge only against a rich background of varying contexts, that phenomenal material is required to discern phenomenal structure. The difficulty in obtaining objective knowledge is then not how to connect with (or "hook onto") the world, but to liberate ourselves from particular tasks or involvements so as to gain an overview of (or higher-order insight into) its structures. Accordingly, on Heidegger's nondualist picture, purposive and predicative awareness are simply two different ways of being in genuine contact with our surroundings, two different ways of being "in the truth" (*SZ* 221).

An implication of this view is that the traditional dualist distinction between "for us" and "in themselves" collapses. On the one hand, while ready-to-hand Equipment clearly does not have an agent-independent way of being (it is "for us"), nothing prevents us from knowing our tools just as they are ("in themselves"). On the other hand, as just explained, while Heidegger's conception of present-at-hand Things preserves their independence (they are "in themselves") and allows that we can gain undistorted knowledge of them, his foundation thesis implies that this independence and the possibility of "objective" knowledge are intelligible only in relation to a background of purposive activity (i.e., to the "for us"). Thus both ready-to-hand Equipment and present-at-hand Things can be known just as they are ("in themselves") but retain a constitutive link with access conditions ("for us").[5]

In place of the "for us"/"in themselves" distinction Heidegger sees agents as having different modes of "access" to their world and relies on what might be called an access-condition-relative notion of entities (e.g., as ready-to-hand or present-at-hand).[6] This can be seen as an intermediate, nondualist notion which allows to idealism that entities are defined by a particular mode of access while denying—in realist spirit—that this entails any break with how things are "in themselves." Of course, Heidegger's talk of "access" (*Zugang*) is not supposed to suggest that Dasein could be altogether out of touch with, or lack access to, the surrounding world. Rather it reflects a recognition of the methodological need for a phenomenological conception of knowledge to cohere with an account of the kinds of experience in which knowledge is acquired.[7] In this respect "access" corresponds to the act side of Husserl's act-object duality, while avoiding any hint of a subjectivist or psychological notion of consciousness that the latter might carry.

To talk of an "access-condition-relative" notion of entities invites the objection that this seems ill-suited to natural objects, which are putatively individuated in "access"-independent (e.g., spatiotemporal) terms: Isn't one forced to admit that the individuation of rocks, atoms, and so on, is no more "access" dependent than it is "mind" dependent? Here it is important to appreciate that there is a close link between Heidegger's notion of access and that of individuating entities. For, as just pointed out, he thinks of the individuation of entities in certain ways as a correlative of a certain mode of access. Indeed, independently of Heidegger, the two are plausibly linked: to know how something is individuated is to know, at least in principle, how to access it; conversely, to know how to access something requires knowing, at least in part, how it is individuated. This suggests, however, that within the framework of Heidegger's thought to say that entities are access-dependent is in effect a tautology, since all this means is that entities are necessarily related to some mode of individuation. Taken this way, it also seems undeniable, since to claim that there are entities without access conditions—that is, individuated features of the world without individuating conditions—no longer makes any sense. In that case, however, natural objects present no challenge, because Heidegger's talk of access conditions requires only that such objects be understood in relation to some (e.g., spatiotemporal) individuation conditions.

There is one further complication to Heidegger's views, which might be thought to reveal a new variant of idealism based on his distinction between being (*Sein*) and that-which-is (*Seiendes*). The basis for this thought is provided by a number of passages in which Heidegger talks of the "dependency of being, [but] not of what is, on understanding of being," and "the dependency of reality, [but] not of the real, on Concern."[8] Part of the point of these claims is clearly to acknowledge that if humans did not exist most of what there is in the universe would be unchanged—that "what is" (*Seiendes*) is independent of Dasein. This acknowledgment is often read as confirmation that Heidegger's position is a form of "realism"—such as "hermeneutic realism," "ontic realism," or "empirical realism."[9] Conversely, his claim that being depends on Dasein, is sometimes interpreted as meaning that he remains an "idealist" about being—as an ontological idealist perhaps or a "transcendental idealist about being."[10]

What kind of "idealism" this might be clearly hinges on how Heidegger's claim that being (*Sein*) depends on Dasein is interpreted. However, that is clearly a central question of Heidegger scholarship, one that I cannot attempt to pursue in detail here.[11] Instead I want to indicate briefly why this matter can be ignored for my purposes here. One consideration is that it is very unclear what it might mean to describe Heidegger's position as an "idealist" one: usually "idealism" marks a link with subjectivity, whereas Heidegger categorically rejects the notion of a subject along with that of objects. Nor can the subject's role be simply transferred to Dasein. Given that Dasein cannot be defined independently of its world, neither being nor the world can depend on Dasein in the (unidirectional) way idealism is thought to depend on subjectivity—indeed it is arguably unclear what sense of Dasein-dependence could be in play here.[12] Similarly, given that Heidegger's position collapses the "for us"/"in itself" distinction, it is unclear what the point might be of calling his position an idealist one, in particular what kind of "realist" (?) position it might be supposed to contrast with. A second consideration is that, whatever Heidegger intended to say in claiming that being depends on Dasein, it is unlikely to bear any specific relevance to a discussion of the disclosive function of language. For his claim is intended to be highly general in scope, aiming at transcendental conditions of possibility for anything to be an entity. And in the context of *SZ* it is clear that this encompasses intelligent nonlinguistic behaviors, rather than telling us anything specifically relating to the role of language in revealing the world.

To conclude, for my purposes here there are two main lessons to be drawn from Heidegger's treatment of the Reality Problem. The first is that Heidegger saw not only the alternatives of realism and idealism, but also the questions these purport to answer as misguided.[13] Traditionally the realism/idealism issue has addressed the relationship between subjects and objects, or between things-for-us and things-in-themselves, and the general constellation of appealing but conflicting "intuitions," with their respectively problematic consequences, that results from this picture of the epistemological situation. Given Heidegger's rejection of the ontological presuppositions on which these contrasts rest, it is difficult to see what the realism/idealism issue might still be about. The second main lesson is that Heidegger's emphasis on Dasein's embeddedness in its world is such that the idea of a principled difference between the way the world appears "for us" in the perspective of our projects and the way the world is "in itself" breaks down. On this view Dasein is instead in unbreached cognitive contact with its surroundings, which are genuinely revealed to it through purposive engagement with the world. Indeed, the thought that the idea of disclosure is in tension with Heidegger's conception of world—the supposed contrast between genuine revelation and mere interpretation—appears to be simply another product of dualist presuppositions of the kind Heidegger rejects.

2 Linguistic Contact with the World

What would it be for a phenomenological conception of language to have realist or idealist implications? Any answer to this should respect its emphasis on the embeddedness of language-in-the-world and the concomitant rejection of the idea that language has an "inside" and an "outside." Given the parallels between this view of language and Heidegger's notion of being-in-the-world, *SZ*'s treatment of idealism and realism should provide some guidance in developing such an answer. Accordingly, in this section I will take up its lesson that Dasein is always in cognitive contact with its world, and consider what kind of "contact" we have with objects through language.

It will help to begin with a contrast. Traditionally the realism/idealism debate has drawn its poignancy from the worry that a subject's contact with its world is inherently problematic. On the Cartesian picture, most notably, representations—the "veil of ideas"—both stand between and link subjects and objects. This picture of representations *intervening* between (i.e., literally coming between) knowing and what is known, portrays a subject's contact with its surroundings as indirect. Because of this, particularly with the need for mediation by such representations, it suggests that our awareness of objects may be systematically refracted or occlusive, and even that we may be radically deceived concerning the existence of external objects.[14] One might be tempted to think of language as functioning in an analogous way. Thus in the present case emphasizing the involvement of words, via their presentational and pragmatic sense, in constituting awareness of the world might be taken to suggest that language too plays the role of intervening representation, reintroducing familiar concerns about a basic disconnect between the way objects are "for us" and how they are "in themselves."

The task here is therefore to understand the disclosive function of language while avoiding the influence of this picture of intervening representations. I approach that task in two complementary steps. On the one hand, we need to understand in what sense the use of words is constitutive of the understanding in which they are involved. On the other hand, we need to understand how this involvement fails to generate the possibility of systematically distorting, refracting, or occluding the world in the way the intervening-representations picture does. These two steps will allow us to see language not as standing *between* us and real objects, but as a way of being *in contact* with them.

The phenomenological conception of language developed here combines two ways in which the involvement of words is constitutive of understanding.[15] The first is Merleau-Ponty's view that "thought tends toward expression as toward its completion" (*PdP* 206), which, I suggested, should be interpreted as claiming that what a thought is can be understood only in terms of its being expressed. At one level this is a causal claim to the effect that words—as the ontologically enduring linguistic entities—are quite literally the means, the forms and materials, with which we fashion thoughts, and in the absence of which it is difficult to see what would count as having (many) thoughts. However, its point is primarily criterial in the sense that the only general way of understanding and stating what the content of a thought is is in relation to its linguistic expression. In this dual causal-criterial sense the involvement of words is constitutive for the "completion" of thought in the way Merleau-Ponty describes. There is also, second, the Wittgensteinian insight that the use of words in certain ways is constitutive of many forms of practice. This kind of constitution, corresponding to the use of words as means for specific tasks, is again both causal and criterial. It reflects not only the fact that words are actually used to do these things, but also the tautological sense entailed by the internal link between language use and practice: to use language in such and such a way just is to engage in the corresponding language-game, whereas to use language differently would be to act differently.

To be "constitutive" in either of these ways thus has a dual aspect. On the one hand, the claim is to describe the way language is actually (causally) involved in our interaction with the world. On the other hand, there is a criterial aspect in that what it is to be a particular thought or a particular practice involves reference to language. With either mode of constitution, presentational or pragmatic, it could be said that language in some sense "makes possible" the expression of thoughts and content.[16] However, I want to emphasize that rather than attributing to language some kind of transcendental status, to be "constitutive" in these ways is a prosaic, nuts-and-bolts affair, akin to the way that using a hammer is constitutive of hammering a nail into a wall. Accordingly, to say language has a constitutive function is a descriptive claim rather than some kind of explanation, it is to describe how language is part of certain phenomena, rather than to identify language as a distinguished entity—so to speak, semantic ether—within which meaning arises. This modest sense in which language "constitutes" meaning is perhaps best conveyed by talking—in a way hinted at by Merleau-Ponty—of "conditions of actuality" rather than "conditions of possibility" (cf. *PdP* 48, 74). That is, it is a claim not about what or how linguistic meaning *has to be*, what *could be* meant, but about how it *is* and how language *is actually* involved in generating meaning.[17]

Given the two respective modes of linguistic meaning constitution, presentational and pragmatic sense can be thought of as explicating different aspects of how the perspective under which objects are linguistically addressed is built up. In view of this, the phenomenological conception of language can be usefully summed up as a kind of perspectivism, provided that several points are kept in mind. First, to someone accustomed to thinking in terms of the distinction between objects-for-us and objects-in-themselves, a perspectivist view might look like idealism. However, within a Heideggerian framework, I have argued, that distinction is inoperative, such that linking the concept of an object with access conditions neither entails a loss of genuine contact with the world nor has "idealist" implications. Indeed, this is precisely what the visual metaphor of perspectives should lead us to expect: for taking up different viewpoints on a particular object or scene in no way suggests that one is failing to see that object itself, or that the scene in question is being misrepresented. On the contrary, since it relies on the idea of being situated in a space of possible alternatives, to talk of one perspective implies the possibility of adopting others and with this the possibility of correcting idiosyncrasies of any single standpoint. A perspective is simply one of many possible direct takes on the world itself.

Second, it might be thought that to talk of perspectivism involves reference to an agent or group that bears such perspectives. While it is, of course, true that linguistic perspectives are borne by and to some extent shared by groups of speakers, the point of presentational and pragmatic sense is to identify specific factors that make up such perspectives and their differences or similarities. The fact that linguistic perspectives are shared (or not shared) is of no importance in understanding the role of these perspective-constituting factors. Accordingly, by focusing on features of public phenomenology, rather than appealing to cognitive or motivational states of speakers, the pragmatic-presentational perspective proposed here respects the need for linguistic sense to be impersonal.[18]

Third, to talk of perspectivism does not imply any kind of constraint or limit on knowledge. Not only is expansion or improvement of a linguistic perspective possible—specifically by developing new practices or new means of expression—but it is quite plausible for such processes to be driven by the epistemic aims of acquiring better or more comprehensive knowledge of the world.[19] So to say that a phenomenological conception of language is perspectival does not imply that human understanding or knowledge is constrained in some principled sense by the "limits of one's language." Rather than identifying some kind of transcendental or necessary limit on what can be said, observations about the limits of our linguistic perspective are straightforwardly technical and contingent in character, hinting at what we have yet to learn to say: not having a way to talk about ... is a limit just as the fact that no one has yet built a solar-powered family car is a "limit of mobility."

Finally, acknowledging the possibility of improving or expanding perspectives does not entail that such processes either have or are guided by the ideal of some final, absolute or unconditioned terminus. Rather the pursuit of more expansive and detailed perspectives is better thought of—in Merleau-Pontian manner—as an open process. The reason for this is not simply that no one has any clear conception of what such a terminus might be, nor even that the ideal of full determinacy might—as Wittgenstein thought (see chapter 6, section 4)—not be a coherent ideal. It is simply that nothing in real discourse requires it. Conceiving of possible improvements to a perspective no more requires directedness toward an "absolute" conception than aspiring to run 100 m faster requires an absolute conception of speed.[20]

So much then for the first step identified above of understanding the constitutive function of language. The second step is to understand why such perspectival and constitutive modes of linguistic contact should not be thought of—as with the intervening representations model—as systematically refracting or occluding awareness of our surroundings. I want to suggest three reasons for thinking this is not the case. The first is the embeddedness of language in other phenomena, which contrasts with the traditional picture of representations' being sandwiched between the thinking subject and external objects. Such embeddedness, reflected in my talk of "language-in-the-world," implies that knowledge of language is not sui generis, but is intrinsically bound up with more general awareness of the world. On this picture linguistically constituted knowledge cannot be thought to consist of knowledge of language and knowledge of the world attained through language as two disjunct elements. If any clear sense can be given to this distinction, their relationship would be as two aspects of an integrated whole.

A second reason for not thinking of language as a refractive medium on the model of intervening representations is its artifactual ontological status. If we stood in relation to language as we do to natural objects, as features of the world that exist prior to and independently of our awareness of them, then it might be reasonable to wonder whether language systematically refracted or put limits on our access to the world. In that case language might be presenting the world to us in a way that we are ignorant of. So, for example, if my eyes' lens had always been tinted (rose colored, say), I might remain unaware that colors are always being presented to me in this refractive manner. This is clearly how philosophers have often thought of mental representations. But it is not how we stand in relation to language. Rather, as something produced by human activity, language has never been completely beyond the pale of human understanding. Moreover, it is something that has been attuned and refined in interaction with the environing world. So although one might be unaware of various nuances and implications of the language one uses, it is implausible to think anyone could speak a language without being constantly aware of its relation to other aspects of the world. It is therefore difficult to see how language—in contrast to mental representations—could stand in an undiscovered, and systematically refractive, general relation to what we understand in using it.[21]

Finally, it is important to realize that there are other models in terms of which linguistic constitution of meaning can be conceived, so that there is no need to assimilate it to the intervening representations model. To appreciate this, it will be helpful to distinguish two different ways that something might function as an intermediary, or as mediating some effect. To think of representations as intervening in our awareness of the world is to think of them as effecting what I will call *material mediation*. The defining feature of material mediation is that whatever functions as an intermediary either forms part of, or leaves some characteristic imprint on, the result. Understanding either the process or the result of material mediation consequently requires some mention of the nature of the material intermediary involved. For example, in describing how either a pair of glasses or mental representations affect our awareness of the world, it seems necessary to take account of the optical properties and so on of the respective material intermediary.[22]

An alternative model for thinking about mediation is provided by the chemical function of catalysts, these being substances that play a role in either initiating or facilitating some reactive sequence, but that precipitate out to comprise no part of the overall reaction's end product. By definition a catalyst is something constitutively involved in the (chemical) process, but not the product, of what it brings about. Such mediation—*catalytic mediation*—thus contrasts with material mediation, in which the intermediary is implicated in both the process and product.[23] It is important, I suggest, to realize that instruments, being typically used to bring about states of the world of which they comprise no part, often function in a manner that contrasts in the same way with material mediation. To use the customary example, though a hammer is used to hammer in nails, the result is ("hammer-free") nails in the wall—the hammer is part of the process, but not of its product. In such cases there is surely no temptation to think of instruments as standing *between* us and the world, since in using them we are obviously in direct contact with our surroundings. So part of the point of emphasizing the instrumental character of linguistic signs, as I have been doing, is to suggest both that language is in this kind of direct instrumental contact with the world, and that its mode of mediation is catalytic rather than material. That is, thinking of awareness of the world as linguistically mediated does not mean that this awareness remains tacitly dependent on one's grasp of language as such; instead the latter is immaterial to the result, though not the process of arriving at such understanding. Once this catalytic mode of mediation is recognized, however, it is not only clear that there is no need to model the constitutive function of language on intervening representations. Rather, given the embeddedness and artifactuality of language, it is also difficult to see any reason to do so.

At this point some clarification is required. The claim that we are in unbreached, undistorted linguistic contact with the world might appear to imply some form of infallibility, as though in using language we could not possibly be mistaken about our surroundings or in our views about language itself. This would clearly be an unacceptable implication. Not only would it be obviously wrong to suggest that using language prevents us from being mistaken about the world. It would also undermine the coherence of a phenomenological conception of language in suggesting it is impossible for opposing philosophical positions to misunderstand language and in denying such a conception the capacity to correct conflicting views.

However, the point of saying that we are in contact with our surroundings is merely to deny that in using language our understanding of the world is subject to principled philosophical constraints that systematically and unavoidably shape it. The claim is that the use of language does not systematically distort our awareness of the world on ontological or metaphysical grounds, due to the "nature of language" or a supposed "gap" between language and its referents. Further, it is that our use of language does not block or limit our access to the world "itself" by erecting an edifice of ontologically distinct "appearances" between us and "real things." These denials are necessary because such assumptions have often been made by philosophers transferring traditional post-Cartesian patterns of thought to the case of language. Yet they do not amount to an infallibility claim. The point is rather that when we are wrong about the world, or in our understanding of language, there is no profound philosophical reason that makes such errors necessary. Instead we are failing to understand something that lies open before our eyes, a failure for which we bear the responsibility rather than having an underlying ontological or metaphysical excuse.

There are nevertheless many, more prosaic, ways in which the understanding of the world we develop in using language or our views about language itself may be incomplete or mistaken. We might, for example, lack sufficiently broad experience, never have been attentive to specific features; or we may exaggerate or overgeneralize certain aspects, perhaps due to inherited concepts or a theoretical bias. In most cases such limitations or errors are (quite rightly) unlikely to appear of general philosophical significance—that is, they would not lead us to conclude there is something systematically problematic about our linguistically mediated contact with the world. Indeed, we are most likely to come to such a conclusion when attempting to reflect explicitly on language itself. For in doing this we are particularly prone to overgeneralize the features of a limited range of examples, while losing sight of the many ways we use language prereflectively in navigating the world around us. In this respect Wittgenstein is perhaps right to suggest that it is the attempt to philosophize that leads to misunderstandings that could be avoided by more careful attention to everyday language use (e.g., *PU* §§94, 116, 593).

One other kind of case might be thought to show that a more general and hence problematic deception is possible. Imagine a particular civilization—like many ancient and some contemporary cultures—that believes that words have magic powers, such as the power to call forth spirits or elicit supernatural intervention in worldly events. We might think such beliefs about the function of words are mistaken and linked with a fundamental misrepresentation of the world, while nonetheless allowing that they are an integral part of that civilization's everyday life rather than an artifact of attempts to theorize about language. However, apart from these beliefs being more widespread and more integrated in the community's lifestyle, it seems to me that this case does not differ significantly from misdirected philosophical theories about language. In both cases the concern is with reflective views about the function of language, which could be corrected by proper understanding of established language uses. Specifically what is needed in the present example is proper attention to the causal powers (or lack thereof) actually manifested in various kinds of "language-game." Both cases involve contingent and corrigible error, rather than some necessary problem that flows from the nature of language itself.

3 Beyond Realism and Nonrealism

The second main lesson of Heidegger's discussion of realism and idealism is that he saw both as bad options resulting from inadequate ontological assumptions. All the same, it is possible to persist in asking whether Heidegger is an idealist or a realist. And if one does this, the chances are that his position will be taken—or rather mistaken—for a realist or idealist one. This is hardly surprising. To begin with, because Heidegger's position aims to combine the advantages of both without reducing to either, it is bound to look like a realist position in some respects and an idealist one in others. It is also unsurprising because, as Collingwood (1960, 149) memorably puts it, "When ideas are dead their ghosts usually walk." That is, in the present case, other philosophers are likely to continue asking what they take to be good questions. The aim of this section is to prevent the phenomenological conception of language developed here from being confused with either realism or nonrealism—and thus haunted by the ghost of Cartesian dualism—by clarifying how it differs from familiar forms of both.

Returning first to Putnam's metaphysical realist, I pointed out at the start of the chapter that the position developed here conflicts with its assumption of a "fixed totality of mind-independent objects." One clear reason for this was my rejection of an ideal or final determinacy. However, with regard to mind-independence we can now see that the situation is more nuanced than it initially appeared. In this respect I suggested that a phenomenological view of language should take up the lessons of Heidegger's treatment of realism/idealism. Heidegger had little time for the idea of mind-independence, objecting that to be intelligible (metaphysical) realism would need to explain what this independence is. Nor, however, did he accept the opposing idea of mind-dependence, and instead reconceived the Reality Problem in terms of access conditions (individuation conditions). This enabled him both to define the sense of independence involved in moving from readiness-to-hand to presence-at-hand and in this way to acknowledge the possibility of objective knowledge. Furthermore, by collapsing the for-us/in-itself distinction Heidegger allows language use to be seen as a way of being in pervasive contact with our surroundings rather than only indirectly or imperfectly connected with them. Thus a Heideggerian approach of the kind proposed here is able to accommodate the kind of independence the realist values without suggesting a philosophical standoff between the perspectives of "for-us" and "in-themselves" or conflating different modes of access with the issue of what is and what is not "real."

Putnam characterizes metaphysical realism in terms of two further theses, namely, that there is "exactly one true and complete description of 'the way the world is'" and that "Truth involves some sort of correspondence relation between words or thought-signs and external things and sets of things" (Putnam 1981, 49). Given its perspectival character, my phenomenological conception of language is at odds with the first of these, the idea that there is only "one true and complete description" of the world. Clearly the vagueness of this suggestion makes it difficult to assess—what, for example, is "complete" description? However, it at least intimates a contrast with the idea that a language embodies a certain perspective on the world that is subject to change or differences. Yet if the optical metaphor is taken seriously, it seems fairly obvious that pictures of the same scene, produced from different viewpoints, using different palettes, or under different perspective conventions, and so on, can look quite different without being "wrong." Further, once this kind of representational pluralism is acknowledged, it seems reasonable to think that the—again, admittedly vague—idea of correspondence could be invoked to describe the correctness of different pictures of the same scene. So, by this analogy, there is no obvious reason to suppose that a perspectivist conception of language in fact conflicts with the metaphysical realist's third thesis of truth as correspondence.[24]

It is perhaps by now unsurprising that the relationship with the "internalist" position Putnam sees as contrasting with (and improving on) the "externalist" perspective of metaphysical realism is no more straightforward. This internalism is again characterized by three theses. First, for the internalist, "*what objects does the world consist of?* is a question that it only makes sense to ask *within* a theory or description" (Putnam 1981, 49). Second, internalism allows that there might be "more than one 'true' theory or description of the world" (Putnam 1981, 49). Third, on this perspective, truth "is an *idealization* of rational acceptability," something justified under "epistemically ideal conditions" (Putnam 1981, 55).

In some respects, the phenomenological position I have been developing is closer to this internalist position than to the externalist-metaphysical one. In relating objectivity to "theory or description," internalism clearly attempts to meet the requirement I have discussed in terms of access or individuation conditions. In acknowledging that there are "experiential *inputs* to knowledge" (Putnam 1981, 54), it also seeks to accommodate the fact that we are in contact with our surroundings and that not all descriptions or theories are equally good—that the world can, so to speak, resist being described in certain ways. This acknowledgment, together with the commitment to some notion of truth as a norm of inquiry, underlies Putnam's conviction that the internalist perspective can be construed as a kind (internal) of realism. At the same time, the internalist's allowance that there might be more than one acceptable description of the world seems congenial to the perspectivist view advocated here.

Nonetheless—disregarding the obscurity of an appeal to "epistemically ideal conditions"[25]—my position is more saliently opposed to Putnam's so-called internal realism, with its invocation of interiority, than it is to metaphysical realism. This interiority is defined in relation to sets of concepts, or "conceptual schemes," with internal realism accepting that "'Objects' do not exist independently of conceptual schemes" and that objectivity is "*objectivity for us*" (Putnam 1981, 52, 55). However, since they are supposed to be instantiated in language use, it is difficult to see how the notion of conceptual schemes can sustain inside-outside imagery. For the whole point of my talk of language as "language-in-the-world" is that phenomenological accountability requires conceiving language as ontologically and functionally embedded in the wider world, so that its use cannot create a schism between objects-in-themselves and objects-for-us.[26] Indeed, to describe the way certain linguistic or conceptual means determine access to objects in the world in terms of an inside-outside topology seems simply to misappropriate the traditional picture of mental representations intervening between thinking subjects and the physical world. And while it may be plausible with phenomena such as perception, transferring the intervening representations model to linguistic phenomena is—as the preceding section argued—both unnecessary and misleading.

Although Putnam thinks of the internalist perspective as a form of realism, in holding that knowledge deals only in "objectivity for us," this position has obvious proximity to positions otherwise labeled "idealism."

Thus the identification of an "internal" aspect of representation that conditions and constrains agents' cognitive access to the world is reminiscent not only—as Putnam (1981, 60–64) himself emphasizes—of Kant's transcendental idealism, but also of the nonsolipsistic type of linguistic idealism sometimes associated with the late Wittgenstein. This line of thought arises from Wittgenstein's evocative talk of the "limits of language" and the supposed role of "grammatical propositions" in delimiting sense from nonsense.[27] As Bernard Williams (1981, 152–153) describes it, the basic gambit of such linguistic idealism is to treat the "fact that our language is such and such" as a "transcendental" fact that results in a form of idealism insofar as "everything can be expressed only via human interests and concerns."

Now, while it is plausible that language should have such a transcendental status on the calculus view, with the autonomy of language it supposes, I have argued here that in later writings the predominance of the language-game analogy led Wittgenstein increasingly to an embedded view of language that erodes the supposed autonomy on which a transcendentalist reading of his views relies.[28] Exegetic considerations aside, however, two reminders should suffice to show that the phenomenological conception of language advocated here is not to be interpreted as linguistic idealism of the kind Williams characterizes. First, the view of linguistic constitution outlined in the last section does not attribute transcendental feats to language. Rather, as I put it, on the phenomenological view language functions as a condition of actuality, not possibility, for meaning. Second, given the openness to continually developing or expanding the linguistic perspective we have on the world, the idea that language imposes principled "limits" on what can be said is unconvincing. Like the limits of scientific or technical knowledge, such supposed limits are simply a reflection of what can be done here and now (i.e., contingently and empirically) rather than principled ones.[29]

The fact that my phenomenological conception of language cannot be straightforwardly aligned with either of the perspectives Putnam distinguishes, or with linguistic idealism, is a reminder of the underlying fundamental difference. Both Putnam's alternatives and the idea of linguistic idealism conspicuously operate with the same basic coordinates of internal and external, things-for-us and ("mind-independent") things-in-themselves, as the Cartesian tradition.[30] Throughout this book, however, I have been emphasizing that a phenomenologically adequate conception of language should treat language as not having an inside-outside topology. The failure to match up with the familiar alternatives discussed in this section is one consequence of that approach. For, just as Heidegger's notion of Dasein, its rejection of a presupposed basic dualism makes it a hybrid seeking to reconcile the desirable aspects of each of the traditionally opposing "internalist" and "externalist" positions. A more profound consequence is that without an inside-outside topology the traditional realism/idealism contraposition loses its point. The conclusion to be drawn from this situation, paralleling Heidegger's own attitude, is that to ask of a phenomenological conception of language-in-the-world whether it is idealist or realist is simply not an illuminating question: it is neither or both, depending on how the question is meant. Or rather, the lesson to be drawn is that language is not itself the kind of thing that has realist or nonrealist implications, because commitment to phenomenological accountability results in a conception of language that is beyond the realism/idealism alternative.

9 Phenomenology and Semantics

Having discussed Heidegger's view that predicative awareness is founded in prepredicative awareness of entities, I suggested in chapter 1 that a distinctive feature of the Heideggerian framework is that both these modes of awareness can be linked with language use. To allow better understanding and assessment of these claims, chapter 2 then examined the idea of prepredicative foundation, suggesting that this is concerned with factors underlying propositional content. What makes this idea of prepredicative founding particularly interesting is that it gives a specific focus to the differences between the phenomenological conception of language developed here and more mainstream philosophical approaches to language, particularly by challenging the primacy of propositional meaning usually assumed in post-Fregean philosophy of language. In this chapter I return to this foundation claim to consider in more detail both how it should be interpreted and how the phenomenological approach developed here, particularly the prepredicative factors it identifies, contributes more generally to philosophical understanding of the articulatory and disclosive feats of language.

The chapter begins by clarifying how the two approaches distinguished in chapter 2—the phenomenological and the semantics approaches—differ and introduces various options for the interpretation of Heidegger's foundation claim. The second section of the chapter considers several ways of accommodating a phenomenological conception of language from the perspective of the semantics approach, each of which implies a founding relationship that is too weak to impact on the semantics approach's supposed ability to provide a philosophically adequate conception of language. The third section argues for a stronger claim—a moderate functional foundation claim, as I call it—on behalf of the phenomenological approach based on differences in the functioning of prepredicative factors and semantic properties. The following two sections then seek to clarify further and to highlight the philosophical interest of this foundation claim by situating it in the context of a debate between Hubert Dreyfus and John McDowell about the role of concepts in prereflective agency. Thus the fourth section sets out how my position both parallels and differs from the idea of absorbed coping defended by Dreyfus, while the fifth section considers how it relates to—in particular, whether it is challenged by—McDowell's thesis of pervasive conceptualism. Finally, the sixth section draws on an analogy with developments in twentieth-century physics to outline some overall implications of the moderate functional foundation claim for the relationship between the phenomenological and semantics approaches.

1 Two Approaches to Language

In chapter 2 I introduced a contrast between two general approaches to language. The first is a phenomenological approach of the kind I have been developing through parts I–III of this book.[1] The second, the semantics approach, might reasonably be considered the standard approach in contemporary philosophy of language. The first task in assessing the relation between these two approaches is to get clearer about how these two approaches differ and about how Heidegger's foundation claim might be interpreted.

The semantics approach—or simply semantics—will be taken here to be characterized by four minimal commitments that might be described as the "common sense" of post-Fregean philosophy of language.[2] The first of these is its working vocabulary: language is conceived of in terms of "propositions" and "concepts," which are "expressed" in sentences and words; the workings of language are discussed in terms of "reference" and (perhaps) "meaning" or "sense," and distinguished as "semantic" rather than "causal" in character. The relationship between propositions and concepts is characterized, second, by assuming the notion of propositional content—rather than, say, word meaning—to be explanatorily primary; and, third, by thinking of concepts (along with singular terms) as the building blocks of which propositions are composed. Fourth, the semantic properties of propositions and concepts are assumed to be linked in a systematic way, with the functioning of concepts being understood in terms of the contribution made to (the explanatorily primary) propositional content. In Frege's view this systematic link is represented by means of a predicate calculus, based on the idea that concepts are analogous to mathematical functions and take various objects (or other functions) as their "arguments" to make up propositions and thus yield a "value"—according to Frege (1994a), a truth value. Although the details of its use may vary, it will be assumed in the following that the semantics approach is characterized by the belief that the kind of systematic link holding between concepts and propositions can be modeled using some version of a predicate calculus.

The semantics approach has attractions that are widely held to be obvious and powerful. Above all, by construing language in a quasi-mathematical manner it seems to hold out the prospect of a systematic theory that will exhibit the pervasively rational internal workings of language. With this intimation of systematicity and its close link with formal logic the semantics approach exemplifies modern philosophy's craving for a *mathesis universalis*.[3] At the same time, the semantics approach is clearly theory-driven: its central notions—and the systematicity they augur—are primarily guided not by the goal of accurately describing the way speakers experience language, but by the availability of a certain model and the prospect of revealing an initially nonmanifest ("deep") structure. In this respect it not only paints a somewhat different picture of language from that developed in the preceding chapters, but is precisely the kind of "reconstructive" approach traditionally opposed (often quite rhetorically) by "phenomenological" positions.

Against this background, the question as to the relationship between the semantics approach and the phenomenological approach developed here becomes particularly poignant. However, as pointed out in chapter 2, the fact that the two are prima facie at odds does not entail genuine dissent, since not only do different "approaches" not necessarily lead to different conceptions, but apparently "different" conceptions may not be in genuine conflict. On the one hand, that an approach is theory-driven, or "reconstructive," is not in itself an objection, and may be fully compatible with phenomenological facts. On the other hand, the predominance of the semantics approach cannot count against the relevance of phenomenological facts to a philosophically satisfactory conception of language. The situation is rather that if a systematic semantic theory is to be possible, a condition of its adequacy will be that it should ultimately cohere with a phenomenological conception of language. In this light, examining this relationship in some detail will be a means of probing the philosophical relevance of *both* the phenomenological approach developed here and the semantics approach with which it—in some sense—contrasts.

How then do the conceptions of language yielded by the respective approaches in fact differ? One way they do this is with regard to the ontological priorities they suggest—that is, which items they posit as central to a philosophical conception of language. My phenomenological view has focused on words and acts of speech (utterances) as the basic features in terms of which linguistic phenomena are to be conceived. In this perspective propositions are derivative entities, or an abstraction based on commonalities between sentences, with concepts, as the constituents of propositions, a further abstraction. By contrast, the semantics approach treats propositions and concepts as fundamental posits, in relation to which sentences and words stand in a derivative relation of "expression" (e.g., "Snow is white" and "Schnee ist weiß" are said to express the same proposition). Although not always overtly platonist, there is at least an intimation on this approach that propositions exist in some way independently of their instantiation or expression in sentences. For the assumption is that the way propositions and concepts function can be understood or modeled independently of the role of actual words and sentences—that is, those embedded in and shaped by a historically evolved horizon of forms and sense. Thus it seems that, whether or not they do, propositions and concepts might in principle exist in some way antecedent to their expression in linguistic acts.

Another way the phenomenological and semantics approaches differ is in how they see language functioning. It might be said, somewhat generically, that both approaches are concerned with, but provide different answers to, the question of what meaning is or what it is for language to be meaningful.[4] In the terms I have used here, the two approaches provide different general views of the articulatory processes at work in linguistic disclosure, and so of what Taylor (1985, 256) describes as the "speaking activity" or "what is going on in language." On the phenomenological approach this is understood with regard to the public phenomenology of language use, with the suggestion here being that linguistic signs are instruments in a dual sense—for presenting objects and for specific tasks—whose articulatory feats can be thought of in terms of the structure of the sign system and practices. On the semantics approach the functioning of language is likened to that of mathematical functions, the picture being, roughly, one of set-theoretical mappings of entities and a hierarchical network of functions onto truth values.

This divergence over the functioning of language can also be characterized by saying that the two approaches offer different views of the subpropositional—that is, of the structural and functional makeup of propositional content. To keep this difference in focus, some terminological distinctions will be convenient in the following. The phenomenological approach, I will say, is concerned with—presentational and pragmatic—*sense*; structurally relevant influences on or features of sense will be termed *factors*, with the "sense" due to such factors being something that can be grasped either prepredicatively or predicatively.[5] Conversely, the semantics approach deals with—propositional and conceptual—*content*; features bearing on content will be referred to as (semantic) *properties*. The relationship between the two approaches can then be seen as the relationship between sense and prepredicative factors on the one hand, and content and semantic properties on the other.

Having distinguished the two, the question naturally arises of how the phenomenological and semantics approaches are related. Chapter 2 introduced the thought that this question is given the requisite focus by Heidegger's claim that propositional content is founded in prepredicative awareness. This claim, in effect that the idea of "founding" should apply to the relationship between the phenomenological and semantics approaches, might be interpreted in several ways. To begin with, corresponding to the differences just mentioned, it may be interpreted as either ontological or functional foundation—that is, as viewing propositions and concepts either as existing only via sentences and words, or as something the semantic properties of which presuppose the operation of prepredicative factors.

The first thing to be said of the ontological foundation claim is that it is true, perhaps even trivially so. For it is obvious that the actual presence of language must be conceived of in terms of linguistic acts, the sentences uttered in these, and the words of which they are composed. So if the obscurities of platonism are to be avoided, the ontological priorities at work in language are straightforwardly elicited. If they are to be considered part of the ontological picture, propositions must be thought of in terms of classes of (synonymous) declarative sentences, and concepts linked with the existence of words as their bearers. Moreover, since sentences—utterances—are transient events, words (and sublexical elements) have some claim to primacy as more enduring items in a linguistic ontology.[6] These considerations serve as a reminder—paralleling the ontological realizability condition that Heidegger had relied on against Husserl (see the introduction, section 2)—that the way language is thought to function must be able to be instantiated by the entities assumed in a conception of language. In this way, they provide some justification for thinking both that propositions ought not to form the centerpiece of a philosophical conception of language, and that any satisfactory conception should reflect how the actual functioning of words is involved in linguistic meaning.

Even so, it is not clear that the ontological foundation claim suffices to establish genuine tension between the phenomenological and semantics approaches. To begin with, as long as word-sentence relationships are thought to function in the same way as those between concepts and propositions, such foundation can readily be acknowledged by the semantics approach. Nor does it mean there is no point in talking of propositions. For it may still be necessary to characterize the function of words—so to speak, teleologically—in relation to their role in propositions, and simultaneously dispute that doing this requires propositions to be thought of as reified in a platonist manner.

The more interesting and more involved claim is that of functional foundation. A preliminary difficulty is how to construe such a claim in contemporary terms, given that Heidegger's views are formulated using the traditional idiom of subject-predicate logic. Two considerations suggest, however, that this is fairly straightforward. First, Heidegger's antiformalism is presented generically and does not hinge on the details of the "formalist" approach he opposes. Second, while providing a framework for a contemporary formalism, Frege himself (1994c, 67, 72n) describes functional concepts as "predicative" in the general sense of not yielding a value without an argument (being "unsaturated"), and makes no provision for anything that might be described as "prepredicative" contributions to or formation of propositional content. With these two points in mind it seems natural, in a post-Fregean context, to construe Heidegger's claim simply as being that the notions deployed in the semantics approach are functionally founded in those identified by the phenomenological approach.

A further complication is the ambitiousness or intended strength of the functional foundation claim. A strong foundation claim would be that semantic properties are effectively epiphenomenal and can be explained (reductively) in terms of prepredicative factors. A weak foundation claim would acknowledge that prepredicative factors underlie semantic properties, but deny that this in any way compromises the explanatory adequacy of the semantics approach. Between these two opposites lies the further possibility of a moderate foundation claim that semantic properties presuppose prepredicative abilities, while allowing that the two levels function in mutually irreducible ways, so that an account of both is required for a satisfactory philosophical understanding of linguistic articulation. The task for the next two sections will be to consider which claim best represents the relationship between the semantics approach and the phenomenological approach.

2 Weak Functional Foundation

I want to begin by considering how a phenomenological conception of language might look from the perspective of the semantics approach. In particular, this section will briefly review three ways the semantics approach might respond to such a conception and suggest how this would be compatible with a suitably constrained version of the functional foundation claim.

The first option for the semantics approach would be effectively to ignore the phenomenology of language, an attitude that, it seems to me, is more common than one might naively expect. In the most extreme cases this might mean that literally no attention is paid to the connection between a preferred semantic theory and actual linguistic phenomena.[7] In other cases some sketch of linguistic phenomena might be offered, either as an afterthought or an illustration of a proposed theory. This is surely the case with the radical translation scenario discussed by Quine (1960, 26–79), which is clearly tailored to his behaviorist assumptions and which—even if it were assumed plausible as far as it goes—focuses on extremely exceptional circumstances that can hardly be taken to be typical of linguistic phenomena. It is also plausibly the case for Davidson's views on radical interpretation, which, despite claiming to set out a framework for serious empirical study of semantics, provide no more than programmatic hints as to how to correlate the truth theory for a language with actual linguistic practices.[8] One other such example—though no doubt there are many more—would be Brandom's peripheralization of the "material aspect" of the linguistic practices in which scorekeeping activities are assumed to be realized (see chapter 7, section 2).

That many philosophers of language pay little attention to the phenomenology of language no doubt reflects a conviction that predicate calculus provides a powerful theoretical tool which captures all that is significant (philosophically) about the functioning of language. However, that conviction is equally consistent with other strategies that might be used to accommodate a phenomenological conception of language, in particular the prepredicative factors it identifies, within the broad framework of commitments defining the semantics approach. Here I will consider two such strategies.

A second option for the semantics approach would thus be to see a phenomenological conception of language as contributing a theory of speakers' abilities. To see how this might be done, consider the tripartite general architecture proposed by Dummett (1976), according to which a theory of meaning should encompass a "theory of reference," a "theory of sense," and a "theory of force." The core of such a theory of meaning is the theory of reference, which aims to characterize the meaning of all assertoric sentences of a language in terms of one semantic concept that is assumed to be basic (in particular, truth or justification). A theory of force is to expand the theory's account of assertoric meaning to cover sentences differing in mood, such as interrogatives, imperatives, and optatives. Of particular relevance here, however, is the theory of sense. For Dummett this is to "lay down in what a speaker's knowledge of any part of the theory of reference is to be taken to consist," thus relating the theory of reference to "the speaker's mastery of his language," correlating "his knowledge of the propositions of the theory of truth with practical linguistic abilities which he displays."[9] Given this general architecture, it would be natural to interpret the phenomenological approach developed here as comprising such a theory of sense, with the views drawn from Heidegger, Merleau-Ponty, and Wittgenstein providing a phenomenological description of the abilities of speakers in which semantic properties are manifested. In saying this, it should be highlighted that understanding sense in terms of speakers' abilities is not to understand it in terms of mental or psychological states, or anything else "in the head." Rather phenomenological descriptions of abilities linked with presentational and pragmatic sense rely only on publicly available features of signs and the practices in which they are used. They can thus be considered to explicate how the perspective under which objects are linguistically addressed is built up, while accommodating the objective perspectival character that Frege saw as characterizing sense without resorting to Frege's own obscure platonism.[10]

The simplest way of construing their complementarity in this Dummettian framework would be to think of the phenomenological and semantics approaches as addressing the same processes in different terms. Their respective conceptions of language would then be *functionally continuous* (i.e., function in the same way), so that the semantic properties of expressions correspond 1:1 to their roles as described (or at least describable) within a phenomenological perspective. There is, however, a problem with assuming complementarity of this straightforward kind, due to the key feature of the semantics approach: its claim to capture how the meaning of sentences can be thought of as composed of the meaning of its parts. This is not to presuppose that the semantics approach has the right answer to this question. It is open to debate whether there has to be, or even could be, a unitary notion of sentence meaning, such as predicate calculus assumes, that explains the various disclosive feats of sentences satisfactorily. It is also quite possible that alternative accounts of how sentence meaning is built up by words could be offered. Nor is it beyond question that there is a need for a "systematic" account of this relationship between words and sentences. Perhaps this supposed need is bound up with the conviction that propositional meaning is primary and with the perceived ability of predicate calculus to provide such an account. Nonetheless, it is true that the semantics approach articulates an answer to the question of how sentence and word meaning are related, whereas the phenomenological approach, so far as it has been developed here, does not. Hence there is an asymmetry in that there is a key feature of the semantics approach to which nothing in the phenomenological approach here corresponds, meaning that the complementarity of the two cannot be one of straightforward correspondence.

This has the consequence of ruling out the strong functional foundation claim: for if nothing in the phenomenological conception corresponds to the combination of words in sentences, then there is nothing to strongly found a semantic account of the relationship between propositions and concepts. So while Dummett's notion of sense points in the right direction, the functional foundation claim that can be made on behalf of the phenomenological approach must instead be a partial one. On the one hand, word meanings should, I suggest, be thought of as presupposing and corresponding to the kind of factors—presentational and pragmatic sense—identified by a phenomenological "theory of sense." On the other hand, however, the specific relationship between sentence and word meaning remains an aspect of linguistic functioning addressed only by the semantics approach. Hence, even if it is indispensable, the phenomenological conception offers only a partial account of the abilities speakers have, a necessary part perhaps, but not the whole of a Dummettian theory of sense.

A third, and somewhat more specific, option in conceiving the relationship between the phenomenological and semantics approaches is suggested by John Searle's idea of "Background" as the foundation for intentional (or "representational") states. Whereas intentional states are defined in terms of satisfaction conditions, Searle explains, "Background is a set of nonrepresentational mental capacities that enable all representing to take place"; it comprises not propositional "'knowing-that,'" but "certain kinds of know-how: I must know how things are and I must know how to do things."[11] The principal role of Background, which encompasses "deep" human biological and "local" enculturated levels, is to provide the operational context for literal or metaphorical propositional content, which "requires for its understanding more than the semantic content of the component expressions and the rules for their combination."[12] The dependency, Searle emphasizes, is such that the Background *"permeates"* intentional states; it is "enabling not determining"—that is, it provides "necessary but not sufficient conditions for understanding" (Searle 1983, 151, 158).

Two qualifications should be mentioned. First, although Searle (1983, 16) presents his as a "non-ontological approach to Intentionality," he nonetheless characterizes the Background as mental. Given Heidegger's rejection of the notion of subjectivity (*SZ* 46), this might seem to suggest incompatibility with a Heideggerian framework. For the present purposes, however, this difference can be disregarded, since in describing the functioning of language, any relevant Background abilities can be (in Wittgensteinian spirit) thought of as characterizable in terms of an impersonal public phenomenology, so that the label "mental" entails no obviously problematic assimilation to a mind-based ontology. Second, Searle sometimes explains the role of background in terms of background assumptions, such as those made when ordering a meal in a restaurant (cf. Searle 1979, 127; 1992, 180). Presumably to accommodate this feature, he came to include in Background "unconscious intentional" states, understood to be "dispositional states of the brain" of the right form—that is, genuinely intentional and with aspectual shape—"to produce conscious thoughts and conscious behavior."[13] Although such states might be "background" in the sense of being presupposed or implied by conscious (i.e., occurrent) intentional states, their inclusion in it means that not all Background is "preintentional." Because Searle had already distinguished intentionality from consciousness, this modification has the consequence of making it difficult to see what Background is to be distinguished from. For my purposes here, however, this modification can be disregarded, as the concern will be only with Background in the sense of capacities or factors that are pre- or nonintentional according to Searle's definition—that is, prior to and not characterized by semantic content.[14]

With these two qualifications, the present suggestion (going beyond Searle) is that the Background has a specifically linguistic component. Accordingly, the accounts of presentational and pragmatic sense developed here identify—as a partial theory of speakers' abilities—Background features that shape and pervade semantic content. This suggestion is possible because the prepredicative factors these kinds of sense rely on are presupposed in the same "preintentional" way as Searle's Background. For the abilities to use words to mark distinctions or to instantiate correct use in the context of a language-game, corresponding to presentational and pragmatic sense respectively, cannot be traced back to the ability to manifest in language an awareness of corresponding conditions of satisfaction.[15] Rather, this kind of awareness presupposes the former abilities, as precisely the kinds of knowing how things are and how to do things that define Background for Searle. Moreover, because to exercise these abilities requires a grasp of the structure of both a language's sign system and the relevant language-games, the notions of presentational and pragmatic sense can be seen to highlight factors that quite literally structure both prepredicative abilities and the predicative abilities that build on these. In this respect, although by definition the Background is not a fully formed ("explicit") anticipation of intentionality, the notions of presentational and pragmatic sense tell us something about the way it is structured.

From the standpoint of the semantics approach, the phenomenological conception of language presented here can thus be thought of as a partial theory of speakers' abilities that explicates specifically linguistic aspects of Background. Perhaps it is worth noting that in this way a phenomenological conception of language would have some role to play even in a semantics-centered conception of language, and would form part of a complete (philosophical) understanding of what is involved in intentionality and propositional meaning. Nonetheless, the three options outlined in this section share the assumption that this role can be treated as peripheral and that the semantics approach captures everything of real philosophical interest about linguistic meaning. This assumption does not rule out construing the relation between the phenomenological and semantics approaches as one of functional foundation, but it does limit such a claim to what I previously described as *weak foundation*. That is, a foundational relation can be acknowledged in the sense that semantic properties would not exist without manifestation in ways described by the phenomenological approach, while considering this foundation to be weak in the sense that it does not compromise the theoretical adequacy of the semantics approach. One way of summing up such weak foundation (in a way that will be helpful below) is as the thought that language can be characterized in terms of a single, uniform functional space, one adequately described by the semantics approach without any need for modification or addition in light of phenomenological considerations.

The idea that the semantics approach is able to capture everything of philosophical interest about linguistic meaning, and hence the weak functional foundation claim, might be defended by relying on two thoughts. The first is that prepredicative factors can be reduced to semantic terms. Thus if the two approaches are functionally continuous, with every feature of prepredicative functioning having a correlate in the semantics approach, it would seem reasonable to suggest that the former can be understood fully in terms of the latter, so that the semantics approach provides a functionally complete picture of language. The second thought is that the semantics approach is functionally autonomous and that the role of prepredicative factors can be classified as mere "causal conditioning." While allowing that prepredicative factors are involved (causally) in the formation of semantic facts, this assumes that semantic characterizations are structurally and functionally independent of those factors. Thus Searle (1983, 143) himself describes the Background as "bedrock," a Wittgensteinian metaphor (*PU* §217) suggesting that our understanding of what concepts are and how they function rests on an impenetrable basis that imposes a principled limit on such inquiry—such that there is "no digging deeper." The suspicion expressed in this metaphor is that, even if the weak foundation claim is true, it is too anodyne to be of philosophical interest: Background is background, and rightly not the focus of proper philosophical analysis. If either (or both) of these thoughts is plausible, as I said, then the weak functional foundation claim would be justified. However, the following section will argue that they are not and that a stronger claim can be made on behalf of the phenomenological approach.

3 Moderate Functional Foundation

In chapter 2, section 2, I suggested that a distinctive feature of Heidegger's view of the prepredicative is the idea of functionally discontinuous foundation, the founding of predicative judgments in something differing in its functional nature and so not to be understood in terms of predication. This feature suggests that a somewhat stronger claim should be made concerning the foundational role of prepredicative factors in relation to semantic properties than that discussed in the preceding section. The basis for such a stronger claim is that the semantics and phenomenological approaches identify different processes—so to speak, different "mechanisms"—of linguistic articulation, so that an understanding of prepredicative factors becomes an integral part of a philosophically adequate conception of language.

For this to be the case, two conditions need to be fulfilled, corresponding to the two thoughts—mentioned at the end of the last section—that potentially justify the weak functional foundation claim. First, the phenomenological and semantics approaches must differ in their functional topology (i.e., identify factors that function in significantly different ways), so that there is no 1:1 mapping between them. Meeting this condition would challenge the thought that prepredicative factors can be reduced to semantic terms, or indeed that either of the two approaches is reducible to the other. The second condition is that, despite differing functionally, the factors identified by the phenomenological approach must be semantically relevant. That is, to avoid being dismissed as semantically inert, so that semantic relationships could be considered explanatorily closed, prepredicative factors need to be seen to have some role in the formation of semantic content. Meeting this second condition would rule out the thought that the semantics approach is autonomous or independent of prepredicative factors. If either of these conditions were not fulfilled—that is, if either semantic and prepredicative functioning corresponded perfectly, or prepredicative factors were not relevant to semantics—it would be reasonable to think of their relationship in terms of weak foundation, so that semantics tells us everything a philosopher might want to know about the way language works.

The second condition is, I suggest, straightforwardly met. For in describing factors involved in the actual phenomena in which linguistic articulation occurs, pragmatic and presentational sense pertain directly to the meaningful use of language and hence to the formation of concepts. The basis of this claim (set out in parts II and III) is phenomenological accountability—so that these descriptions are more reminders than discoveries, acknowledging that the functioning of language lies open to inspection. It has also been indicated how both pragmatic and presentational sense can be exploited in predicative contexts, and so shape inferential properties. In this way they exhibit a key characteristic of Searle's Background, the whole point of which is to allow that nonrepresentational abilities generate propositionally expressible commitments.[16] And while it may be plausible to consider some generative factors to be irrelevant to the disclosive function of language, as the bedrock metaphor implies, this is not the case for the form of linguistic signs and practices linked with the notions of presentational and pragmatic sense, respectively. Indeed, an important advantage of distinguishing these two kinds of sense is to reflect the involvement of formative processes in semantic content, rather than presenting the latter as a functionally independent superstructure.

The key question therefore relates to the first condition: What basis is there for thinking that the phenomenological and semantics approaches identify different functional topologies? In my initial discussion of how claims about prepredicative factors are to be interpreted (chapter 2, section 2) I suggested a need to distinguish their functioning, in terms of appropriateness or otherwise, from that of propositional truth. I now want to expand on this by suggesting three ways in which presentational and pragmatic sense, as characterized in the meantime, differ in their functional topology from semantic content.

(i) Basis of Structuration

The first difference lies in the features taken to underlie linguistic articulation, or their basis of structuration. According to the views developed here, both presentational and pragmatic sense reflect structural influences on language. In the case of pragmatic sense, the internal link between practice and language highlighted by Wittgenstein means that language is structured by language-games, the form of which determines both the degree of differentiation required and which aspects of the world need to be marked linguistically. In the case of presentational sense, the lexical and sublexical structure of language is used to map the world in a schematic or typological way. And while the phonetic basis of linguistic signs is ultimately arbitrary—there is no temptation here, as perhaps there is with pictures, to suppose natural resemblance—the system of signs develops in nonarbitrary ways through iterative use, acquiring intelligibility as a system of relatively motivated features. Being simultaneously the result of such processes and an operative factor involved in the ongoing evolution of language, the morphological array of linguistic signs is part of the basis of language's structuration. In contrast, the basis of structuration for the semantics approach lies in the inferential or truth-preserving role of words in sentential contexts, which identifies semantic elements and rules for their combination. These different bases of structuration—expressed in Heidegger's talk of a change in "*as* structure"—underlie familiar discrepancies, often dealt with, somewhat tendentiously, by distinguishing "deep" semantic from "surface" grammatical structure.[17] Not only do the words individuated in practice differ from the "semantic structure" of a language—which has historically given rise to the thought that real languages fail to exhibit their logical form adequately in the way a philosophically "ideal" language would. Rather, with the involvement of sublexical form, the question of the sign system's morphological form corresponding to the basic structures identified by the semantics approach simply cannot arise (because the latter takes as basic the level of Saussurean signs, which are already composed of sublexical forms).

(ii) Nondiscreteness

The second way the functional topologies differ concerns the question of discreteness. A central assumption of the semantics approach is its modeling of propositional and conceptual operations as discrete functions, usually bivalent ones such as the truth or falsity of propositional contents or the satisfaction or nonsatisfaction of predicates. This tidy picture of discrete and determinate content "valencies" contrasts with both presentational and pragmatic sense. Presentational sense is nondiscrete due to the semantic openness of linguistic expressions, the fact that they are constantly open to extended or new uses. Such openness is characterized by continuous variability rather than a fixed number of values—not least because the difference between simply extending the use of a term and a new use will often be marginal. Similarly with regard to pragmatic sense: drawing on Wittgenstein's later work, I have suggested that an ideal-free conception of the rules inherent in linguistic practices should be generally understood as having the shape of empirical or statistical rules—that is, as characterized by greater or lesser variation about a mean. On this picture the notion of falling under, or complying with, a rule is itself not sharp. Although there will be clear cases of both falling under and not falling under such a rule, there will also be gray areas of transition between the two, so that pragmatic sense too is best thought of in terms of continuous variability rather than a fixed number of values, as continuously rather than discretely articulated.

To some extent the "nondiscreteness" of prepredicative sense merely corresponds to the idea of concepts' "vagueness." However, it is important to be clear that nondiscreteness is a general feature, not a peripheral effect that allows bivalency to be thought of as the default functional mode of linguistic meaning.[18] For while bivalency is a plausible approximation in strictly regulated language-games, the fact that both presentational and pragmatic feats of language are in principle characterized by openness and nondiscreteness suggests that this is the more general topology of linguistic meaning. A second, somewhat deeper consideration leads to the same conclusion: because predicate calculus models the functioning of language in terms of determinately falling under a predicate, being true or otherwise, thinking of bivalency as the *basic* mode of conceptual operation presupposes an idealized fully determinate picture. Yet it is relinquishing the idea of full determinacy, which is shown to be inherently unstable by Wittgenstein's critique of the calculus notion of rules, that leads the phenomenological approach to its emphasis on the openness and nondiscreteness of linguistic functioning. Whereas the first consideration of generality suggests that to think of language as functioning bivalently would be an approximation at best, the second shows that it would be fundamentally misleading. Either way, to think of the functionality of concepts as basically bivalent is to attempt, so to speak, to put round pegs in square holes.

(iii) Language as Acquisitional

The phenomenological approach also differs from the semantics approach in recognizing the acquisitional character of language as an essential aspect of the way it functions. Talking of language as an acquisition—as Merleau-Ponty, adopting Husserl's use, characteristically does—clearly suggests it is something built up over time, which humans have come to possess through processes of discovery or invention. This is most directly seen in the way meaningful use becomes sedimented in the morphology of the system of linguistic signs. The phenomenological view accommodates this fact by seeing the presentational sense of words—due to their place in the differential framework of language and the relative motivations pervading it—as inscribed in the forms of the lexical system. This acquisitional view coheres with both the Wittgensteinian idea of language as an institution or custom and Heidegger's emphasis on the historicity of disclosure, while better doing justice than either of these to the temporality of language. In particular, as seen earlier, by viewing lexical items as type artifacts, characterized by an iteratively evolved situated interpretability, this view explains the ongoing relevance of etymology to word meanings.[19]

The significance of this acquisitional aspect can be brought out by considering the obvious objection that *at any given time* (and perhaps relativized to certain speakers) a language is in principle representable as instantiating a synchronous system of rules. This objection amounts to the suggestion that language can be detemporalized—that is, considered as functioning in a time-independent manner. This is not obviously a coherent suggestion. For it seems to ignore the fundamental fact, emphasized by Wittgenstein, that the concept of a rule makes sense only against the backdrop of repeated use. Yet, even if taken to be coherent, it assumes that the complex means of differentiation provided by a lexical system and its established patterns of use are independent of their history. This thought perhaps draws some plausibility from the fact that we often learn and explain the use of linguistic terms without any recourse to historical or etymological considerations. Nonetheless, it fails to take account of the fact that the techniques and forms that make up the whole of language use are the product of a complex evolution and are no more possible without a history of development than are the abilities to travel to the moon or build a computer. By contrast, the fact that the functional economy of natural language is intrinsically temporal is built into the phenomenological view's acquisitional picture of language. In Heideggerian terminology (cf. *SZ* 329), this view acknowledges that language is temporally ecstatic, standing-out-of-itself in the sense of always having been shaped by its own past and comprising a particular present that conditions its future disclosive possibilities.

Both conditions identified at the start of this section are therefore met. Having previously clarified that pragmatic and presentational sense cannot be dismissed as semantically irrelevant, the three factors just described show that presentational and pragmatic sense function in ways identifiably different from semantic content, so that the phenomenological approach does not map straightforwardly (in a 1:1 manner) onto the semantics approach. Whereas the first condition prevents semantics from being thought of as structurally or functionally independent of presentational and pragmatic sense, the second blocks the idea that presentational and pragmatic sense can be reduced to semantic equivalents.

The fact that these two conditions are met rules out the weak functional foundation claim, discussed in the last section, which views the semantics approach as capturing everything of philosophical interest about the functioning of language. The obvious alternative to this might seem to be a "strong" functional foundation claim. However, I want to avoid this label, because it appears to imply that semantic features are in general explicable in terms of prepredicative factors. This would be misleading in part because, as the last section conceded, the position developed here provides only a partial theory of speakers' abilities. Moreover, this line of thought preserves the idea that there is functional continuity between the two approaches while simply reversing the order of supposed primacy, so that semantic terms become reducible to phenomenological ones rather than vice versa and the semantics approach adds nothing of importance to a phenomenological one. This would simply be an inverted form of weak foundation and is likewise ruled out by the functional discontinuity between the two approaches that I have been highlighting in this section.

Instead I will talk of a *moderate* functional foundation claim so as to distinguish it from both weak foundation and the idea of strong foundation. The point of this claim, based on the thought that the relationship between the two approaches is one of functionally discontinuous foundation, is that the prepredicative Background of semantic content has structure that is (a) not directly accountable for in semantic terms, but that (b) cannot be considered functionally irrelevant to meaning, and (c) can be understood in phenomenological terms as presentational and pragmatic sense. This moderate functional foundation claim not only acknowledges that prepredicative factors underlie the semantics approach, but also recognizes that they add something of functional significance to the latter. Its contrast to the weak foundation claim (and indeed the strong claim) can be summed up by saying that with the moderate functional foundation claim language is no longer thought of in terms of a single, uniform functional space, but as encompassing different functional spaces that are mutually irreducible.

The claim that presentational and pragmatic sense differ in kind from and challenge the primacy of propositional/conceptual content parallels claims sometimes made about the notions of nonconceptual content and knowing-how. A common objection to both the latter notions is that in principle they remain reducible to propositional/conceptual content, so that they fail to identify anything not captured by the semantics approach. In section 5 I consider a sophisticated strategy of this kind (McDowell's) in some detail, but to conclude this section I want to look briefly at one other line of thought sometimes offered in support of this objection.

This thought is that the notions just referred to are conceptually dependent on, and hence not genuinely opposed to, those of propositional or conceptual content. Consider first the idea of "nonconceptual content," which is sometimes appealed to in philosophy of mind to make space for the claim that a state or action can have content without an agent's being "explicitly" aware of it. A common way of describing such content is found, for example, with Adrian Cussins, who distinguishes between conceptual content, defined in terms of "objects, properties, or situations" as determinants of truth conditions, and nonconceptual content, which is to be "canonically specified by means of concepts that the subject need not have."[20] Appealing to "canonical specification" in this way is problematic because it prompts the question of what is nonconceptual about the *content* concerned. Both conceptual and nonconceptual content, on this view, are defined conceptually. The difference between the two cases turns not on the kind of content involved, but on the way content is available to an agent's consciousness. Thus it seems—hence the present objection—that the content of a nonconceptual state can be understood only in conceptual terms, rendering the expression "nonconceptual content" oxymoronic and so misleading at the very least.

Parallel considerations are often relied on in opposing the distinction between knowing-how and knowing-that. In this case the problem arises because knowing-how must be something cognitive, since knowing how to Φ cannot be equated with actually being able to Φ (for example, someone with a broken leg might know how to, but be unable to swim). And yet it seems that whatever it is that a person P knows in knowing how to Φ must be statable in the form "P knows that ...." However, this would mean that knowing-how is to be understood, against its original intention, in terms of propositional knowing-that.[21] Hence it seems not only that reductive strategies to the latter are possible, but that the only way of specifying what "nonpropositional" knowing-how amounts to is in terms of propositional content.

It might be thought that the ideas of presentational and pragmatic sense are susceptible to the same type of objection, such that one can only understand what it is for these to be about "meaning" via the semantic notion of propositional/conceptual content.[22] The fact that the ideas of presentational and pragmatic sense can be—as in the preceding chapters—characterized without the technical terminology of semantics already indicates that this is not the case.[23] This conclusion is supported by noticing what makes possible the conceptual-dependence objection to nonconceptual content and knowing-how. In each case the underlying problem is that the feature in question is not characterized in a way that makes apparent how it differs structurally and functionally from its intended counterpart. By contrast, the significance of the three features identified in this section is to make clear how presentational and pragmatic sense differ in their structure and functional character from the notions of propositional/conceptual content. These differences in functional topology—that is, the basis of structuration, (non)discreteness, and acquisitional character—mean that presentational and pragmatic sense cannot be thought of as mapping in a 1:1 manner onto propositional/conceptual content.[24] It would therefore be mistaken to think of the former as reducible to or even definable in terms of the latter, so that it is difficult to see how presentational and pragmatic sense *could* be thought of as conceptually dependent on the notions of propositional/conceptual content.

4 Intelligent Absorbed Coping

The claims I have been making on behalf of a phenomenological conception of language in some ways parallel those involved in a high-profile debate between Hubert Dreyfus and John McDowell.[25] For the role I have been attributing to prepredicative factors is in some ways similar to that played by what Dreyfus calls "mindless absorbed coping" (Dreyfus 2007b, 353). As a result it might appear to conflict with McDowell's position, which emphasizes the ubiquity of conceptual capacities, and to be susceptible to the challenge McDowell's position presents to that of Dreyfus. Rather than attempting to adjudicate their differences directly and in detail, this section and the next will draw selectively on their debate in order to clarify further and highlight the philosophical interest of my own phenomenological conception of language and its moderate functional foundation claim.[26] In very general terms, I will be trying to show that my own position has the advantage of being intermediate between the two extremes of nonconceptual coping and pervasive conceptualism expounded by Dreyfus and McDowell respectively.

The debate between Dreyfus and McDowell turns primarily on whether our prereflective engagement in embodied action should be seen as conceptually structured. Typical examples of such action are sports activities (playing basketball, swimming, skiing) or everyday skills (opening doors, using tools, driving a car), but they extend—according to Dreyfus—even to paradigmatic intellectual activities such as playing chess. In phenomenological terms these activities have in common that we can feel immersed or totally absorbed in them, and achieve a state of being "in the flow" that is characterized by—and plausibly requires—an absence of reflection or deliberation on what we are doing and how.

Building on this phenomenological fact and his readings of Heidegger and Merleau-Ponty, Dreyfus advocates a "bottom-up" approach to the articulation of meaning based on a distinction between a "ground floor" and "upper stories" of human experience and action (Dreyfus 2005, 47). The ground floor is absorbed coping, a level of engagement with the world that Dreyfus portrays as "nonconceptual and nonminded" (Dreyfus 2007b, 355). The application of concepts and mindedness are to come in only subsequently, at the upper stories of human experience, when we exercise rationality by consciously reflecting or deliberating. As Dreyfus sees it, absorbed coping is the hallmark of mastery. The performances of experts—from baseball players to grand masters—are characterized by the absence of reflection, which would require both disengagement and the inhibition of competence.[27]

By contrast, in order to make sense of the role of appeals to experience in justifying knowledge claims, McDowell defends the view that conceptual capacities are, as he terms it, "operative" in all experience, including apparently "passive" processes such as sense perception.[28] The capacities he has in mind are "conceptual in the sense that they are rationally integrated into spontaneity at large," such that experience is always in the "space of reasons" (McDowell 1996, 58, xiv, and passim). This emphasis on the connection with rationality gives the impression at least that McDowell's position is a "top-down" approach to the articulation of meaning, an approach that focuses exclusively on our highest-level intellectual capacities and seeks to understand the function of lower-level abilities in terms of these. Whether or not this impression—which McDowell disputes—is accurate, this pervasive conceptualism clearly excludes the idea of nonconceptual absorbed coping championed by Dreyfus.

The differences and debate between Dreyfus and McDowell thus concern the general question of whether all experience/action involves concepts, in particular prereflective embodied actions. Despite the apparently clear focus of this debate it is not obvious, as we will see, what exactly using "concepts" means and how this relates to a Heideggerian position. Another useful way of summing up their differences is therefore to say that while McDowell sees all experience as falling within a single, uniform functional space (the "space of reasons"), Dreyfus discerns a functional discontinuity, with experience coming in both nonconceptual and conceptual forms.

In the remainder of this section I want to clarify how my position differs from Dreyfus's, before going on to identify a central feature that our positions have in common and that appears to be challenged by McDowell's position. It will help to begin by highlighting two general features of Dreyfus's position that are problematic. First, despite the fact that this is its supposed main virtue, his description of embodied coping seems phenomenologically wrong. In describing embodied "coping," Dreyfus often makes it sound as though total absorption in the execution of a task is such that we are unaware of what we are attending to in an articulate manner. In "fully absorbed coping," he says, the "coper does not need to be aware of himself even in some minimal way. … If the expert coper is to remain in flow and perform at his best, he must respond directly to solicitations without attending to his activity or to the objects doing the soliciting" (Dreyfus 2007a, 374). Similarly, when ensuring we are standing at a socially appropriate distance from other people "we are not aware of what we are doing"; nor does it follow, he thinks, that "in order to act kindly the kind person must be aware of the situation as a situation calling for kindness" (Dreyfus 2013, 23, 34).

Dreyfus is no doubt right to distinguish the experiences he has in mind from disengaged deliberation (i.e., thinking through what to do or how to do it), which in many cases will plausibly deteriorate an agent's performance.[29] Nonetheless, to characterize prereflective engaged agency in the above ways seems obviously wrong. More specifically, they disregard the important difference between a lack of salience, or attentional focus on some feature, and a complete lack of awareness. Yet it remains an important aspect of (prereflective) skilled performance that we are *aware* of the relevant features of the environment in some way. No matter how smooth the coping feels, for example in driving a car or skiing down the piste, we are surely aware of what we are adjusting our actions to, what we are doing, and that we are doing it: we do not fail to notice the rock we're swerving to avoid, or that we are skiing rather than swimming or running, or that it is me who is driving, not the person in the backseat. Being aware of such features—that is, consumed by this awareness, without distraction—is precisely what it is to be immersed or absorbed in such activities. (Conversely, if that awareness is lost—if our attention to the activity lapses or we fall asleep—then coping ceases, competence disintegrates.) Underlying Dreyfus's claims here is a mistaken view of sensitivity to appropriate performance and of monitoring our surroundings, according to which such feats would (if they were to occur) have the same phenomenological feel as disengaged reflection or deliberation.[30]

A second, related problem is that it is not clear what role, if any, intentional content is supposed to play in relation to absorbed coping. On occasion, Dreyfus appears to deny it has any role: "there is no place in the phenomenon of fully absorbed coping for intentional content mediating between mind and world" (Dreyfus 2013, 28). Elsewhere he suggests that absorbed coping involves its own kind(s?) of content. For example: "motor intentional content ...[,] a kind of content which is non-conceptual, non-propositional, non-rational ... and non-linguistic" (Dreyfus 2007b, 360). In various passages he also talks of "non-mental content," "background content," and "motivational content," and suggests that "the expert in every skill domain, even everyday coping, has a kind of *intentional* content; it just isn't *conceptual* content" (Dreyfus 2007b, 352, 361; 2005, 58, 55). In denying that coping-specific content is conceptual, Dreyfus relies on the thought—which he attributes to Heidegger—that "most of our activities don't involve concepts at all. That is, they don't have a situation-specific 'as structure'" (Dreyfus 2007b, 371).

It is difficult to know what to make of these various claims. What kind of "content" is motor intentional content if it lacks an *as* structure? And why should motor intentional content be thought of as a distinct kind of content rather than a specification of the means by which we are sensitive to or grasp some contents? Against this background Dreyfus also often gives the impression that the content our bodies respond to motorically is somehow ineffable, unable to be expressed in conceptual terms (as though we could never reflectively describe or explain what our own bodies are up to when in absorbed coping mode).[31] It is far from obvious that this is right: a ski instructor or swimming coach seems perfectly able to specify *what* it is that each part of my body should do in order to perform effectively, even if my body fails to do so. To be clear, I do not want to deny that Dreyfus may be right in some way, such that practice or embodied behaviors resist assimilation to propositional/conceptual content. However, to establish this he would need a clear line about the role of intentional content, and in particular would need to specify what it is about motor intentional "content" that resists expression under different (e.g., linguistic and conceptual) modes of presentation.

The two general problems just identified have in common a failure to register the very feature of "coping" behaviors that makes them interesting: the fact that they are intelligent forms of behavior. While I think Dreyfus is right to resist assimilating such behaviors to highly reflective or deliberative acts, by describing them as "mindless" and "nonconceptual" he seems simply to recoil to the opposite extreme, and so to deny himself the means required to describe the intelligence at work in absorbed coping. In particular, his denial that absorbed coping involves concept use makes it difficult for him to characterize the articulate and (so to speak) "contentlike" grasp of our surroundings that such coping surely involves. The underlying reason for this is that Dreyfus conceives the features he denies to absorbed coping—mindedness, concept use, monitoring of our own actions, responsiveness to reasons (rationality), and so on—as being realizable only by a disengaged mind, the mind of an agent who has interrupted practical action to take up a reflective, or even theoretical, attitude. This fact is not lost on McDowell, who justly characterizes this limitation of Dreyfus's position as the "Myth of the Mind as Detached," a residual Cartesian assumption that "rational mindedness is always detached, so that it must be absent from the absorbed coping that occupies the ground floor."[32]

Avoiding the two extremes of failing to register and overstating the intelligence exhibited in prereflective "coping" behaviors requires conceptual resources for describing the intermediate mode of their intelligence. It seems to me that such resources are found, somewhat ironically, in precisely those authors—Heidegger and Merleau-Ponty—that Dreyfus sees as sharing his position. In particular, on my reading, many of the activities Dreyfus cites as examples of absorbed coping are—just as Heidegger's example of hammer use—characterized by purposive awareness, circumspective Setting-out, and hence the *as* structure that individuates features of the world as ready-to-hand (*Zuhandenes*).[33] However, I am not going to attempt to establish here that Dreyfus is exegetically wrong about Heidegger's and Merleau-Ponty's view of intelligent nonlinguistic behaviors. For not only would this require detailed and extended discussion, but this discussion would run at a tangent to my own interest here in the specific case of language.

Against this background, it becomes easier to see how my position both parallels and differs from the claims that interest Dreyfus, and how it does justice to the intelligence inherent in absorbed coping. Despite the difference in focus—on language rather than intelligent nonlinguistic behaviors—there is a clear parallel: in the preceding chapters I have used Heidegger and Merleau-Ponty, along with Wittgenstein, to build up a picture of the foundational function of prepredicative uses of language that, I suggested in chapter 7, culminates in a form of linguistic knowing-how. This can be seen as a natural extension of the "coping" idea, in recognition of the fact (as it surely is) that our use of language typically has the same prereflective character, inconspicuous feel, and effortless fluency as the intelligent nonlinguistic behaviors that Dreyfus focuses on. In addition, the position I have developed here parallels Dreyfus's bottom-up approach in resisting overintellectualization, having been guided by the thought that reflective, theoretical, and conspicuously rational uses of language are no more a general model for *all* language use than they are for intelligent nonlinguistic behaviors.

Nevertheless, in contrast to Dreyfus, through the notions of presentational and pragmatic sense I have at the same time been trying to make clear how these three authors provide the resources to understand the "intermediate" intelligence at work in prepredicative language use. This is perhaps best seen by reviewing briefly how I have suggested a prepredicative grasp of pragmatic sense should be understood—this being the feature of my position that comes closest to Dreyfus's concerns.

First I set out how Heidegger's analysis of disclosure allows for two basic modes of Articulacy (*Rede*), each of which can be manifested in the use of linguistic signs. One of these modes is linked with circumspective Setting-out, in which features of the world take on an "*as* structure," being individuated in a ready-to-hand or purposive manner—that is, with respect to what they are for or what use they serve. We also saw that this mode of awareness picks out entities as tools in a context-relative manner rather than in the context-independent manner that characterizes predicative awareness. As Heidegger put it, that "with which" we are dealing purposively is grasped in connection with the circumstances of use, whereas that "about which" we speak predicatively is grasped in isolation or abstraction from those circumstances.[34] In this way Heidegger defines, at least in outline, an intermediate form of prepredicative intelligence that is linked with purposive intentionality characterized by a particular *as* structure.

I subsequently argued that the late Wittgenstein's conception of language-game rules provides a versatile means for characterizing what we are responsive to in the practical use of language (i.e., pragmatic sense). What I want to emphasize here, however, is that Wittgenstein's view of rule-following—as expounded in chapter 7—can plausibly sustain the kind of experience of prereflective immersion that is taken to characterize absorbed coping. In allowing that rule-following can be achieved on the basis of rule-exemplification abilities rather than requiring rule stating, Wittgenstein allows that our capacity to monitor and correct our behavior is prereflective and continuous with our ability to perform certain acts. That is, rather than requiring the disengagement, distance, or detachment of reflective/deliberative acts, as Dreyfus assumes, "monitoring" our performances is an inconspicuous part of normally engaged sensitivity to the patterns of behavior (e.g., language use) we are producing. Prereflectively monitoring our performance, on this view, is an integral part of knowing how to do whatever we are doing.

The fact remains, nonetheless, that for Heidegger this intermediate form of prepredicative intelligence differs in kind from predicative awareness, which he links with a more general (context-independent) grasp of entities in terms of properties. In other words, as we might put it, Heidegger distinguishes the "space of purposes" from the "space of predication." Further, rather than being a pervasive feature of intelligent human action, predicative awareness is to require a "modification" of the *as* structure in terms of which we grasp the world, a modification Heidegger attributes to the use of language in propositionally expressed judgments (chapter 1, section 4). The thought that there is a need for some such modification, that not all meaningful disclosure of the world can be characterized in terms of a uniform functional space, is a fundamental point of agreement between my position and Dreyfus's (linked in each case with our respective readings of Heidegger). However, it is precisely this point that brings both of us into conflict with McDowell's position, a central feature of which is to deny any such functional discontinuity, so that the intentional realm can be described as a uniform functional space (the "space of reasons") characterized by the operation of concepts.

Before turning to McDowell's position, however, I want to consider an obvious question: Should we, or should we not, say that purposive intentionality of the kind just sketched is "conceptual" or involves grasping some kind of "content"? On the one hand, doing this would be problematic, because these terms are central to the vocabulary of what I have called the semantics approach, and would almost inevitably be taken to be defined in relation to the notion of propositional content. Given Heidegger's opposition to the primacy of propositional meaning, describing his notion of purposive/prepredicative awareness as involving "concepts" would no doubt invite confusion. This is presumably part of what leads Dreyfus to deny that absorbed coping is "conceptual" in character.[35] On the other hand, in some ways it would be helpful to talk here of "concepts," because this would better reflect the intelligent character of the purposive awareness that Heidegger identifies. However, this advantage would come at the cost of requiring constant clarification that the concern is with "concepts" in a different sense from that standardly assumed by the semantics approach. In sum, there is both some rationale and potential for confusion whether or not purposive awareness is labeled "conceptual." This situation might be dealt with in different ways. One would be to say that Heidegger's position allows for two kinds of concepts and content—namely purposive and predicative—or for an extended notion of conceptuality that exhibits two such modes.[36] Another might be to adopt the attitude Wittgenstein expresses in *PU* §79 that it doesn't really matter what we say, as long as we are clear about how matters stand.

One reason for adopting the latter indifferent attitude is that a virtue of Heidegger's position, particularly its emphasis on intelligent prepredicative behavior, is to move beyond the clumsy binary alternative that the nonconceptual-conceptual question presents. Indeed the whole point of his own, sometimes irksome and unorthodox, terminology is surely to provide a more differentiated framework for distinguishing the various features involved in and characterizing human disclosive feats. That is, rather than pressing us to decide between two traditional alternatives—mindedness and concept use, or mindlessness and dispositions (causation)—Heidegger tries to find ways of more accurately describing such feats.[37] Of course, it remains possible to persist in asking whether or not what I have called intermediate modes of (prepredicative) intelligence are "conceptual." But the situation here, as in the case of the realism/idealism issue discussed in the last chapter, is simply that they can be made to look either way, depending on which features of the superseded alternative are emphasized. And while it is possible to persist with the question, doing this fails to register and so obscures the full interest of a Heideggerian position.

5 The Challenge of Pervasive Conceptualism

In the last section I distinguished my claims about the role of prepredicative understanding from Dreyfus's noncognitive notion of absorbed coping as mindless and nonconceptual. My insistence on the intelligent character of prepredicative understanding, in particular its involvement of an *as* structure and purposive intentionality, might be seen as agreeing to some extent with McDowell's approach. Indeed, at the end of the section I conceded that if we are prepared to be flexible about the use of the label "concepts," albeit at the risk of confusion, then prepredicative understanding might be said to involve concepts. However, I also pointed out that my position shares with Dreyfus the assumption of a functional discontinuity linked with Heidegger's view of propositional content. Our interpretation of this discontinuity differs: whereas Dreyfus reads it as a switch from a nonconceptual "ground floor" to conceptual "upper stories" of experience, I have presented it as the transition from prepredicative to predicative forms of understanding. All the same, it is the assumption of some such discontinuity that brings both our positions into at least apparent conflict with McDowell's pervasive conceptualism. To assess whether McDowell's position challenges the moderate functional foundation claim proposed here, I will therefore consider two thoughts—as will become clear, related thoughts—to which McDowell is committed and which might appear to rule out any such functional discontinuity. The first is that conceptual capacities are operative in all experience and action; the second is that all experience and action involve content that is conceptual in shape or form.

One main focus of the debate between Dreyfus and McDowell is the sense in which, as the latter often puts it, conceptual capacities are "operative" in prereflective "coping" actions. As Dreyfus sees it, such actions involve only "ground-floor" abilities, abilities that we plausibly share with nonrational animals, with the "upper stories" of conceptual awareness—in particular, rational abilities—playing no role. Accordingly, he challenged McDowell to explain in what sense conceptual capacities are "operative" in prereflective behaviors despite our being unaware of exercising them (Dreyfus 2007a, 372, 376). In McDowell's view, this challenge is met by seeing the basic perceptual responsiveness exhibited in embodied coping as playing a different role in the human case from that of such responsiveness in nonrational animals. In the human case, he suggests, this responsiveness becomes the background for a wider range of rational abilities that can always be called on and allow us to relate to the world in a distinctively human manner (cf. McDowell 2007b, 343–345; 2013, 42–43). Thus it becomes part of an expanded space of possibilities, with the implication that humans differ from nonrational animals in their relation to such basic responsiveness.

Dreyfus appears subsequently to have accepted that McDowell succeeds in explaining how conceptual capacities are pervasively operative (see Dreyfus 2013, 29). However, it seems to me that he might have been more insistent on this point. Consider, for example, McDowell's explanation that his concern is with content that is "present in the content of a world-disclosing experience in a form in which it already either actually is, or has the potential to be simply appropriated as, the content of a conceptual capacity."[38] While the thought that humans occupy an extended space of cognitive possibilities is well made and plausible, this passage might nonetheless be seen as conceding the point that was important to Dreyfus—that higher-level rational abilities are not being exercised (actualized) in "ground-floor" responsiveness. It is also far from obvious that the additional availability of higher-level rational abilities modifies the actual exercise of "ground-floor" abilities. Suppose, for example, that I learn a new foreign language, such as Norwegian. In doing this I will clearly expand my linguistic abilities and expressive possibilities in some way, but it seems obviously wrong to suggest that acquiring this new capability modifies in any way my previous ability to speak English. Why should matters be any different in the case of the additional rational capacities McDowell emphasizes?

McDowell's answer to this question would probably rely on a further thought that he presents as an implication of the picture just sketched and as underlying his broad application of the term *operative* (McDowell 2007b, 346, 348). In order to think of prereflective experience as being available to higher-level rational abilities, McDowell holds, they must share a common form. As he puts it, all experience—including prereflective experience of "coping"—is conceptual in the sense of being "conceptually shaped" or "conceptual in form": "What is important is this: if an experience is world-disclosing ..., *all* its content is present in a *form* in which ... it is suitable to constitute contents of conceptual capacities."[39] This thought is independent of any concerns about the actual or potential exercise of capacities, and might plausibly suffice to establish the claim that conceptuality pervades even basic responsiveness. It would also appear to rule out the kind of functional discontinuity that both Dreyfus and I have proposed, because it suggests that no features of experience would require a transformation into conceptual form in order to be exploited in high-level rational discourse. Understanding and assessing this thought—that all experience is conceptual in shape or form—are therefore central to assessing both the overall tenability of McDowell's position and its potential impact on my own claims.

So what does it mean to say that content is conceptual in "form" or "shape"? It might seem obvious that such locutions are at the very least saying *something* about the form or shape of the contents referred to. Whatever it is taken to consist in precisely, the notion of conceptual form should at least have some demarcational force, constitute some kind of constraint, such that whatever it applies to has a determinate or specific form, the "conceptual" one as opposed to some other(s).

Although McDowell says surprisingly little to elucidate the notion of conceptual form, several passages suggest that he does think of it as such a determinate constraint on the form of content. For example, in *Mind and World* he describes the content of an experience as corresponding to a *that*-clause: "*That things are thus and so* is the content of the experience, and it can also be the content of a judgement. ... *That things are thus and so* is the conceptual content of an experience" (McDowell 1996, 26). In later work he explicitly rejects his previous assumption that conceptual content is propositional in kind, but continues to describe the ("intuitional") content of experience as being "*in a form* in which it is *already suitable* to be the content associated with a discursive capacity" (McDowell 2009, 264; italics added). Similarly, in his exchange with Dreyfus, McDowell draws on the Kantian background of his position in suggesting that conceptual form entails "categorial unity"—that is, a connection or relation to other possible contents of experience—and emphasizes that his "claim is that when experience is world-disclosing, its content has a distinctive form."[40]

Several observations should be made about these passages. First, if it indeed is a determinate or "distinctive" form, then to say that content has a conceptual form would be an informative claim, something like saying "*x* is triangular in form," which rules out many particular shapes and tells us something specific about the shape of *x* (it has three corners). Further, it would in that case impose some constraint or exclusion, a filtering so to speak, such that anything that is not conceptual "in form" would fail on principle to be captured by experience. It is also significant that McDowell often links conceptual form with the traditional notions of predication, judgment, and propositional content in the way assumed—and opposed—by Heidegger's view of prepredicative foundation. This appears to confirm the suspicion that McDowell's position is a "top-down" intellectualist approach. Of course, this label might be resisted on the straightforward grounds that if the same form or functioning pervades all experience and understanding, then it applies "across the board" rather than being "top down." And McDowell is not guilty, as pointed out above, of thinking that high-level rational abilities are actualized in prereflective activities. Nonetheless, it is difficult to avoid the conclusion that he applies terms designed to describe the semantic functioning of highly rational discursive activities to all intentionality, which is precisely what characterizes top-down approaches of the kind Heidegger, Dreyfus, and I oppose. Despite this, the present concern is not with such specific claims about *which* determinate constraints McDowell assumes, but the more general claim *that* conceptual form/shape is a determinate form of some kind or other.

If McDowell's view is that conceptual form is determinate or distinctive, as so far suggested, then one might expect him to argue that the contents of experience cannot take on a form that is "nonconceptual" in the sense of violating the constraints (whatever these are taken to be) imposed by the notion of conceptual form. An argument serving this purpose is found in *Mind and World*, where McDowell opposes Gareth Evans's attribution of a basic epistemological role to "informational states" that bear a kind of nonconceptual representational content. Evans's view is based on a distinction between conceptual content, linked with the role of states in predicative judgment (as with McDowell) and nonconceptual content ("information"), which is to be instantiated by the structure of perceptual states and due to causal interaction with the environment (see Evans 1982, 122–129). These two kinds of content are to be linked by the possibility of "conceptualization" through which nonconceptual content is taken up into conceptual content, thus entering into conscious experience and becoming "available to" judgment (Evans 1982, 227, 157f.). Evans's position is clearly analogous to the moderate functional foundation claim under consideration here: the nonconceptual-conceptual distinction parallels the prepredicative-predicative distinction, just as the idea of "conceptualization" parallels Heidegger's talk of a "modification" of the hermeneutic *as* of circumspection into the apophantic *as* of judgment.[41] To put it another way, in contrast to McDowell's view that all intentionality can be described in terms of a single, uniform functional space (the "space of reasons"), Evans thinks there is a discontinuity that requires the assumption of more than one functional space—in his case a "space of information" in addition to the "space of reasons."

Particularly relevant here is the thought that nonconceptual content is distinguished from conceptual content by differing in its form or structure—for example, by being more fine-grained. Evans (1982, 229) illustrates this thought by suggesting that we are able to see and discriminate (i.e., experience) many more shades of color than we have concepts to represent such discriminations. In response, McDowell offers an argument intended to show that such fineness of grain fails to establish that content is not conceptual. Thus in the case of color sensitivity he argues that, if one genuinely has the ability to recognize a certain color over an extended period, "the conceptual content of such a recognitional capacity can be made explicit with the help of a sample" and referred to by "a phrase like 'that shade,' in which the demonstrative exploits the presence of the sample" (McDowell 1996, 57; cf. also 2009, 263). Generalizing somewhat, the argument is that the use of determinable and demonstrative concepts in combination with recognitional capacities suffices to ensure that fineness of grain does not succeed in distinguishing something as nonconceptual. Rather, according to McDowell, by such means any discernible differences (in a fine-grained manifold) are conceptual in the sense he requires of being available for use in rational discourse.[42]

One relatively minor problem with McDowell's argument is that it is difficult to see anything that Evans would have to disagree with, and hence how it counts against the latter's claims. Evans could no doubt agree with McDowell that any feature of nonconceptual content can be "exploited in active judgements." The difference between their positions turns on what it is to be available for such exploitation: for Evans this requires a "conceptualization" of intrinsically nonconceptual content, the imposition of a specific form of organization (defined by the space of concepts actually possessed), whereas for McDowell content intrinsically has conceptual form. Yet the scenario McDowell describes can equally well be seen as both fully compatible with Evans's point about fineness of grain and as a way of explicating the process of conceptualization—as a process in which the use of indexicals and determinable concepts yields ad hoc conceptual abilities.

A second, more important, problem with Dowell's argument is that it does not suffice to show that Evans's information manifold is conceptual in form. For as long as it is assumed that the notion of conceptual form imposes a determinate or distinctive constraint on the *form* of content, the procedure he describes fails to establish that whatever our conceptual capacities (e.g., demonstrative terms) are directed at (e.g., subpersonal information-bearing states) is antecedently conceptual in form. It fails to do this because it cannot be inferred from the mere possibility of applying a certain mode of representation that the "distinctive" form of whatever is represented is being captured. To infer this would involve an illicit projection of the kind Wittgenstein mocks as "predicating of the thing what lies in the mode of representation" (*PU* §104). In other words, it would be simply fallacious: the possibility of applying concepts to any given fine-grained manifold no more implies that this is antecedently conceptual than the existence of digital watches or black and white photographs entails that time lacks continuity or that the world lacks color.[43]

It might seem odd to suggest, as I have so far been doing, that to see the notion of conceptual form as saying something determinate about the form or shape of the contents is an *interpretation*. After all, one might wonder, how else it could possibly be interpreted? To see why an alternative appears necessary, it will help to draw out an implication of McDowell's argument against Evans. Despite the two problems just mentioned, this argument does establish something important. Whether or not one thinks to talk of "conceptualization" is appropriate, the procedure McDowell describes does show that the use of relatively coarse, context-dependent expressions suffices for any distinguishable feature of a differentially structured manifold to be exploited in conceptually articulated rational discourse. This has the implication that concepts are universally applicable in the sense that they can be applied to identifiable features of any shape or form whatsoever. This means, however, that while plausibly recognizing the extreme versatility of our conceptual and rational capacities in latching onto features of the world, McDowell's argument against Evans simultaneously undermines the idea that these capacities can be thought of as attributing a determinate "conceptual" form or structure to their referents.

Perhaps then, as odd as it may initially appear, McDowell's talk of "conceptual in form" might instead be interpreted such that it does not say anything determinate about the form or shape of whatever is referred to.[44] Interpreted in this way, McDowell's pervasive conceptualism would not be attributing a determinate form or functional organization to all of experience or reality, but merely asserting that whatever determinate form the latter have, concepts, as employed in discursive practices, will in principle be able to track it. Rather than imposing any exclusion or constraints on experience, this would simply affirm that conceptual capacities can adapt themselves to any feature presented in our sensory experience. This would clearly be a far less informative claim than that suggested by the first interpretation of "conceptual form," comparable now to saying "*x* has a spatial form"—that is, roughly that it has a shape of some kind (occupies space in some way) rather than a specific shape (e.g., triangle). However, it is important to note that this claim might still suffice for McDowell's overarching aim of securing the justificatory role of experience in knowledge claims, as it leaves intact the thought that nothing lies beyond conceptual determination as a matter of principle, and in this sense avoids postulating a mythical Given.[45]

This second interpretation would, however, be unable to rule out a functional discontinuity, in particular the modification that Heidegger discerns from purposive to predicative awareness. It would be unable to do this precisely because, on this interpretation, McDowell is not attributing any determinate form or shape to whatever rational discourse refers to. All his position implies, on this second interpretation, is that we will be able to apply propositional and predicative abilities to talk about any underlying form of experience or content, such as Heideggerian purposive awareness, that there may be. Taken this way, the claim that experience or features of the world are conceptual "in form" would be somewhat misleading, because it is really a claim about the applicability of concepts rather than their form, amounting to an affirmation of the power of concepts and discourse to capture whatever form the world takes on in our experience.

At first glance, then, McDowell's idea of pervasive conceptualism might appear to rule out the kind of functional discontinuity suggested by a Heideggerian picture and hence challenge my moderate functional foundation claim. The key to understanding whether it really does have these implications lies in the interpretation of McDowell's talk of experience being "conceptual in form." Here I have distinguished two (generic) alternatives and argued that in neither case does McDowell's position present a significant challenge to the kind of modification Heidegger postulates. On the first interpretation discussed above, the "conceptual form" label attributes a determinate form to experience—the "conceptual" one as opposed to others—and *would* conflict with the Heideggerian approach. However, on this interpretation McDowell's argument against supposed nonconceptual forms of content not only fails to establish their impossibility but also undermines the idea that concept application depends on a determinate underlying conceptual form. On the second interpretation, this label does not attribute a determinate "conceptual" form to basic modes of experience or action. This interpretation, which is motivated by the implications of McDowell's argument against Evans, reduces to a claim about the universal applicability of concepts and does not conflict with the Heideggerian approach.[46]

6 Dissolving Bedrock

Against the backdrop of the Dreyfus-McDowell debate discussed in the last two sections, it should be clearer how the phenomenological approach proposed here constitutes an intermediate position. On the one hand, a prepredicative grasp of language and the world has an intelligent character that distinguishes it from Dreyfus's image of mindless absorbed coping. On the other hand, it is not pervasively intellectual or predicative in character in the way at least some of McDowell's comments appear to suggest. Nor, I have argued, does McDowell's pervasive conceptualism succeed in undermining the moderate functional foundation claim.

As I have so far been discussing it, the point of this foundation claim is that explicating the workings of the founding Background is not otiose. To some extent this already followed from the weak foundation claim, but there the philosophical interest of such explication was simply a matter of completeness or illustration, since Background was taken to contribute nothing distinctive to linguistic meaning. By contrast, the message of the moderate foundation claim is that a phenomenological account of Background, of the kind offered here, tells us something distinctive about the functioning of language that is not captured by the semantics approach. However, this foundation claim, being based on the idea of a functional discontinuity, also has the important implication of preserving the idea that predicative understanding is a distinctive feat. Thus it also exemplifies the antireductionist mood that I have claimed should characterize a phenomenological conception of language by suggesting a relationship of mutual irreducibility and some form of complementarity between the phenomenological and semantics approaches. In this final section I want to outline how I think this relationship should be thought of by drawing on an analogy with the change brought to classical physics by the development of atomic physics in the course of the twentieth century.

As the terminology testifies, it was assumed in classical physics that the basic constituents of the physical world are indivisible "elements" or "atoms," in much the same way as the semantics approach often assumes there to be basic units of meaning. The realization that the atom could be split, leading to the development of complex models of the subatomic, clearly made a great difference to our understanding of both physical matter and the status of the theories of classical physics. But what, in very general terms, was the nature of this difference? And how does this analogy shed light on the relationship between semantic properties and prepredicative factors?

First and foremost it became clear that classical physics could not claim to be a comprehensive theory of matter and that an understanding of the subatomic realm would be part of any such theory. Similarly, any ambition the semantics approach might have to deliver a comprehensive theory of linguistic meaning is undermined by the vindication of the moderate functional foundation claim. For this implies that understanding how linguistic meaning functions requires an account of the prepredicative factors in which semantic properties are founded. To insist otherwise would parallel the suggestion that classical physics had nothing to learn from an understanding of the subatomic. This is an important result, implying that the philosophical significance of the semantics approach is considerably less than is widely assumed. To illustrate this I want to mention briefly three of its implications.

The first implication is that it is misleading to think of semantics as a closed or self-sufficient functionality. Semantics is not "autonomous" and such supposed closure cannot justify reassuring dualisms—for example, language/world, normative/causal, conceptual/empirical—that philosophers of language often rely on to delimit their subject matter. Interpretation of the philosophical importance of semantics should instead reflect the fact that linguistic meaning is not only ontologically, but also functionally embedded in the world. This might be summed up by saying that philosophical understanding of language, and indeed of intentionality more generally, should give up Wittgenstein's entertaining but misleading metaphor of a "bedrock" lying below the fertile soil of concepts as one of those "pictures" that have led philosophers astray.[47] Rather than identifying a principled limit on what it is philosophically interesting to understand about language, this metaphor serves at best to flag the limitations of the semantics approach's theoretical vocabulary.

A second implication is that the phenomenon of language should not be overrationalized. A characteristic feature of the semantics approach is the attempt to model language as a pervasively rational phenomenon, as instantiating a system of determinate rules that circumscribe—as the common metaphors have it—our conceptual "mechanisms" or "machinery." This is no doubt a contemporary incarnation of the age-old belief that language use distinguishes humans as rational. But even if it is accepted (in accordance with Heidegger's views) that language is the vehicle of predicative competence, and as such makes rationality of the predicative type possible, this does not mean that language should or can be thought of as functioning homogeneously in a pervasively rational manner.[48] In identifying the formative, foundational role of prepredicative factors, the phenomenological approach serves as a reminder against such convenient oversimplifications. As the preceding discussions of presentational and pragmatic sense should have made clear, prepredicative factors have a kind of intermediate status between straightforwardly mechanical causation and full rationality. Although formed by intelligent behavior, they are not assimilable to the semantics approach's preferred model of what it is to be rational; they are instead inchoately rational, rationally ambivalent, and typically the product of unreflective pragmatic agency.[49]

The third implication, which is linked with the second, is that taking prepredicative factors seriously suggests philosophy of language should give up the aspiration to a tidy "systematic" theory of language. Both the versatility of language in practice and the expressive potentials that lead language, in Merleau-Ponty's terms, to constantly "exceed" or "transcend" itself suggest that thinking of it as a systematically ordered whole would misrepresent the specific modes of intelligence that language use embodies. Rather than projecting onto language use the moribund ideal of rational orderliness, standing firmly on a bedrock of arrested causation, these features should be recognized as its vital element, such that language is more realistically seen as characterized by nonsystematic fecundity, openness, and dynamic fragmentation. An obvious response to this suggestion would be that the aspiration to a systematic overall view is precisely what defines a philosophical approach to language (as opposed, say, to an empirical one). However, if attention to linguistic phenomena attests to such openness and versatility in the functioning of language (as I am suggesting it does), it is difficult to see why the imposition of artificial closure and systematicity should be considered a philosophical gain. What is problematic is not that a philosophical conception of language might give up the dull dream of systematic order, but the phenomenologically groundless demand for simplicity and order that generates mistaken views of language.

Faced with these claims, a proponent of the semantics approach might respond that it is intended in the spirit of a scientific hypothesis, as the best available model of, and a good approximation to, how language works. The analogy with physics gives this suggestion some plausibility: after all, macroscopic physics did not become obsolete with the discovery of subatomic properties; rather it remained useful despite being strictly speaking "false." Nonetheless, the suggestion is unsatisfactory. Apart from the fact that it seems inaccurate to describe (philosophical) semantics as an empirical field of inquiry, an appeal to the idea of approximation is problematic. With scientific theories the claim to be an approximation can be justified and it is standard practice to consider and explicate constraints on a theory's applicability. This might be done quantatively (error calculations) or by specifying the range of applicability or purposes in which a given theory is properly deployed. Admittedly, it is true that semantics *is* a good approximation in the context of what I have called strictly regulated language-games, those in which most word-use patterns are covered by well-defined rules. But the semantics approach itself is oblivious to—that is, does not factor in or make explicit—the circumstances that regulate its applicability. This means that to assume that semantics generally approximates well to linguistic phenomena lacks justification, resting instead on the prejudgment (or a set of "intuitions") that language just is rule-governed, determinate, and fundamentally bivalent in its operation. Yet this obliviousness leaves intact the suspicion that the semantics approach is characterized by omissions that are principled, and which lead it to misrepresent the way linguistic disclosure actually works.

Despite this, finally, the analogy with physics also provides a model for thinking of the way the phenomenological and semantics approaches complement each other. Classical physics does not "approximate" to subatomic physics, but it does succeed in modeling some aspects of the behavior of matter, behavior exhibited at a different level, due either to statistical aggregation or the fact that atoms, for example, have (emergent) properties in virtue of their organization as a whole that might not be predicted on the basis of the properties of their parts taken in isolation. So the development of subatomic physics quite literally took understanding (of matter) to a new level. It was discovered that the subatomic realm functions in ways not evident from the top-down macroscopic perspective of classical physics, yielding a gain in understanding by explaining and predicting phenomena that were beyond the scope of classical physics. At the same time, such insights did nothing to impair the reliability or validity of most macroscopic theories, provided their level and range of applicability are respected. I want to suggest that the phenomenological and semantics approaches complement each other in an analogous way. They do not correspond to one another in a straightforward, functionally continuous manner—so to speak, telling the same story from different perspectives. Rather, due to their functionally discontinuous relationship, their complementarity is such that they each model different (mutually irreducible) aspects of the phenomenon of linguistic meaning. While this is consistent with the antireductionist outlook I have been advocating, in the current context of post-Fregean philosophy of language the take-home message is that the semantics approach is of far more limited philosophical importance than is usually assumed—that its theoretical ambitions should be reined in just as those of classical physics were in the twentieth century. By contrast, as with the development of subatomic physics, a phenomenological account of prepredicative factors should be seen as taking understanding of language to a new level, enabling understanding of phenomena that are beyond the pale of the semantics approach.[50]

10 Phenomenology and Beyond

In the introduction I suggested that my approach in this book would be phenomenological in a minimalist sense—defined by a commitment to phenomenological accountability—of aiming to describe accurately the way language is experienced by speakers. I went on to characterize this approach metaphorically as mapping the experiential surface of linguistic phenomena, due to its reliance on noninferential recognition of its validity as opposed to systematic or reasoned explanation of how language works.

The thought naturally arises—and is encouraged by the surface/depth metaphor—that a phenomenological conception of language of this type might not be enough. This is not, of course, to deny that describing the experiential surface of language is important, because this conviction underlies any phenomenological approach. But if it tells us only about the "surface" of linguistic phenomena, surely philosophers are entitled to ask what is going on below this surface and to demand a "deeper" explanation of language. Indeed, some might even wonder why we should care about a phenomenology of language at all, since (as was once suggested to me) we might be mistaken about how language really is. This is a strange suggestion, and one that I think—for reasons that will emerge below—is based on and reveals a fundamental misunderstanding of the kind of entity language is. But even if that suggestion is rejected, it still seems plausible to think that a minimalist phenomenology of language is not the whole story (i.e., does not tell us everything of philosophical interest about language), and that an account is needed of what underlies or explains the experiential surface that I have so far been focusing on. The question, in that case, is: What kind of explanatory perspective ought to be adopted to complement the phenomenological conception of language set out in the first three parts of this book?

Before going any further I want to underline that precisely its indifference to explanatory claims is a key advantage of a minimalist conception of phenomenology. It is central first of all to fulfilling the task that phenomenological conceptions have traditionally claimed to address of defining the subject matter and explananda for theoretical explanations (see the introduction, section 1). Phenomenologists sometimes further link this task with specific kinds of cognitive feat (e.g., eidetic insight or aprioristic interpretation), or a particular division of labor (with phenomenology investigating features presupposed by and perhaps unavailable to scientific thought). The important point here, however, is simply that phenomenology claims to identify conditions of adequacy for the respective areas of inquiry, in the present case identifying features that any putative explanatory theory of language would need to account for. Yet clearly, to play this role of identifying explananda, a phenomenological approach must itself be explanatorily neutral.

From a strategic point of view an indifference to explanatory claims also helps to maximize the potential appeal of a phenomenological approach. To some extent this was true already of my original characterization of a minimalist phenomenology, which identified core commitments that would be acceptable to a range of phenomenological positions, ranging from Heidegger to typical usage of this term in analytic discourse. Yet this indifference to explanatory claims further allows that in principle different kinds of explanatory framework or theory might be developed so as to complement a minimalist phenomenological approach and yield a more complete and satisfactory philosophical understanding of language. Indeed, on the basis of what I have said so far, a very open answer is possible to the question of what explanatory perspective on language to adopt, namely, that any kind of explanation that makes sense of the various features set out by a minimalist phenomenological conception of language would in principle be acceptable.

Despite this, some features of the phenomenological conception of language developed here suggest that not all explanatory approaches would be equally suitable. Most obviously, the matter is complicated by my criticism in the last chapter of the semantics approach. For in rejecting the primacy of propositional content and the idea of a fully determinate rule system that might define a teleological functional perspective, and in suggesting the need for an account of founding prepredicative factors, the message is clearly that the semantics approach does not provide a satisfactory basis for a philosophical theory of language. The aim of this chapter is therefore to identify a kind of explanatory approach that is particularly suited to taking philosophical understanding of language beyond (minimalist) phenomenology. The first section of the chapter briefly reviews two more ambitious (nonminimalist) phenomenological strategies and introduces an alternative cognitive science approach—"4e" cognitive science—that emphasizes the interactive embedding of agents in their surroundings. Although the former are plausibly closer in spirit to main figures in the phenomenological tradition, some generic reasons are offered here for preferring the latter. The second section sets out more fully how 4e cognitive science shares key commitments with the phenomenological approach to language developed here, both in its overall metaphysical outlook on language and in sharing a more specific focus on prepredicative factors. The chapter's third section argues that a minimalist phenomenology cannot simply be eliminated in favor of 4e cognitive science, as might be suspected, and that the two approaches should be thought of as complementary and mutually illuminating.

1 Below the Experiential Surface

One obvious possibility for developing a deeper explanation of linguistic phenomena would be a more ambitious phenomenological approach. Indeed it might be thought that the minimalist conception of phenomenology I have so far relied on sells phenomenology somewhat short by ignoring the fact that authors in the phenomenological tradition have typically seen themselves as transcendental philosophers. In the Kantian usage, on which this tradition relies, transcendental knowledge is "concerned not only with objects, but with our way of knowing objects altogether, insofar as this is to be a priori possible" (Kant 1983a, 63 [B 25]). Accordingly, phenomenology has often not contented itself with describing prima facie or surface features of experience, but has sought to extract structural principles or rules of phenomenal manifestation that constitute "conditions for the possibility" of objective knowledge in a broadly Kantian sense. One complicating factor is what this Kantian formula is taken to mean. If it means simply that the function of phenomenological descriptions is to define the subject matter or explananda for scientific theories, then the minimalist approach I have advocated here might also be labeled a "transcendental" one. However, I argued earlier (introduction, section 2) that minimalist phenomenology should be thought of as lacking the transcendental touch, and that in order to make transcendental claims a phenomenological position will require further assumptions or methodological provisions. I now want to consider briefly two—historically influential—ways of attempting to do this and so to develop a (nonminimalist) phenomenological position that gives the Kantian formula a more specific and stronger sense.

The first and most obvious is Husserl's conception of transcendental phenomenology. With Husserl the claim to be investigating the a priori conditions for objective knowledge is based on the idea of describing the realm of "pure phenomena" or, in his view equivalently, of transcendental subjectivity.[1] Access to this realm of pure phenomena is first ensured by "bracketing" all assumptions about the existence (or nonexistence) of objects through the so-called phenomenological reduction. Through a process of "free variation" (of act type, object types, and so on) Husserl believes it is then possible to explore systematically and hence describe the structure of all—that is, actual and possible—intentional acts and objects.[2] These two steps are intended to distinguish phenomenology from empirical psychology and to underpin Husserl's claim to be carrying out an a priori investigation into the constitution of objects of knowledge, just as Kant requires, thus making phenomenology "transcendental."[3]

In this perspective it is easier to see the significance of the two features of Heidegger's and Merleau-Ponty's views highlighted in the introduction (section 2). It was suggested there that Heidegger and Merleau-Ponty both emphasize the importance of lived experience—actual experience rather than imaginative variations—and reject the notion of a pure transcendental subject as an unintelligible construct. In doing this, they deny precisely those features of Husserl's position that were supposed to make phenomenology a transcendental discipline. For without the extension of experience beyond the real into the realm of "pure" phenomena, a view such as Heidegger's or Merleau-Ponty's cannot claim to be investigating conditions under which objective knowledge is possible in a completely experience-independent (a priori) manner. Instead their positions remain empirically conditioned ones—that is, retain an underlying bind to facts. In view of this Husserl came to dismiss Heidegger's position as merely a form of anthropology, objecting that the idea of a "transcendental philosophy that remains standing on the natural ground" is simply "nonsense [*Widersinn*]" and "misunderstands the most profound sense of the intentional method or that of the transcendental reduction."[4]

At this point some arbitration seems necessary. On the one hand, it seems to me that Heidegger and Merleau-Ponty are right—in the spirit of phenomenological accountability—to insist on the primacy of lived or actual experience and to reject the notion of transcendental subjectivity. The most obvious reason for this is that the extended notion of experience

Husserl relies on is clearly untenable. As soon as one disregards ("brackets") the difference between the actual and possible, real and imagined, experience simply collapses into imagination and loses the ability to function as a constraint on knowledge or thought. On the other hand, Husserl has a point in emphasizing the connection between the extension of experience he assumes and the capacity of phenomenology to make transcendental claims. Whether or not one agrees with him, Husserl clearly has a developed view of what steps are required to investigate conditions of a priori possibility. So although Heidegger and Merleau-Ponty continue to present their claims as though they were practicing a form of transcendental philosophy, it is far from obvious that they are entitled to do so.[5] At the very least, their positions need a different account of what makes their claims "transcendental."

The minimalist phenomenological approach I have been advocating therefore seems incompatible with at least one—historically the most influential—kind of transcendental phenomenology. For if one thinks (as I have claimed) that a commitment to describe experience accurately is the principal characteristic of a phenomenological approach, then one must give up the claim that phenomenology is transcendental in a Husserlian sense—that is, in any sense which relies on an appeal to pure phenomena and the corresponding method of reduction.

An obvious place to look for an alternative to Husserl's transcendentalism is to Heidegger, whose hermeneutic version of phenomenology might be seen as providing an alternative way of developing explanatory claims according to a phenomenological method. Heidegger's interpretive method—as practiced in *SZ* itself—involves several steps. The first is to ensure the right focus on or approach to phenomena, so as to avoid distorting one's understanding from the start. In *SZ* this requirement is taken to be met by analyzing Dasein's everyday existence and instrument use, which are intended to avoid overlooking the essential link between humans and their surroundings.[6] Once a suitable phenomenal focus has been found, the task begins of interpreting the relevant phenomena by identifying their various structural features. For Heidegger this interpretive process moves through different stages, with an initial interpretation being subjected to iterative reinterpretation(s), guided by the aim of attaining greater unity and comprehensiveness—referred to collectively as "originality."[7] Thus *SZ* starts (in Division I) by identifying existential structures making up Dasein's everyday disclosure, interpreting these first as belonging to the unifying structure of care, and then attempting (in Division II) to reinterpret those structures in (more original) temporal terms.

Although, as might be expected, somewhat closer in spirit than Husserlian transcendentalism, this method seems ill-suited as an explanatory complement to minimalist phenomenology for two reasons. First, it is not clear that this hermeneutic method yields genuine explanation of any kind. It is surely right that structural interpretation and reinterpretation can provide new insights and perspectives on the phenomena under consideration, and in this sense expand or improve our understanding of them. Yet Heidegger seems to think that the interpretive process provides explanatory insight into conditions of possibility in a Kantian sense. However, although it might be allowed that this process is aprioristic (i.e., nonempirical), it is far from clear in what sense more abstract or unified levels of structure "make possible" or explain the phenomena they are taken to subsume. Instead it seems that Heidegger runs together the possibility of redescribing or reinterpreting some phenomenon with explaining it.

Second, the commitment to phenomenological accountability seems too weak in the interpretive process Heidegger advocates and employs. It is true that phenomenological accountability plays a significant role in the first component of his method—that is, in the choice of how to focus fully and appropriately on the target phenomena. It is also plausibly at work in the initial identification of phenomenal structures. Yet Heidegger seems to believe that choosing the right "phenomenal ground" to be bound by suffices to distinguish "disclosing the a priori" from "'aprioristic' construction" (*SZ* 50n). In the subsequent process of (re)interpretation he becomes more concerned with systematizing his analysis in a programmatic manner than with showing how the various structures he proposes are motivated by the complexities of experience. To some extent this is unsurprising, because it is implicit in the aspiration to full unity and generality of explanation that higher-order structures cannot be motivated by phenomenal differences. Nonetheless, it seems fair to suggest that Heidegger both fails to recognize the inevitable tension between striving for unity and phenomenological accountability, and underestimates the full methodological implications of the latter.[8] As a result, despite its importance in determining the initial stages of Heidegger's inquiry, the commitment to phenomenological accountability gets lost along the interpretive way.

Taken together these two considerations—the apparent assimilation of explanation to reinterpretation and the waning influence of phenomenological accountability—suggest that Heidegger's method runs the risk of drifting into unconstrained speculation. While I do not want to consider here whether this is an advantageous or disadvantageous feature of Heidegger's philosophical views in general, it is certainly difficult to reconcile with the minimalist phenomenological approach I have advocated here, which emphasizes descriptive accountability to experience precisely in order to avoid a speculative, theory-driven approach to language.[9]

Having rejected two options for a more ambitious explanatory version of phenomenology, the remainder of this chapter aims to show that a minimalist phenomenological approach is most naturally complemented by recent approaches in cognitive science that emphasize the importance of the embodiment, embedding, enaction, and extension of cognitive processes. One of the core features of 4e cognitive science—as it commonly called—is opposition to earlier cognitivist or representationalist conceptions of cognition as processes of symbol manipulation taking place in the human brain/mind. This symbol-processing model of cognition lent itself naturally to associations not only with serial CPU-based computing architecture, but also with the Cartesian picture of the mind as a neatly delineated internal unit that communicates with the outside world via transducers (the senses), the reliability of which is open to doubt.

The various *e*'s of 4e cognitive science mark different nuances and degrees of contrast with that older representationalist approach and can be seen as a progressive decentralization of cognitive processes. The basic thought behind the claim that cognition is *embodied* is that various parts of the body outside the brain play a constitutive role in cognitive processes. This may be because cognitive processing extends beyond the brain (e.g., is shaped by sensory abilities) or because cognition is attuned to the physical constitution of bodies (e.g., the weight distribution and shape of limbs).[10] An emphasis on embodiment already marks a significant departure from the idea, shared by representationalist cognitive science and Descartes, that operations of the mind (as "software") can be understood independently of the physical body (as "hardware") with which they are linked.

The claim that cognition is *embedded* goes slightly further by suggesting that cognitive processes cannot be understood independently of the body's surroundings. This thesis—one clearly reminiscent of Heideggerian being-in-the-world—implies that various entities, such as tools or other people, have some function that must be considered in order to understand cognition properly, so that attention can no longer be limited to the abilities of "individual" agents or organisms.

Insisting on the embodied or embedded nature of cognition might be seen as a fairly conservative outlook, because both these claims are consistent with the thought that cognitive processes as such do not extend beyond a single agent or organism.[11] The idea that cognition is *enactive* goes further still by taking cognition itself to consist in the interaction between an embodied agent and its surroundings. On this view, cognitive functions do not simply aim to map an objective environment, but are geared to the specific kinds of action—for example, modes of perception and movement—performed by particular types of body in their surroundings.[12]

Finally, viewing cognition as *extended* means that features of the environment that have a constitutive function in cognitive processes are considered parts of the cognitive system itself, so that an agent's mind spreads out over its physical and social surroundings. On this view, equipment such as (pen and paper) notebooks, computers, social institutions and networks, and so on, that we use to offload cognitive effort become freely interchangeable, but functionally integrated parts of an "individual's" cognitive system.[13] The idea of such cognitive extension presents the most radical challenge to both representationalist cognitive science and Cartesian mindedness, because it appears to give up the idea that cognitive systems have a fixed and determinate locus correlated with particular agents' brains or even bodies.

There are several generic reasons for thinking of 4e cognitive science as complementing the phenomenological position developed in the first three parts of this book. The first is that the four e-theses have in common the tendency to afford greater attention to the role of things outside the head in understanding cognition. While there is room for debate about the precise implications of and relations between these theses, this general tendency suggests a parallel with the view I summed up using the label "language-in-the-world." This is not really surprising in view of, second, the anti-Cartesian tenor that 4e cognitive science shares with Heidegger. Indeed, classical phenomenological authors are widely recognized as having anticipated and have in some cases influenced these recent developments in cognitive science.[14] Third, although its literature often focuses on more novel examples than the familiar but challenging example of language, the latter is obviously a paradigm case of the kind of environmentally embedded apparatus that 4e cognitive science emphasizes.

Finally, from a methodological standpoint minimalist phenomenology and cognitive science share a sense of accountability to experience. In cognitive science the factors or structures thought to explain cognition must be motivated empirically by features of the phenomena they seek to explain. This contrasts, to take the example above, with Heidegger's hermeneutic phenomenology, in which an initial sense of remaining true to experience gets left behind in the interpretive quest for greater comprehensiveness and unity. To put it another way, a connection with 4e cognitive science seems to be not merely a coincidence but a natural consequence of a minimalist phenomenological approach. For if descriptive accountability to experience is taken to be phenomenology's basic commitment, then consistency seems to require that any "deeper" understanding should also be properly informed by experience—as is the case with an empirical science. Put simply, minimalist phenomenology and cognitive science are two aspects of a more general commitment to understanding experience.

2 A Shared Outlook

Against the background of the generic considerations just listed, this section aims to bring out more specifically how 4e cognitive science converges in outlook with the phenomenological conception of language developed here. To this end I briefly discuss two characteristic themes of 4e cognitive science—the use of so-called scaffolds and the role of representation—so as to show how it shares the various commitments previously summed up under the label "language-in-the-world" (see chapter 2, section 1). By doing this it should become clearer that, and how, 4e cognitive science is particularly suited as an explanatory approach that complements the phenomenological conception of language presented here.

The distinctive tendency of 4e cognitive science is, as I put it above, to pay greater attention to the role of things outside the head in understanding cognition. This tendency is epitomized in the notion of "scaffolding," which refers to the use of features of the environment as a tool that either enables or simplifies cognitive processes (see Clark 1997, 45–51 and passim). Perhaps the most straightforward examples of scaffolding are the use of notebooks as a memory aid, or of pen and paper as a means to keep track of and manipulate our thoughts (e.g., when writing an essay or solving a mathematical problem).[15] Somewhat less obvious is John Haugeland's elegant example of the reliance of our navigational abilities on built-in features of the human environment. In answer to the question of how he gets to San Jose, Haugeland (1998, 234) pithily responds: "I pick the right road (Interstate 880 south), stay on it, and get off at the end." The point of this otherwise deadpan comment is that the ability to get to San Jose does not require representing where San Jose lies, being able to read a map, or indeed any complex navigational skills. Instead all the information required for successful navigation is literally laid out in the road system, so that the navigation task reduces to that of following road signs correctly within the road network.

This simple example illustrates effectively how scaffolds allow cognitive effort to be offloaded and reduced. First, the overall task is broken down and distributed between agents and over time/space, so that individual agents do not have to solve the complete, potentially complex, problem by themselves. The road system itself is an intelligently structured environment, being built to accommodate and coordinate complex traffic-flow requirements, with relevant information cues distributed (and often repeated) throughout to guide its users. As a result the entry-level demands on user competence are relatively low, and there is no need for us to master the system or understand, for example, how such an environment must be constructed to function well. As Haugeland (1998, 235) emphasizes, what holds for the road system also applies to many other contexts, such as tool use, institutions, and social or corporate structures. Indeed, as his richly suggestive example intimates, we pervasively navigate a human world that is more intelligently structured than we are required to realize. Another powerful example of this is provided by Hutchins's (1995) case study of the navigation work (again!) on a naval ship, which documents in detail how the task of tracking and planning a ship's course is broken down into subtasks distributed over multiple agents and pieces of equipment, each with specialized roles that are coordinated in real time. Through devolution and distribution of the overall task the cognitive demands on individual agents are reduced in such a way, as Hutchins (1995, 224) puts it, that "one can be functioning well before one knows what one is doing."

Closely linked with this breaking down and distribution of the overall task is the fact that the road system, as an intelligently structured environment, is designed for use by situated and responsive agents. It takes account of the fact that agents (drivers) are able to access and exploit features of the environment in real time—just as fluidly as they might access anything "inside" their minds. Road signs located before and at junctions, for example, prompt and enable actions at the right time. Further, the road system's design requires, structures, and facilitates context-specific responses: if you don't know where to leave the highway, you just keep checking the road signs, choosing between simplified options ("Do I turn off here or not?") in real time. Thus scaffolds allow cognitive effort to be offloaded both by requiring less work of individual agents and by structuring and prompting their real-time decision making.

There is clearly scope for disagreement as to how the role of scaffolds should be interpreted. At one extreme is the so-called extended-mind thesis, according to which environmental features used as scaffolds literally comprise part of the mind, part of the physical supervenience base for mental processes. The other extreme is to insist that such cases should be understood as (embedded) agents with minds making use of (external) equipment. Through these disagreements a number of important questions arise. One concerns the kind of connection or interaction there must be between users and the environment for scaffolds to count as an integrated part of a user's cognitive system. In the road example, say, there is arguably no genuine interaction, and the road system seems to be a passive resource that agents access in a one-way manner. This asymmetry makes it easier to deny that the road system becomes an integral part of an extended-mind system. A possible improvement on this is therefore to require that for scaffolds to be considered as integrated with an agent, as forming a single cognitive system, there must be reciprocal influence or feedback—"continuous reciprocal causation" as Clark (1997, 165) calls it. But even then, given that reciprocal feedback processes are common in physical and biological systems, the question remains of how to distinguish cognitive from noncognitive cases.[16]

These are good questions that sustain rich and interesting debate within 4e cognitive science, and promise to scientifically deepen our understanding of the human-world relation. What I want to highlight here, however, is that once such questions are formulated, whether or not the "mind" is said to be "extended" is of no great importance. For once it is acknowledged that various bits of the environment can be co-opted as scaffolds for cognitive processes, whether one talks of "cognitive systems" or "extended minds" reduces to no more than a terminological issue.[17]

What does matter, however, is the insistence—one basic to 4e cognitive science—that in order to understand human thought properly it is necessary to consider the constitutive role played by features of the physical and social environment that function as scaffolds. Once this is accepted all the important questions—such as those just referred to about the constitution and functioning of cognitive systems—will be set up in a manner indifferent to the inside-outside distinction. Indeed, I would suggest that the extended-mind thesis is best interpreted as an expression of this indifference. Thus in the view of Clark, probably the most vocal advocate of the thesis, the barrier between the "inside" and "outside" of cognitive systems is fluid, transient, and ad hoc. As he puts it, "profoundly embodied" intelligent agents make "maximal problem-simplifying use of an open-ended variety of internal, bodily, or external sources of order" and are "able constantly to negotiate and renegotiate the agent-world boundary itself" Clark (2008, 42, 34). Such indifference to an inside-outside distinction is also inscribed in the working principle that is often used to weaken unreflective ("intuitive") resistance to the extended-mind idea, namely, that we should not preclude processes from being cognitive simply on the basis of their not taking place "in the head."[18]

Against this background, the convergence between 4e cognitive science and the commitments that I suggested characterize a conception of language as phenomenological begins to emerge. The first of these commitments, to recall, was the rejection of any inside-outside topology within which intentional relations are contained, together with the suggestion that intentionality instead be seen as a having a distributed topology. Both these moves have obvious parallels in 4e cognitive science. First, the distinctive tendency of the 4e approach is to break with the preconception that cognition is something that goes on (exclusively) in the head. As this tendency unfolds, as just outlined, it leads to indifference concerning the inside-outside distinction, which becomes irrelevant in determining the range of phenomena focused on and the ways cognition is conceived. Second, the 4e approach also embraces an "overall vision ... of cognition [as] distributed among brain, body, and world" (Clark 2008, 137). Accordingly, cognition is conceived of as ontologically and functionally spread out both over features of the physical or equipmental environment that function as scaffolds, and over multiple agents who share the load in the division of cognitive labor.[19]

To bring out further aspects of its convergence with a phenomenological approach to language, I turn now to the role of representation—or rather to a paradigm shift in the role attributed to representation—as a second theme characterizing 4e cognitive science. The basis for this paradigm shift is a general acceptance in more recent cognitive science of the view that not all representation is linguistic in form. This acceptance is directed against a thought central to older representationalist cognitive science, according to which language provides a basic model in understanding all cognition. One variant of this thought is the idea that, once properly understood, the abstract structure of mental representations—the "language of thought"—would turn out to be language-like, a symbol system with a combinatorial syntax and semantics of the kind usually attributed to natural languages (by what I have been calling the semantics approach).[20] A second variant is the idea found in classical artificial intelligence that cognition is essentially algorithmic—comprising "rules and representations"—and so is reducible to sequences of explicitly formulated commands. Although not necessarily formulated in natural language, such commands would again approximate linguistic statements in their form.

The idea that language should have such a distinguished role has been challenged in two basic ways in more recent cognitive science. First, inspired by the structure of biological neural networks, connectionism developed forms of representation that are plausibly nonlinguistic in type. Second, attempts to explain cognition using the more abstract mathematical methods of dynamic systems theory have led to claims that—in many cases at least—the notion of representation can be dispensed with in explaining cognitive processes. As these two challenges are well known and widely discussed in the literature, I will not rehearse them here.[21] The important point for my present purpose is simply that in their wake 4e cognitive science in general has moved away from the idea that language—understood as a system of propositions—is the central, exclusive, or basic paradigm of representation.

Beyond this general shift it would no doubt be misleading to suggest that a single view of representation is distinctive of 4e cognitive science. Rather, as one would expect, widely varying views persist about the nature and role of representation, including—as just mentioned—eliminativist views that see representation as having no role in explaining cognition. To illustrate the potential for convergence with my position, I will focus in the following on a less radical response to the above challenges that is exemplified in the work, among others, of Andy Clark and Michael Wheeler. For these authors the success of dynamic systems theory in explaining examples based on relatively simple physical systems is not to be overgeneralized to the claim that representations have no explanatory role in cognitive phenomena. The conclusion they draw from the rejection of language as a central paradigm is simply that language has a more peripheral role as a specific kind of representation, one based on the use of (scaffolded on) established sign systems and forming part of a cognitive environment distributed over many speakers.[22] Against this background they see a role for various modes of representation, with linguistic representation and representation-free behaviors as two extremes allowing a range of intermediate possibilities.[23]

Both Clark (1997, 149ff.) and Wheeler (2005, 195ff.) particularly emphasize the importance of a strongly context-dependent genre of representation that they call "action-oriented representation." As the label suggests, the basic idea is that the information represented by cognitive systems will be attuned to the needs of specific tasks. Beyond this, however, such representation reflects the kinds of cognitive feat that situated agents might be expected to perform under the conditions that motivate 4e cognitive science—for example, when it is assumed that information is freely available in the environment, that agents can move around and interact with that environment in real time, and that features of the world function as scaffolds which reduce cognitive load by structuring and eliciting context-specific responses. Such conditions reduce the amount and modify the types of information that agents store, with the result that action-oriented representations will typically capture the agent's surroundings only partially, transiently, and from a particular spatial location and orientation.

The role of action-oriented representation is often illustrated using the example of mobile robots designed to perform specific tasks while avoiding obstacles in an unknown environment.[24] Two features—pioneered by Rodney Brooks—have been experimentally found to be particularly expedient to such tasks. The first is for the robot continuously to sample certain kinds of information while moving through its surroundings and to adjust the direction of its movement accordingly. This ongoing process of monitoring and response exploits the presence of its surroundings as an "online" resource, reducing the amount of information gathering and processing required.[25] The second feature is for the robot to be controlled by the interplay between distinct subsystems, known as "layers," each of which is capable of generating a certain kind of behavior, such as obstacle avoidance, random wandering, or fixing on a distant target. Each of these layers detects information from the surroundings and is independently capable of driving the robot's behavior, but the overall behavior of the robot is determined by the various layers inhibiting or overriding each other, as determined by the robot's current situation. In this respect the robot consists, in the words of Brooks (1991, 149), of "a collection of competing behaviors." The information processing in systems with this kind of architecture is distinguished by the fact that (a) each layer represents only information relevant to its respective activity (e.g., the obstacle-avoiding layer does not represent distant targets), and (b) there is no unified representation or procedure that determines the robot's motion. This contrasts with older representationalist approaches, which would typically try to build up a complete model of relevant features of the robot's environment, before computing and then carrying out a complete sequence of movements to attain the goal. The innovative feature of the newer, action-oriented approach is thus to get by without either a complete model of the environment or a complete plan of actions.

It may be wondered whether such robots tell us anything about cognition in general, rather than simply illustrating some neat problem-solving techniques used in engineering. In this respect it is worth noting that the choice of tasks such robots were designed to address—negotiating a real environment in real time—and some of the techniques deployed are modeled on relatively simple forms of biological intelligence.[26] Further, in their behavior these newer robots exhibit responsiveness, flexibility, and robustness of a kind usually associated only with intelligent biological systems, and which proved elusive for traditional artificial intelligence. Facts such as these encourage the thought that intelligent biological systems also use action-oriented representation in manipulating information about their environment.[27] However, independently of any extrapolation of results from experimental robotics, a similar view has been proposed by the so-called sensorimotor theory of perception.[28] According to this view, human perception and cognition are attuned to action and constant exploitation of the surroundings, so that action-oriented representation is plausibly central in the human case too. In this light, the action-oriented nature of representation illustrated by simple robotics systems is at least a live option for a more general view in contemporary cognitive science.

It might also be suspected that action-oriented representations are not proper representations, that their attunement to a practical and/or partial perspective entails that they fail to represent the surrounding world accurately. This would amount to a cognitive scientific version of the concerns discussed in chapter 8 in relation to Heidegger. However, here as there, the effect of a practical or partial perspective is not to introduce a potentially distortive intervening representation. Although they are strongly context dependent, and hence selective and spatiotemporally indexed, action-oriented representations are nonetheless determined by objective factors—including features of the environment, the agent's (robot's) location and motion, as well as the task being carried out. As Wheeler (2005, 197) nicely puts it, "how the world is *is itself encoded in terms of* possibilities for action."

Even this brief look at action-oriented representations allows 4e cognitive science to be seen as sharing the remaining three general features I identified as characterizing a phenomenological conception of language. The first of these was *antireductionism*, the idea that to understand what or how language is requires consideration of the processes and phenomenal relations it is involved in. Although the example just discussed does not relate immediately to language, it does make clear that 4e cognitive science shares a commitment to antireductionism insofar as action-oriented representation is strongly context dependent. Although agents will represent some information neurally (or electronically), they will do so only selectively, as a function not only of the environment but also of the context of action and the agent's physical and functional makeup. It follows that the represented information can be made sense of only nonreductively, in relation to the agent's embeddedness and embodiment. This antireductionism lends plausibility to the thought that 4e cognitive science, second, shares an *antiformalist* stance, according to which focusing simply on the formal or relational properties of a system of signs is an uninformative and hence inadequate way of thinking about language. Indeed the focus on symbol systems in early cognitive science seems to embody precisely the kind of formal system Heidegger intended to oppose. Hence the emphasis on embodiment, embeddedness, and enaction in 4e cognitive science, which is largely in response to the inadequacies of the older symbol-system approach, can be seen as corresponding to Heideggerian antiformalism. Third, and perhaps most obviously, 4e cognitive science shares the commitment to what I called a *process view*, as opposed to thinking of intentionality as a static or detemporalized system. For action-oriented representation is clearly calibrated to the in situ requirements of mobile agents performing temporally unfolding actions.

Beyond the general features just highlighted, a parallel can also be seen between 4e cognitive science and the more specific focus on prepredicative factors I identified as part of the Heideggerian framework. This can be appreciated by noticing how two features of 4e cognitive science correspond to Heidegger's discussion of the role of Statements and the transition these effect from purposive to predicative awareness. The first is the rejection by 4e cognitive science of the idea that language is the paradigm of representation, which implies that propositions (since they are linguistic in form) are no longer to be thought of as the basic vehicles of meaning. This corresponds to Heidegger's emphasis on the founding of Statements in purposive awareness, the aim of which was clearly to deny explanatory primacy to propositional content. The second is that a defining feature of action-oriented representations, as characterized by Clark and Wheeler, is that they are context-dependent modes of representation. Again a parallel is found in Heidegger's discussion of Statements. For the shift from purposive to predicative awareness is characterized there in terms of the attainment of context independence (the "cutting off" of instrumental relations), so that purposive awareness corresponds to action-oriented representation as basic context-dependent features.[29] Given these two parallels, it seems reasonable to suggest that 4e cognitive science shares an interest in information- or content-relevant factors that are—in Heideggerian terms—prepredicative.

By now it should be fairly straightforward to see why 4e cognitive science is particularly suited as an explanatory perspective to complement a minimalist phenomenological conception of language. In addition to a sense of accountability to experience, the two approaches share both a general outlook and an interest in prepredicative factors. More specifically, it is at least plausible to conjecture that action-oriented representations of the kind emphasized by both Clark and Wheeler might provide a cognitive scientific explanation of the prepredicative grasp of the world that I have claimed forms an important aspect of a phenomenological conception of language. Although (obviously) not directed to the specific view of language set out here, this suggestion draws some support from the fact that Wheeler has worked out a detailed account of what he describes as a "structural symbiosis" between embodied-embedded cognitive science and Heideggerian phenomenology, correlating phenomenological awareness of action-oriented representations with cases in which instrumental engagement with the world breaks down and our tools become "unready-to-hand."[30]

I want to finish this section, however, by clarifying that its aim has been more modest than the wide-ranging discussion above might seem to suggest. The intention has not been to claim that 4e cognitive science, as it stands, provides a ready-made explanation for the various features I have presented as forming part of a phenomenological conception of language. For example, although the idea of action-oriented representations appears plausible and explanatorily powerful—in other words, a good hypothesis meriting further investigation—it might turn out empirically to have no role in explaining human cognition. However, this would not impact on the *phenomenological* conception of language developed in this book—which is intended to stand or fall on the basis of its adequacy in describing speakers' experience of language. Rather, in that case the result would be a need to find an alternative explanation of phenomenologically identifiable features of language, another way of explaining the phenomenologically defined explanandum. The intention in this section has instead been simply to argue that 4e cognitive science seems to be the right kind of approach for providing such an explanation by showing that it has strong parallels with the phenomenological commitments that define the Heideggerian framework used in this book. It is this natural and deep complementarity that makes 4e cognitive science the best approach to explaining the experiential surface mapped by a minimalist phenomenology of language.

3 Embedding and Extending Phenomenology

It might be thought that recognizing 4e cognitive science as an explanatory approach is unwise, amounting to a Faustian pact in which phenomenology has aligned itself with the greater power of science only by sacrificing its soul. One possible concern is that if 4e cognitive science is recognized as an explanatory activity, then there is no longer any useful task for phenomenology. Another is that recognizing cognitive science in this way is contrary to the whole point of phenomenological inquiry. The aim of this section is to show that these concerns are out of place by first explaining how the supposed problems arise, before going on to argue that a need for phenomenology—specifically a phenomenology of language—is inherent in 4e cognitive science.

The two concerns just mentioned—that phenomenology becomes redundant, and that its specific claims have been neglected—stem from the basic conviction that phenomenology and science are philosophical methods that both differ and are in competition with one another. They differ because phenomenology is a descriptive study of the structure of how the world (or certain aspects of it) becomes manifest to agents, while science is an empirical study of a certain domain of entities and the causal processes governing them. The supposed competition emerges when both phenomenology and some branch of science address the same subject matter, raising the question of which approach is to be considered philosophically primary. Such competition is usually expressed in the question of whether phenomenology can be "naturalized"—that is, whether or not a scientific explanation can be given of the features (of intentionality or language, say) described by phenomenology.

Once this question is raised there seem to be two broad possibilities. The first is that there is no inconsistency between phenomenological and scientific conceptions of the relevant subject matter. This is the case in which it might seem that phenomenology is redundant or has no useful role to play. For in this kind of case, it might be claimed, empirical science gives us a deeper and objective understanding of the relevant phenomena, while phenomenology just reports superficially the standpoint of subjective experience. The second broad possibility is that the conceptions of the relevant subject matter provided by phenomenology and science are somehow in conflict. In this case philosophical *bon sens* might seem to require that the scientific conception is to be preferred. At the very least, it is likely to be thought incumbent on phenomenology to explain why science should answer to phenomenology rather than vice versa.

One of the traditional defining features of phenomenology, however, is precisely the claim that science *is* accountable to phenomenology. Indeed, opposition to attempts to naturalize intentional phenomena (qua "psychologism") was Husserl's foremost aim in developing phenomenology as a transcendental discipline.[31] Without transcendental phenomenology—as pointed out above—Husserl believed science would remain "naive," lacking proper understanding of both the objects it investigates empirically and its own epistemic values (cf. Husserl 1965, 18ff.). Hence the second concern mentioned above that recognizing the validity of cognitive science's explanatory claims appears contrary to the aims of phenomenology.

Although Husserl's task—persuading us that he has a method that transcendentally grounds science—is certainly forbidding, it might be thought that the minimalist phenomenological approach I am advocating is in a worse position still. After all, Husserl at least has a story—centering on the reduction and the extended field of "pure" phenomena—about why science answers to phenomenology. However, having rejected transcendentalist claims of the kind Husserl advocated, it might seem that no justification remains for insisting that phenomenology adds anything significant to scientific explanation. According to this line of thought, while accepting cognitive science as an explanatory approach is a consistent move, minimalist phenomenology does this because it has no choice—because it has simply missed the whole point of phenomenology.

The potential challenge to my position can be summed up as follows. Having claimed that 4e cognitive science is the right approach in seeking deeper explanations of the disclosive function of language, the question arises of why a (minimalist) phenomenological conception of language is still needed. If the latter is consistent with the findings of 4e cognitive science, a phenomenology of language seems superfluous; should they not agree, however, then minimalist phenomenology seems to have deprived itself of the means to claim philosophical importance in the face of cognitive science. In response to this challenge I want to argue in the following that the specific character of 4e cognitive science is such that in the case of language—and indeed cognitively relevant artifacts more generally—the two approaches, although different, should be thought of as complementary. More precisely, my claim will be that the requirement to articulate and to converge with a minimally phenomenological conception of language is inherent in 4e cognitive science itself.

To begin with, it is important to keep in mind some key differences between the case of language and many standard debates about the possibility of "naturalizing" phenomenology. Such debates often concern the relation between consciousness or introspective experience and physiological or neurological processes that are taken to underlie them. As long as these debates are thought of in terms of processes internal to an agent it remains plausible to reformulate the relation between phenomenological and scientific views in certain ways. For example, phenomenology can then be linked with the "first-person perspective" as opposed to the "third-person perspective" of science.[32] Similarly, thinking of perception in terms of internal representations can easily suggest that while phenomenology is concerned with appearances ("how things seem to us"), science is concerned with objective reality ("how things really are"). These two reformulations encourage the thought that phenomenology is eliminable, rather than offering a distinct take on the relevant features of reality. However, no matter what one thinks about such agent-internal cases, disanalogies are again to be expected in the case of linguistic phenomena, given that these take place in a public space. To begin with, the relation between a phenomenology of language and a candidate cognitive scientific explanation cannot be portrayed in terms of a distinction between first- and third-person perspectives. For language does not have a functionally relevant first-personal aspect, so that any phenomenological conception of language will focus on so-called third-personal factors. Similarly, as argued in chapter 8, section 2, it is difficult to see how appearances and reality could come apart in a principled way in the case of language, so that this distinction also cannot be used in thinking about the relation between the phenomenology and cognitive science of language.

Why, then, does 4e cognitive science need phenomenological description of our experience of language? The short answer to this is simply: due to the role it attributes to scaffolds—that is, features of the environment in which cognition is embedded and enacted. To see how the need arises, it will help to distinguish two aspects of this claim.

The first is what might be called the *ontological aspect*. A basic implication of saying that cognition is embedded and enacted is that cognitive processes make use of features of or entities in an agent's environment: parts of the material surroundings in public (non-first-personal) space. But such "scaffolds" are clearly not individuated naturally; rather, in order for any particular feature to be used as a scaffold, it must be picked out by an agent specifically for that use. This means, on the one hand, that individuation of the cognitively relevant ontology is already constitutively dependent on the experience of agents. On the other hand, because scaffolds must be visible in the experience of agents, phenomenological description of that experience has a role to play in identifying the constituent parts of embedded/enactive/extended cognitive systems.

The second is what might be called the *functional aspect.* An inherent peculiarity of scaffolds is that a given feature of the environment will be able to perform a particular cognitive function (i.e., serves as a scaffold) only if agents experience and understand it as having that function. At the very least, an agent must have some idea of what the scaffold is for, or which task(s) it is simplifying. To put it negatively, if we don't know what task a given scaffold is supposed to perform, we will be unable to integrate and exploit it in our actions and thinking, with the consequence that it would fail to perform its intended cognitive function. Note that this does not mean that an agent must grasp *all* the possibilities for cognitive scaffolding a particular piece of equipment affords.[33] A pocket calculator, for example, might be capable of computing many advanced mathematical functions of which I remain ignorant. But in making use of its simpler functions, say to calculate loan repayments, the intended easing of cognitive load can occur only if I understand which tasks the calculator is taking over (e.g., compounding interest rates). Similarly, looking back to Haugeland's example, the road system may be designed to simplify the cognitive demands of navigation, but for it to do this drivers must realize that road signs and so on serve this purpose, and integrate this realization into their overall pattern of behavior.

The two aspects just described can be summed up by saying that it is implicit in the use of environmental features as cognitive tools (scaffolds) that it will show up in an agent's experience *which features* are doing cognitive work and *which work* they are doing. The constitutive function of scaffolds thus has two sides in much the same way as Merleau-Ponty's conception of lived sense has both a phenomenological and a biological side (cf. chapter 4, sections 1–2). Accordingly, while their cognitive role and causal operation can be studied from the perspective of cognitive science, it is a basic functional requirement of cognitive scaffolds that they will have an irreducible experiential side that can become the object of phenomenological description. Moreover, such phenomenological description is not merely possible, but seems necessary in understanding the cognitive function of scaffolds. For in order to understand how features of the material or social environment are exploited in cognitively relevant ways, it is necessary to understand which environmental features are being exploited and how they are functioning as part of an agent's cognitive equipment. Yet this leads back to the perspective of agents' experience, and so to the need for phenomenological description of the experience of scaffolds. In this way the requirement for a minimalist phenomenology arises at the heart of 4e cognitive science, as a consequence of the role it attributes to cognitive scaffolds. The nature of embedded-enacted cognition is such that if the phenomenological side is neglected, explanation in 4e cognitive science will appear incomplete *by its own lights*.

The fact that the need for phenomenology is implicit in the 4e cognitive science approach has three notable consequences. The first is that it sheds further light on why phenomenology is able to play the role traditionally attributed to it of defining the explanandum for empirical science (see also the introduction, section 1). This is basically because a phenomenology of scaffold use will be describing the same external objects and the same cognitive process that 4e cognitive science aims to explain within an empirical scientific framework—convergence and mutual constraint are to be expected because they target the same phenomena. However, there is no need for phenomenology (nor indeed cognitive science) to claim some overall philosophical "primacy," because 4e cognitive science is antecedently committed to recognition of a phenomenological description of its explananda. A second consequence is that because there is no attempt to eliminate phenomenological description or to "reduce" it to a scientific theory, phenomenology can still be seen as resisting dissolution into a scientific perspective—as resisting "naturalization," so to speak.[34] It is also worth noting, finally, that the above considerations show why it makes no sense to think that we might be generally mistaken about how language really is. This is not simply implausible as a matter of how it "subjectively" appears; rather it is functionally excluded in relation to cognitive scaffolds such as language. For if we were not aware of language as it really is—which spoken or written aspects of the world are language, and what cognitive function they are performing in the midst of the world—then language would be unable to perform the function(s) it does and so would not exist qua language.[35]

I want to underline that the above argument is limited in scope: it is specific to the claims made by 4e cognitive science about the function of scaffolds, and not a generic defense of phenomenology against "naturalization" in all cases. There are nonetheless many cases beyond that of language in which it will apply, because the need for (minimalist) phenomenology in 4e cognitive science arises wherever scaffolds are in play.[36] The need for a phenomenology of the many scaffolds exploited in extended and enacted cognition is, so to speak, embedded in 4e cognitive science itself. Nevertheless, a significant disanalogy remains to more standard debates about the possibility of naturalizing what might be called nonextended phenomenology, where the concern is with agent-internal—"first-personal," "private," or "in-the-head"—experiences. The difference is simply that in the case of nonextended phenomenology (unlike that of scaffolds) we do not need to identify, and understand the function of, the subpersonal systems that underlie conscious experience for them to perform their cognitive function. This fact should not be taken as showing that external scaffolds lack a cognitive function. It does, however, mean that the claim that the need for phenomenology is implicit in scaffold use cannot be carried over to the case of nonextended phenomenology.

The upshot of the considerations of this final section is that the attempt to move to a deeper explanatory perspective does not eliminate the need for a minimalist phenomenology of language. Instead, although phenomenology and 4e cognitive science remain different approaches, they emerge as complementary rather than in competition. One aspect of this complementarity is that the two approaches each seem to anticipate and require the other. While minimalist phenomenology appears to call out for a deeper explanatory perspective, the role of scaffolds in 4e cognitive science refers us back to the descriptive task of minimalist phenomenology. A second aspect of their complementarity is that the two different approaches should (in the long term) provide convergent conceptions of their shared subject matter. Thus, while the conceptions of language provided by a minimalist phenomenology and 4e cognitive science might still differ in focus and content, they should not only not contradict, but also recognizably accord with one another—as outlined for the case of language in the preceding section. The result, to end on an optimistic note, should be a relation characterized not by principled conflict and myopic claims to philosophical supremacy, but by a convergence of efforts and mutual illumination.

Notes

Introduction: A Phenomenological Approach to Language

1. These terms all occur in Heidegger's description of the *Ge-stell* in the essay "The Question Concerning Technology": cf. Heidegger 1954, 19, 24 (*herstellen*); 22, 25 (natural science "stellt der Natur ... nach"); 11, 18, 26 (*feststellen*); 27 ("sicherzustellende Bestände"). In "Science and Contemplation" Heidegger tersely characterizes calculative modern science as a "nachstellend-sicherstellendes Vorgehen" and explains that "Das nachstellende Vorstellen, das alles Wirkliche in seiner verfolgbaren Gegenständigkeit sicherstellt, ist der Grundzug des Vorstellens, wodurch die neuzeitliche Wissenschaft dem Wirklichen entspricht" (Heidegger 1954, 53, 51–52).

2. Heidegger 1954, 24–25. For this view of *Vorstellen* cf. Heidegger 1977, 91ff. It is likely Heidegger also had in mind a connection with positing (of entities), as "stellen" is a standard German translation of the Latin *ponere*.

3. Curiously, Quine (1969a, 80) distinguishes his interests from what he disdainfully refers to as "ordinary unphilosophical translations."

4. For example, despite advocating phenomenology as "a mode of doing philosophy," Dermot Moran (2000, xv, xiv) somewhat awkwardly concedes that "phenomenology cannot be understood simply as a method, a project, a set of tasks; in its historical form it is primarily a set of people."

5. For instance, by Zahavi (2005, 4–5), who also underlines that these uses are predicated on a problematic "divide between inside and outside" (ibid., 223n2).

6. This is the point of Husserl's distinction between the "flow of experience" (*Erlebnisstrom*) and reflective acts in §78 of his *Ideas I*. Though he emphasizes that it is "through reflectively experiencing acts alone that we know anything about the flow of experience" (Husserl 1992c, 168), the knowledge attained is knowledge of the original flow of prereflective experience.

7. Cf. Heidegger's characterization of phenomenological "description" as avoiding determinations that are not directly exhibited (*SZ* 35). Husserl's talk of "evidence" and "intuition" similarly relies on the idea that phenomenological data are noninferentially accessible (e.g., Husserl 1992d, 58 [§24]).

8. Brentano, in many ways a precursor of Husserl's phenomenology, saw the task of "descriptive psychology" as grounding empirical psychology by providing "an analysing description of our phenomena, i.e. of our immediate experiential facts" (Brentano 1995, 139). See also *SZ* 8–11 (§3) and Merleau-Ponty's critique of "objective thinking" in the extensive "Introduction" to *PdP* (9–77).

9. Cf. Husserl 1992d, 70ff. (§34) and 60–61 (§25); Husserl 1992c, 16 (§4).

10. On the reduction see, for example, Husserl 1992c, 61–66 (§§31–32), and 1992d, 21–23 (§8). The term *neutralization* is used in Husserl 1992c, 248 (§109).

11. In his 1927 lectures Heidegger briefly suggests a reinterpretation of the reduction geared to the question of being (Heidegger 1975, 29), but he makes neither use nor mention of the reduction in *SZ*. Merleau-Ponty makes occasional references to the reduction and avoids rejecting it, but he interprets it in a way alien to Husserl, as indeed his own view of humans and intentionality requires (e.g., *PdP* v–ix, 58, 60). I discuss Merleau-Ponty's treatment of the reduction in more detail in Inkpin (forthcoming).

12. Cf. *SZ* 229. The above formulation of Heidegger's challenge is found in correspondence with Husserl during their 1927 collaboration on the *Encyclopaedia Britannica* article (Husserl 1968, 601–602).

13. *PdP* 51, cf. 172: "Experience of the body forces us to recognize an imposition of sense which is not that of a universal constituting consciousness."

14. Husserl 1992d, 74 (§34). Similarly in *Ideas I* he claims that the "old ontological doctrine that knowledge of 'possibilities' must precede that of actual realities [*Wirklichkeiten*], is in my view, as long as it is correctly understood and made use of in the right way, a prodigious truth" (Husserl 1992c, 178 [§79]).

15. *SZ* 143f. Pöggeler (1989) argues powerfully that disagreement over the nature of modality was central to Heidegger's break with Husserl from 1927 onward.

16. *PdP* 74. The world's "ontological contingency," says Merleau-Ponty, is "radical": "The world is the real, of which the necessary and the possible are merely provinces" (*PdP* 456; cf. similarly Merleau-Ponty 1964, 148).

1 The "Place" of Language

1. In translating Heidegger there is an acute tension between—in Ashton's (1971, 22) apt characterization—two forms of literalism, the first a standard "word-for-

word" approach, the second a "radical one of digging up the roots of words, no matter what their dictionary sense and current usage." Sympathetic readers of Heidegger are often drawn to the latter.

2. To provide just one example, Heidegger's analysis of Dasein deploys a number of terms (e.g., *Erschlossenheit, Sinn, Bedeutung, Bedeutsamkeit*) that might be translated as "meaning." However, because each has a specific role in Heidegger's analysis, to equate any one of them with meaning—quite apart from that term's own rich ambiguity—would be precipitous and potentially misleading.

3. According to Heidegger, this way of conceiving entities is a traditional legacy of ancient ontology informed by the idea of (artisanal) production—as such production is guided by an antecedent notion (the "essence") of what is to be made (cf. Heidegger 1975, 108–172).

4. *SZ* 68. As Heidegger puts it, "Zeug ist wesenhaft 'etwas, um zu ...'" I am rendering this *um zu,* as well as Heidegger's talk of an entity's *Wozu* (e.g., *SZ* 78), as what an entity, qua Equipment, is "for" because this more idiomatically renders Heidegger's—equally idiomatic—terms than more awkward renditions such as the "in order to," the "to which," or the "toward which."

5. This term is usually translated as "references." I prefer "instrumental relations" here to avoid confusion with the more common semantic notion of reference.

6. Heidegger's frequent talk of how things appear to us "zumeist und zunächst" might appear to support this thought. But note that this locution usually indicates how things strike us, so to speak *inadequately,* as inauthentic Dasein.

7. Cf. Ryle's (1949, 25ff.) distinction between dispositional "knowing how" and propositional "knowing that." Incidentally, given Ryle's early awareness and eventual adoption of many of Heidegger's views (cf. Murray 1973), it seems likely that Ryle's distinction was influenced by Heidegger's view of purposive and predicative awareness (see also chapter 7, section 2 below).

8. Cf. Kant's definition of the world as the "sum total of all appearances" (Kant 1983a, 336, 408 [B 391, 446]).

9. Cf. Heidegger 1979, 352. In German, *Befindlichkeit* is to capture "daß es sich ... in seiner Geworfenheit befindet" ("that it finds itself in its thrownness"; *SZ* 135) and "wie es sich (dabei) befindet" (roughly: how Dasein finds it to be the way it is).

10. Cf. Kant 1983a, 202 (B 199). Heidegger distances himself explicitly from the debate over the distinction of understanding and explanation at *SZ* 143. For a survey of this debate see Schnädelbach 1983, 138–171.

11. *SZ* 151, 148, 149. "Something as something" is the only idiomatic rendering of "Etwas als Etwas" in English, which is slightly unfortunate given Heidegger's

emphatic distinction between Equipment and Things. The general point he is making concerns the individuation of what is understood (whether as Equipment or Things).

12. *SZ* 149. The term usually translated as "explicit" (*ausdrücklich*) suggests the use of an expression (*Ausdruck*). That translation, which is constrained by the use of "expression" for (*sich*) *aussprechen/(Hinaus)gesprochenheit* (e.g., *SZ* 28, 161, 220), somewhat obscures Heidegger's distinction between being "spoken out" or verbally performed and explicitness or having an expression for (cf. on this Heidegger 1962, 190n).

13. *SZ* 149. "Circumspection," "Umsicht," is Heidegger's term for the kind of understanding that guides practical action (cf. *SZ* 69). This presumably alludes to the Aristotelian practical reason, or *φροόυησις*, which he also translates as "Umsicht" (Heidegger 1989a, 255; 1992, 21).

14. Heidegger (1979, 348) talks straightforwardly of a "foundational nexus" (*Fundierungszusammenhang*). This view is echoed in his characterization of "reality" on *SZ* 201.

15. The main reason for this being that historically "Discourse" tends to suggest a link with predicative awareness. According to Kant (1983b, 205), for example, "the specific nature of our understanding [or intellect, *Verstand*] consists in thinking of everything discursively—that is, through concepts, consequently also through nothing but predicates."

16. The link with Understanding (*Verstehen*) is also reflected in Heidegger's talk of *Verständlichkeit* ("intelligibility," "understandability") as what is "articulated in Understanding disclosing" (*SZ* 151, cf. 161).

17. Lafont interprets *SZ* similarly as exhibiting an inchoate recognition of the constitutive function of language that was to mature and to overcome the instrumental view fully only in Heidegger's later writings (see Lafont 2001, 11). On the link between language and "inauthentic" Dasein see *SZ* §35 and chapter 3 below.

18. Carman 2003, 221, 228; cf. 220–232. Carman links Guignon 1983 with the linguistic model. The pragmatic model is found in Dreyfus 1991, 215ff.; Blattner 1999, 67ff.; and Wrathall 2011, 119–137. It is also hinted at in Haugeland 1989, 64–65.

19. This thought is often relied on to deny that "knowing how" is a kind of knowledge distinct from "knowing that"—since if whatever it is one knows in knowing how to do something can be stated in language the distinction appears to be undermined (e.g., Stanley and Williamson 2001, 425ff.).

20. For an illuminating discussion of tigers' purposive behavior see chapter 5 of Okrent 1988, especially 137, 147, 148.

21. As he puts it a little later: "λόγος as Articulacy [or talk, *Rede*] means what we understand by 'Language' [*Sprache*], but it simultaneously means more than the whole of the vocabulary, namely the basic capacity to be able to talk and hence speak [*das Grundvermögen, reden und demzufolge sprechen zu können*]" (Heidegger 1983, 442).

22. Note also that while Dreyfus's (1991, 215) suggestion that *Rede* is like "telling" the time or "telling" the difference between ... conveys well his own intended meaning, the German term does not have this connotation.

23. Seeing the linguistic model as too narrow and—given the need to distinguish expressive from other features of action—the pragmatic model as too broad, Carman (2003, 234) interprets Discourse (Articulacy) as "the normativity specific to interpretation," whether linguistic or nonlinguistic. Whatever the independent virtues of this approach to expressive phenomena, it is difficult to square with both the clearly linguistic characterizations above and the fact that in presenting it as "equiprimordial" with Understanding, Heidegger sees Articulacy as prior to interpretation (i.e., Setting-out).

24. Wrathall (2011, 127–128). Elsewhere he describes *Rede* in *SZ* as having a "nonlinguistic" character, and contrasts "linguistic understanding" with "practical conversance" (Wrathall 2011, 124; cf. 130–131, 112).

25. To provide just a couple of illustrations: Heidegger paraphrases *Aussage* as *Urteil* at *SZ* 153 and 214 (cf. also Heidegger 1978b, 153, and 1983, 487). As he explains in his lectures on *Logic*, "according to traditional opinion the proposition [*Satz*] or the judgement [*Urteil*] is the proper locus of truth" (*Logik* 127). Note that *Aussagenlogik* is the standard term for propositional logic in German, while *Satz* can be either "proposition" or "sentence" (a well-known source of translation difficulties).

26. "Spoken-out-ness" (*Ausgesprochenheit*) is the same—specific and distinctive—locution as Heidegger uses in defining language (*SZ* 161; cf. section 2 in this chapter).

27. *SZ* 158. Cf. 149, where *as* structure is characterized as the expressness or explicitness of what is understood ("Ausdrücklichkeit des Verstandenen").

28. *Verweisungsbezüge der Bedeutsamkeit*; Heidegger 1979, 361, 370. These views recur in §17 of *SZ*, where signs are described as a kind of Equipment that make us circumspectively aware of a context of action (*SZ* 79–80).

29. Heidegger advocated a far broader conception of "logic," which he thought—as the "science of Articulacy" or "phenomenology of Articulacy, i.e. of the λόγος"—should have a more general concern with allowing things to be seen (*Logik* 6; Heidegger 1979, 364).

30. *Sic*: Heidegger's use of scare quotes here ("sentences") is due to the ambiguity of the German *Satz* between "sentence" and "proposition" qua logical judgment. The point of the scare quotes is precisely to express the claim that *sentences* of the kind listed are not propositions (i.e., predicative judgments).

31. Though Heidegger does elsewhere make the Aristotelian point that not all sentences concern truth (*SZ* 32; cf. Aristotle 1938, 121 [17a5]).

32. In this respect Heidegger's view parallels Dewey's pragmatist and relational conception of objective knowledge for which "*The* object is an abstraction" that "designates selected relations of things which ... are constant within the limits practically important" (Dewey 1988, 190, cf. also 175–176).

33. "kann einer besorgenden Überlegung Ausdruck geben" (*SZ* 360–361, cf. 157).

34. This is part of the explanation for Heidegger's characterization of *Gerede* ("idle talk") as a superficial mode of language use (see chapter 3, section 3).

35. *SZ* 79. Apophantic use of language, as Heidegger later puts it, has the point "solely" or "only" of pointing out something about its referent (Heidegger 1983, 448, 453). It seems to me that Carman (2003, 234) conflates the above distinction in equating "interpretation" (*Auslegung,* Setting-out) with "demonstrative practice," meaning that he cannot distinguish activity in which articulate understanding is manifested from activities with the specific point of drawing attention to articulate understanding.

36. Although not in the above sense, Heidegger sometimes talks of *Rede* as having different "ways of being" or "modes" (*SZ* 167, 269).

37. The suggestion is thus that the "content" of Heidegger's *Rede* can be nonpredicative, rather than bound to the traditional discourse-concepts-predicates triad (cf. note 15).

38. See also Heidegger's insistence that Articulacy is not to be equated with (predicative) judgment (*SZ* 32), and his discussion of language as belonging to the hermeneutic *as* (*Logik* 151–152).

39. They are what Heidegger calls "deficient modes" of the corresponding existential structures, an idea he often makes use of (cf. *SZ* 20, 57, 73, 104, 120, etc.).

40. As Humboldt (1995, 48–49) points out, linguistic representation, in his view an exteriorization of the mental, objectifies and enables us to gain reflective distance toward our thoughts.

41. Mark Okrent (1988, 147–149) further proposes a verificationist argument to the effect that without the use of language there would be no evidential basis to distinguish predicative awareness from purposive awareness. Although this is in itself plausible and would support the above claims, it seems to me rather generous to attribute this argument to Heidegger.

42. Heidegger's later thinking might be seen as emphasizing the "constitutive" function of language. However, while his accounts of poetic experience of language (Heidegger 1959, 157–216), for example, are extremely subtle from a literary point of view and exemplify the proximity between poetry and thinking that Heidegger describes, it seems to me that they do little (albeit perhaps intentionally) to *elucidate* the supposed constitutive function of language.

2 Phenomenological Commitments

1. Cf., e.g., *SZ* 53, 161, 206, 65. The aspects of unity and completeness are combined in Heidegger's characterization of "originality" at *SZ* 231.

2. This parallels Austin's (1979, 182) "linguistic phenomenology": "When we examine what we should say when, what words we should use in what situations, we are looking not *merely* at words ... but also at the realities we use the words to talk about: we are using words to sharpen our perception of, though not as the final arbiter of, the phenomena."

3. The distinction between the formal and fulfillment results from Heidegger's adaptation of Husserl's theory of meaning (see chapter 3, section 2).

4. Emphasis on the processual character of language is the point of Humboldt's often-cited distinction between *ergon* and *energeia*: "Language, grasped in its real essence, is something persistent and at every moment transient. ... [Language] itself is not a work (*ergon*), but an activity (*energeia*). Its true definition can therefore be only a genetic one" (Humboldt 1995, 36).

5. Cf. respectively Davies 1981, 135; George 1990, 291; Chomsky 1986, 25–26. George makes the obscure, yet supposedly "uncontroversial," claim that "when I die, my own language, being an abstract entity, will not cease to exist and will almost certainly never be possessed by anyone again."

6. Heidegger consistently opposed algorithmic approaches to modeling content. Cf., e.g., *PIA* 32; Heidegger 1959, 263–264; and Heidegger 1989b, 24–25. Incidentally, Heidegger was to some extent aware of Frege's work, suggesting in a 1912 review of research in logic that Frege's importance for a "general theory of concepts" was still to be realized. Yet later in the same text he more characteristically claims that "logistics," or formal logic, is simply mathematics and is not "capable of pressing forward to the actual logical problems" (Heidegger 1978a, 20, 42; cf. Heidegger 1959, 116).

7. I discuss some ways this might be done in chapter 9, section 2.

8. Aristotle 1938, 115 (16a13); *Logik* 140. The discussion of *Logik* 127–161 is particularly relevant to the treatment of predicative statements in §33 of *SZ*. The same ideas are also discussed in Heidegger 1983, 456–460.

9. *Logik* 141; *SZ* 159. Aristotle presumably thought this occurred in the "soul" (Aristotle 1938, 115 [16a4]), but this can be ignored for the present purpose.

10. In a phenomenon "which in itself is combining and dividing and lies prior to relationships between linguistic expressions and their attribution and denial" (*Logik* 141).

11. Husserl 1985, 1–230. As will be seen, Merleau-Ponty also took up this interest in the prepredicative.

12. *Logik* 147. This comment parallels Heidegger's thinly veiled criticism of Husserl's reliance on the idea of pure intuition at *SZ* 147.

13. This should apply even if one thinks it possible to have an "implicit" grasp of properties. If such a grasp cannot be seamlessly made "explicit" in the sense of being involved in one's reasoning, it is difficult to see what an "implicit" grasp amounts to.

14. Cf. Aristotle 1934, 331ff. (1139b14), and particularly Heidegger's 1924–1925 lectures (Heidegger 1992, 14–64).

15. See, for example, Tugendhat 1970 and Dahlstrom 2001.

16. This is not intended as an exegetic claim. Nevertheless, although obviously diverging from Heidegger's position in *SZ*, it is presumably consistent with his late concession that it had been incorrect and misleading to call "*ἀλήθεια* in the sense of the Clearing 'truth'" (Heidegger 2000, 77).

17. This represents a significant difference from Husserl, who rather curiously often discerned directly corresponding features of predicative judgments in "prepredicative" perception (cf. Husserl 1985, 97, 104, 111).

18. Heidegger makes a corresponding move by describing Equipment in terms of "expediency" (*Dienlichkeit*) or "suitability" (*Geeignetheit*; *SZ* 83). However, as already noted, the distinctness of this prepredicative functionality is less easy to appreciate where expediency is treated as part of the "original phenomenon of truth."

19. It might be objected that predicative awareness of objects is possible without the overt use of signs. If this is possible, it is clearly of no relevance to the relationship between signs and what they stand for.

3 The Disclosive Function of Linguistic Signs

1. That is, "das schlicht hinsehende Vernehmen der einfachsten Seinsbestimmungen des Seienden als solchen" (*SZ* 33).

2. It is "aufweisendes Sehenlassen" or "Vernehmenlassen": "die Grundbedeutung von λόγος ist Rede," which is "απόφανσις" or "aufweisendes Sehenlassen"; the

"Funktion des λόγος" lies "im schlichten Sehenlassen von etwas ..., im *Vernehmenlassen* des Seienden" (*SZ* 32, 34).

3. See for example Husserl 1992d, 13 (§5). One significant difference is that Heidegger thinks Husserl's notion of "pure" intuition or intuition of essences is founded in the purposive seeing of circumspection (cf. *SZ* 147, 149).

4. As Austin (1962, 14–19) and Stroll (1988, 152–159, 161) have emphasized, without some such contrast with "indirectness" it would make no sense to speak of "directness."

5. How this mediation that leaves no trace can be understood is discussed in chapter 8, section 2.

6. "Das Man-selbst ... artikuliert den Verweisungszusammenhang der Bedeutsamkeit" (*SZ* 129).

7. *SZ* 169, 155. Heidegger elsewhere tersely describes this as "Wortdenken, Hörensagen, Angelesenen" (Heidegger 2002, 274)—that is, merely verbal thinking, hearsay, things we have read.

8. Cf. Putnam 1975, especially 247ff. and 228.

9. *SZ* 168. Incidentally, this provides a good example of Heidegger's exploitation of morphology. His terminology is supposed to make manifest that *Gerede* (routine language use) is a form of *Rede* (Articulacy), and also that *Gerede* focuses on *Geredetes* (what is said). The latter similarity might presumably be invoked in support of Heidegger's—nonetheless hardly plausible—claim that the term *Gerede* is not intended pejoratively.

10. *PAA* 168, 8; cf. 171. Heidegger even described the problem of concept formation as "the philosophical problem in its origin" (*PAA* 169).

11. The main sources of which are *PIA*, *PAA*, Heidegger 1995 and 1996a. He seems to have commenced such an account once, but abandoned it following complaints by students questioning its relevance in lectures ostensibly on the phenomenology of religion (cf. Heidegger 1995, 55–65; Kisiel 1993, 149–150). Awareness in the literature of the importance of formal indication has grown following the publication of Heidegger's early (1919–1923) lectures. Dahlstrom 1994 and Imdahl 1994 offer particularly helpful surveys of the main sources and motifs; much of the relevant source material is summarized in Kisiel 1993.

12. *Sinnbelebter Wortlaut*, Husserl 1992b, 44. Unlike Frege, Husserl famously considers the terms *Sinn* and *Bedeutung* synonymous (cf. ibid., 58).

13. Husserl 1992b, 44, 47. Here there is obvious proximity to Grice's (1989, 215ff.) distinction between natural and intentions-based nonnatural meaning,

although Grice provides greater detail about the kind of intentions supposed to confer meaning.

14. Husserl 1992b, 39, 40. Derrida (1993, cf. 46) has challenged Husserl's distinction by claiming that signs cannot function as expressions without being indications, further suggesting that the expressive function depends on a problematic notion of transcendental subjectivity's full presence to itself. Whether or not these criticisms of Husserl are sound, Heidegger's conception of signs avoids these objections, as will become clear, by being based on the indicative function and a different view of meaning conferral (or expressiveness).

15. For this contrast, see Heidegger 1995, 3–18; 1993a, 141–144, 235–237. On Heidegger's view of ontic-positive sciences, cf. *SZ* 9–10, 51–52.

16. *PIA* 1–2. They have the character of *Vollzug*, from *vollziehen* (to carry out, complete, or perform). As will become clearer in the sequel, Dahlstrom (1994, 782n) is right to suggest that in "'Vollziehen' there is a sense of executing, carrying out, and performing but also a sense of accomplishing, perfecting, and fulfilling."

17. *PIA* 23, *PAA* 179, 180; cf. also 186; Heidegger 1987, 24. Such talk of originality once again echoes Husserl (e.g., 1992a, 246).

18. He sees the "idea of definition" as "nothing but the formal interpretation of the full sense of knowledge" (*PIA* 54).

19. *PIA* 32; Heidegger 1979, 112. This idea is retained in *SZ*, where "indication" is described as the basic structure of the appearance of "all indications, representations, symptoms, and symbols" (*SZ* 29).

20. *PIA* 20 ("'*formal*' anzeigend" means "der 'Weg,' im 'Ansatz.' Es ist eine gehaltlich unbestimmte, vollzugshaft bestimmte Bindung vorgegeben"), 34.

21. Dilthey 1990, 235ff. (cf. 157, 176). Heidegger refers to this triad—*Erlebnis, Verstehen, Ausdruck*—explicitly at *PAA* 169. Many of Heidegger's key notions (e.g., existentiales, significance, temporality, historicity) are prefigured in Dilthey. For a good survey of such themes see Guignon 1983, 45–59.

22. Dilthey 1990, 263–267. This idea was later relied on by Collingwood's (1958, 150; 1994, e.g., 115, 172, 203) "expression theory" in aesthetics and the philosophy of history.

23. Thus grounding experience, as performed understanding, plays the same role as Dilthey's notion of *Erlebnis* (experience). See Gadamer's (1990, 70ff.) insightful discussion of this role as attempting to model methodology in the humanities on that of empirical natural science.

24. Heidegger repeatedly opposes the "prejudice" that phenomenology's emphasis on direct insight means grasping things "in a flash" without interpretative effort (Heidegger 1979, 36–37; cf. *SZ* 36–37).

25. *PIA* 72, 141. Heidegger often talks of "deformalization" (e.g., Heidegger 1996a, 24; *SZ* 35).

26. *PIA* 144. In this spirit, Heidegger also suggests that grammatical relationships are grounded in phenomenal relations (*PIA* 82–83).

27. Indeed Heidegger experimented with the label "expressive concepts" (Heidegger 1993a, 240) to reflect this Diltheyan interdependence of experience, understanding, and expression. Even in *SZ* "*ausdrücklich*" might be construed in terms of such expressiveness (cf. note 12 in chapter 1).

28. This process, which Heidegger once aptly termed "diahermeneutics" (Heidegger 1993a, 262), features in *SZ* (153, 314–315) as the "hermeneutic circle."

29. Not surprisingly therefore §68d of *SZ*—on the "temporality of Articulacy [*Rede*]"—is conspicuously brief. Though Heidegger reassures us that "Articulacy is in itself temporal" and "grounded in the ecstatic unity of temporality," he has nothing specific to say about the temporality of language (either *Rede* or *Sprache*) as such. For Heidegger, as he himself glosses it, the "temporality of Articulacy" is just the temporality "of Dasein altogether" (*SZ* 349).

30. Heidegger 1995, 63. This tripartite view of sense is also found, for example, at *PIA* 53–54; *PAA* 60; Heidegger 1996a, 22; Heidegger 1993a, 261.

31. For example, from 1921 to 1922, *PIA* 33–34. Indeed even in the above 1920–1921 discussion Heidegger seems unable to uphold his own distinction: "because formal determination is completely indifferent in content [*inhaltlich indifferent*] ... it prescribes a theoretical relational sense ..., covers up the performative aspect [*das Vollzugsmäßige*] ..., and directs itself one-sidedly toward the content [*Gehalt*]" (Heidegger 1995, 63).

32. Plato 1926, 142–149 (425d–427d).

33. Heidegger talks literally of "formal indication" (*formale Anzeige*) or cognates at *SZ* 114, 116, 117, 231, 313, 315; of temporary indication at *SZ* 14, 41; of formal sense at *SZ* 34, 43; and of indication of a formal concept at *SZ* 53. These references might seem sparser still in English translation because Macquarrie and Robinson miss, and so do not render, the terminological significance of *indication* in their 1962 translation of *Being and Time* (Heidegger 1962); with the exception of p. 315, this oversight is remedied in Stambaugh's revised translation (Heidegger 1996b).

34. It further requires that "the *whole* of the thematic entity" is treated and that the "*unity* of [its] structural aspects" is found (*SZ* 231–232).

35. The clearest examples of this use are found at *PIA* 33–35, but cf. *PIA* 41, 60, 62–63, 73. For example: "'Formally indicated' means ... indicated in such a way that what's said is of the character of the 'formal,' improper [*uneigentlich*]"; Formal

indication "provides the way to try out and to fulfill what's improperly [*uneigentlich*] indicated," so that it comes to be proper (*zum Eigentlichen kommen*) (*PIA* 33). The opposite of improper indication is "proper possession [*das eigentliche Haben*]," "the specific being of what's respectively performed" (*PIA* 31).

36. Cf. *SZ* 28. Kisiel (1993, 46) talks of a "properizing event" and accurately captures this reciprocity: "It is my proper experience because it appropriates me and I, in accord, appropriate it. I am It, I am of It, It is mine."

37. *SZ* 32. Cf. 315 and 35, where it is emphasized that in phenomenology, the "character of description," "the specific sense of the λόγος," should be fixed by the nature of the subject matter (*Sachheit*) to be described.

38. Although the terminological proximity here to Langer's (1951, 79–102) interesting contrast between discursive and presentational forms of symbol is no coincidence, I do not intend to follow the (often problematic) details of her account. In particular, I am suggesting that language has presentational aspects, so that the discursive-presentational distinction cannot coincide with that between linguistic and nonlinguistic forms in the way Langer often seems to imply.

39. An analogy might be an infrared photo of a familiar face, which presents its features in an unfamiliar, reduced way so that recognition perhaps requires "interpretative" effort.

40. Thus prior to *SZ* Heidegger appears to have overlooked Husserl's warning that for theories that define meaning solely in relation to intuitive fulfillment the phenomenon of "symbolic thinking" remains an "insoluble puzzle" (Husserl 1992b, 72).

41. *SZ* 82. As Heidegger claims (*SZ* 77) to be discussing signs so as to clarify the character of instrumental relations (*Verweisungen*), it might be objected that taking this as bearing on language is an interpretative assumption. However, unless language—the "spoken-out-ness of Articulacy"—is thought not to involve the use of signs, the assumption is surely justified.

42. Even the apparent elucidation "Das Gesagte wird [im Gerede] zunächst immer verstanden als 'sagendes,' das ist entdeckendes" is part of explaining how routine language use is in fact a covering up—that is, is "initially" wrongly taken to be discovering or "saying" (*SZ* 169).

43. For a suggestion of this kind see *SZ* 5.

44. *SZ* 79. Given the expression "vorzügliche Verwendung," and the nature of the distinction he suggests, it is quite possible that Heidegger is here exploiting the relationship of *vorziehen*—the literal sense of which is to bring or draw forth—and *vorzüglich* (distinctive, excellent): the distinctive use of signs being to bring to the fore instrumental relations.

45. This multifactorial constitution also provides a basic motivation for Heidegger's antireductionism because it implies the function of linguistic signs cannot be reduced to either formal or pragmatic factors alone.

46. This instrumentalist-constitutive contraposition is assumed, for example, by Taylor (1985, 1995), Guignon (1983), and Lafont (1999, 2001). In her illuminating discussion of the role of language in the German philosophical tradition Lafont (1999, 13–18) traces this distinction back to Humboldt.

4 Language as the Expression of Lived Sense

1. My exegetic strategy differs from the common approaches of either periodizing Merleau-Ponty's views on language (e.g., Fontaine—de Visscher 1974, 17–18; Silverman 1981) or approaching them from the perspective of his late works (e.g., Dillon 1997; Madison 1981). Briefly this is because (a) Saussure's views introduce the most important systematic development to Merleau-Ponty's otherwise continuous view of how language functions; (b) despite attempts to reconceive the ontological picture, the text and working notes of *The Visible and the Invisible* provide no indication of a significant modification to this view.

2. "The Frenchman says 'l'appétit vient en mangeant' and this empirical principle remains true when one parodies it and says 'l'idée vient en parlant'" (Kleist 1990, 535).

3. *PdP* 206. In one passage Merleau-Ponty goes as far as to suggest that "my [own] acts of speech surprise me" (*S* 111).

4. Though the limitations of such terminology were astutely criticized by Jean Beaufret as early as 1946 (cf. Merleau-Ponty 1996a, 102–13), Merleau-Ponty himself seems to have realized the need to develop an ontological picture appropriate to his views only much later (cf. note 21).

5. Waldenfels (1976, 1981) brings out this continuity well with regard to the notion of structure.

6. The notion of behavior was taken as central to *SdC* precisely due to its neutrality between classical oppositions such as that between subject and object (*SdC* 2–3; cf. 137).

7. Cf. in particular *PdP* 120ff., 181ff., 228–229. What Schneider lacked was the ability to "situate himself in the virtual," "the power to project before him a sexual world, to put himself in an erotic situation," or quite generally to "polarize" the world in terms of significance (*PdP* 126, 182, 130).

8. Indeed, the reliability of even that case might be questioned because some of Schneider's symptoms are alleged to have been faked (see Gardner 1977, 142–151).

9. Taylor (1989, 4–8) emphasizes and illuminates the importance of such bodily orientation. Baldwin (2004, 12) correctly notes that bodily disorders such as Schneider's are "not intelligible without reference to the body's contribution to experience." However, the underlying point of Merleau-Ponty's discussion is that "normal" experience too is intelligible only with this contribution.

10. *PdP* 117, cf. 271. In the sense that it mediates the link between intellectual concepts and the sensory domain the "schéma corporel" might be viewed as analogous to Kant's schematism (Kant 1983a, 187ff. [B176]). Merleau-Ponty's notion of "orientation" is also reminiscent of Kant's appeal to subjective factors in geographic, mathematical, and logical orientation (Kant 1983c, 270n).

11. For example, *PdP* 162; cf. also 161, which similarly talks of the body "haunting" its environment.

12. This formulation of *SdC* 133 anticipates the distinction between "having" (existential) sense and "being" (predicative) sense of *PdP* 203n.

13. *PdM* 23. Given the content of *SdC* and *PdP*, Barbaras's (2005, 210–211) claim that Merleau-Ponty's use of "living" has the dual meaning of living and experiencing (i.e., *Leben/Erleben, vivre/vivre quelque chose*) is clearly plausible. My talk of "lived sense" is intended to accommodate this ambiguity.

14. *PdP* 459; cf. similarly 221 and *PdM* 15, 58.

15. *PdP* 219. Cf. *PdM* 43 and 156, which talks of the "illusion of an absolutely transparent expression."

16. *PdP* 214. Merleau-Ponty's talk of "gestural meaning"—as "immanent within speech" and the basis of "conceptual meaning" (*PdP* 209)—is clearly intended to be synonymous with "existential meaning" (*PdP* 225).

17. *SdC* 136. The basis of this view is Merleau-Ponty's critique of behaviorism, centering on the idea that the individuation of environmental stimuli (perceptual "givens") depends on an organism's internal functioning (cf. *SdC* 1–64).

18. This feature—to be discussed further below—makes the mode of gestural expression appropriate to the "principle of indeterminacy" that in Merleau-Ponty's view pervasively characterizes human existence (*PdP* 197).

19. *PdP* 217. Merleau-Ponty requires: "What one calls an 'idea' is necessarily linked to an act of expression and owes to this its appearance of autonomy" (*PdP* 447; cf. 212 and Merleau-Ponty 1964, 196ff.).

20. Particularly Rousseau's view that "the first languages were singing and impassioned"—a view also hinted at by Condillac (1971, 180–181)—and only later acquired grammatical structure (Rousseau 1990, 67, 73). The claim that certain (linguistic) sounds naturally express feelings, passions, or emotions is common

not only to Condillac (1971, 172, 228–229, 237) and Rousseau (1990, 114, 126), but also—at least with words for things lacking characteristic sounds—to Herder (1966, 57).

21. Including explicit rejection of the "tacit cogito" (Merleau-Ponty 1964, 222, 227). While it seems correct to claim that this development was necessitated by certain aspects of his thinking—for example, his treatments of expression or nature (cf. Barbaras 1993, 2005)—it seems to me that the need extends back to *SdC*'s attempts to clarify the relationship between nature and consciousness using an ontologically undetermined notion of behavior.

22. The distinction between "authentic speech"—which "gives rise to a new sense" (*PdP* 226)—and "secondary expression" occurs first (*PdP* 207n). Authentic speech is then referred to as "original speech" (*PdP* 208n), which in turn is counterposed with the "constituted speech" of "everyday life" that presupposes "the decisive step of expression" (*PdP* 214; cf. 253). Authentic speech is further linked with "speaking speech" ("the significative intention finds itself *in statu nascendi*") as opposed to "spoken speech" (*PdP* 229). Further contrapositions between "secondary speech" and "original speech," as well as between "empirical speech" ("the word as an acoustic phenomenon") and "transcendental or authentic speech," are found at *PdP* 446 and 448 respectively.

23. Merleau-Ponty's terminological variations could be seen as reflecting his view of indirect sense, since "truly expressive speech ... feels its way around an intention to signify" (*PdM* 64), so to speak circling in on a thought, rather than purporting to capture it directly with univocal terminology.

24. *PdM* 195. For Merleau-Ponty the initial impression of possessing a common language is approximative, superficial, or perhaps even illusory: "We speak and we understand each other, at least at first sight" (*PdM* 32).

25. *PdM* 85; cf. 84 and *PdP* 209. See Frank 1999a, 1999b, for an interesting development of the nature and philosophical relevance of (ultimately) individual style that, despite not drawing on it directly, is very much in line with Merleau-Ponty's thinking.

26. Cf. *PdM* 30, 21. Also the curious suggestion that to "say something important" implies saying something original that "imposes" its sense (*PdP* 445; cf. 460).

27. Cf. *S* 113–114. Thus coinciding with the idea that thought is realized in linguistic expression.

28. *S* 112. Merleau-Ponty talks of "excès" at *PdP* 447, *PdM* 9, 24, and *S* 104, 112; and of "transcendence" at *PdP* 449.

29. *PdP* 213; cf. also 217 and *PdM* 20. It is sometimes suggested that Merleau-Ponty himself saw the relationship between established and creative use of language as a

priority-free "reversible" one (Hass 2008, 191; Baldwin 2007, 98–100). While I agree that this ought to have been Merleau-Ponty's view, and that it is perhaps hinted at by the above concession and some comments in *The Visible and the Invisible* (Merleau-Ponty 1964, 199–201), the earlier works under consideration here consistently present creative expression as primary.

30. On the former see *PdM* 60–61 as well as *PdP* 217, 462–463; on the latter *PdM* 160, 162, 165, and *PdP* 448–449. Emphasizing the fact that language expresses something that did not antecedently exist as a "paradox" or "mystery" might be interpreted as urging a pious attitude toward the ineffability of the "silence" underlying language (e.g., by Kwant 1966, 184–191).

31. These implications are brought out well by Charron's (1972, 69–138) insightful exposition of Merleau-Ponty's views on language.

32. For Husserl's view of "adequate evidence" as the consummation, or complete sum, of relative experiential "evidences," see Husserl 1992d, 16 (§6).

33. Merleau-Ponty 1968, 12; *PdP* 254, 465. As highlighted by Waldenfels (1976), this emphasis on the openness of structure and human behavior goes back to *SdC* (cf. 125).

34. The view that language should be modeled on perception, realizes (or "constitutes") thought, and is hence perspectival in character, was later (1960) echoed by Gadamer (1990, 451ff.). For a lucid contemporary discussion of intentionality as perspectival see chapter 1 of Crane 2001, especially 4–8, 18–21.

35. *S* 112, cf. 54. Note that this is not simply about limits of "representation," but a constraint on the mode of presence of what is "represented." A photo of a house, for example, is limited in the sense that it depicts only one view of it; the indirect objects of language are akin to the painting of a fictive house, which is not only depicted from one viewpoint but exists in no other.

36. *PdM* 41; *S* 112 (italics added), cf. also 114. On the "inadequate" character of linguistic expression see also *PdM* 52–53, 79.

37. "It always seems to us that the processes of experience codified in our language follow the very articulations of being because it is through [our language] that we learn to direct ourselves to [being]" (*PdM* 38).

38. In this respect the terms *indirect* and *lateral* correspond to the earlier emphasis on *ambiguity*, which, apart from an occasional distinction between "good" and "bad" ambiguity (e.g., Merleau-Ponty 1960, 14; 1962, 409), disappears after *PdP*. Perhaps this is because, as Sapontzis (1978, 542) has pointed out, in presupposing determinate meanings between which it prevails, the idea of ambiguity is ill-suited to Merleau-Ponty's thinking.

39. Though he continued to talk of linguistic "gestures," after *PdP* Merleau-Ponty no longer refers to them as "emotional." Moreover, it is precisely the view of modern art as self-expression that his theory of expression sought to overcome (cf. *PdM* 76ff. and chapter 5, section 3 in this volume).

40. Collingwood (1958, 122) similarly holds that "a person who expresses something thereby becomes conscious of what it is that he is expressing." He also parallels Merleau-Ponty in likening language to gestures and in his (similarly misleading) account of linguistic expression as founded in emotional meaning (see Collingwood 1958, 243, 225ff.).

41. See Carnap 1935, 27; Tormey 1971, 128; Wollheim 1992, 12ff.; Goodman 1976, 85, respectively.

5 The Art and Science of Indirect Sense

1. See Watson 1983, 210, for a useful catalog of Merleau-Ponty's references to Saussure.

2. *PdM* 37, 34. On the historically plausible claim that Saussure liberated linguistics from etymology, see Culler 1986, 65–85.

3. The published text of the *Cours* was compiled from student notes of three courses held from 1907 to 1911. For a useful survey of the editorial difficulties see chapter 3 of Harris 2001. Bouquet (2004) has argued that linguistics of *parole* was more important to Saussure than the published *Cours* suggests—which would potentially narrow the gap between Saussure and Merleau-Ponty's interpretation of him.

4. As Dastur (2001, 63) delicately puts it, Merleau-Ponty's reading of Saussure "peut sembler peu rigoureuse"; Schmidt (1985, 105) describes it as "so idiosyncratic that it makes his notoriously loose readings of Husserl look like models of hermeneutic chastity." For helpful commentary on Merleau-Ponty's relation to Saussure, see Lagueux 1965 and Schmidt 1985, 105ff.

5. Saussure 1972, 25, 124. Madison (1981, 322n) ponders whether Merleau-Ponty is confusing Saussure with Pos and Wartburg. But, as Schmidt (1985, 195–196) reasonably points out, it "is difficult to believe that Merleau-Ponty would be this confused in the early 1950s after having taught several courses on Saussure."

6. In his 1949–1950 course, for example, Merleau-Ponty more accurately distinguishes *parole* as "what one says" from *la langue* as "a system of possibilities" (Merleau-Ponty 2001, 84).

7. Merleau-Ponty 2001, 82. For Saussure (1972, 45; cf. 44ff.) it is the "spoken word" that "alone" constitutes the object of linguistics.

8. Cf. *PdM* 41, 47, as well as the examples of 41ff. Particularly Saussurean is Merleau-Ponty's claim that phonemes are the "true foundations of speech" (*PdM* 47).

9. *S* 110. This might seem to echo Saussure's (1972, 166) own view, but see below.

10. *S* 109; *PdM* 41. Merleau-Ponty's talk of "use values" further suggests a parallel with Heidegger's discussions of Equipment. Elsewhere he qualifies this parallel by denying that words have a fixed function as hammers and other tools do (*PdM* 64).

11. Saussure 1972, 159–160. A misleading feature, however, of Saussure's analogy with money is the latter's quantitative character, suggesting a uniform principle of comparison—along the lines of more/less—which is hardly germane to language.

12. Saussure 1972, 167; cf. 26, where "a language [*langue*]" is characterized as "a system of distinct signs corresponding to distinct ideas."

13. The point is not merely that we don't know which features to make the basis of distinct units. It is rather that there are no such features. For coastlines are fractals, meaning that "the" exact length (also shape) is a function of the scale at which it is represented (see Gleick 1987, 94–96).

14. By way of analogy, the behavior of a sign might be likened to that of a photon in Young's slits experiments. With photons Heisenberg's uncertainty principle ($\Delta x \Delta p \approx \hbar$) tells us that the precision with which a photon's position is ascertained is inversely proportional to the precision with which its momentum can be known. Analogously, the more precisely a syntactic element (e.g., a morpheme) is focused on, the less determinate are its semantic properties—and vice versa.

15. Corresponding to his earlier distinction between "conceptual meaning" and "existential meaning" (*PdP* 212; cf. chapter 4, section 1).

16. Cf. *PdM* (e.g.,) 41, 161; 43, 46; and 35–36.

17. *PdM* 41n (cf. 53n, 57), 42n, 64. Merleau-Ponty characterizes the basic operation of language as nonconceptual (52, 56–57) and preobjective (42n), and links the two via the notion of style (63n).

18. *S* 108, 54 ("quelque chose comme un être").

19. *PdM* 33; 26–30 deal with psychology, 30–54 with Saussurean linguistics.

20. Saussure 1972, 100–101. Note, incidentally, that Saussure's view of the arbitrariness of signs does not depend, as Benveniste (1939, 24) supposed, on the tacit assumption of referents as a "third term." Saussure's view is rather that arbitrariness attends the pairing of sounds with differentiated thought, not signs with referents.

21. Note also that Merleau-Ponty's notion of *parole* as a productive intentional act, while differing from Saussure's minimalist notion of individual performance, is arguably akin to the latter's treatment of analogy (Saussure 1972, 226ff.).

22. *PdP* 61. This characterization of motives corresponds to the phenomenological concept of foundation described at *PdP* 451. For a clear and detailed discussion of the motivations and their relationship to causes and reasons see Wrathall 2005. Talk of *motivation* as a technical term in the phenomenological tradition again goes back to Husserl (e.g., 1992b, 32 [§2]; 1992c, 101 [§47]), though I leave open here the question of whether Merleau-Ponty uses this term in the same sense as Husserl.

23. Austin uses this argument, about the "natural economy of language," to justify his methodological concentration on ordinary language (Austin 1979, 190, 182).

24. This charge is developed, for example, by Falck (1986) and Dillon (1997, 181–186).

25. Cf. Merleau-Ponty's (1959) essay on structuralist anthropology, in which he insists on the need for "a sort of lived equivalent" to ground formal theoretical structures (*S* 149; cf. also 155).

26. Cf. *PdM* 40. Merleau-Ponty similarly praises Saussure for having freed considerations of meaning from etymology on the grounds that speakers are usually unaware of such factors (*PdM* 33).

27. *PdM* 77. On the preceding, cf. Merleau-Ponty's summary of the task for a modern theory of expression on *PdM* 79. (On the general nature of his view of expression see chapter 4, section 5.)

28. Recall the Husserlian ideal of "adequation" discussed in chapter 4, section 4.

29. *S* 82. For the link with the general system of embodied equivalences—that is, the "schéma corporel" of *PdP*—see *S* 83 or *PdM* 110.

30. Cf. *PdM* 105–106. Merleau-Ponty's Cézanne interpretation discusses this intimate link, arguing that although not straightforwardly explained by facts about the painter's life, "*this work to be done,*" his painterly expression, "*demanded this life*" (Merleau-Ponty 1996b, 26).

31. *PdM* 86, 99. The terms *mise en forme* and *formulation* are found at *PdM* 84–85, 86, and *PdM* 82, respectively.

32. For details, see in particular Gombrich 1960, 55–78, 126–152.

33. Cf. *PdM* 91–92. By "direct copy" I mean here, roughly, like a photo rather than a caricature—for example, there is correspondence without omission, geometric distortion, changes in patterns of color equivalence, a loss of linear-perspectival integrity, and so on.

34. *S* 55. This presentational function is clearly independent of whether the painting has an actual referent. In this respect the presentational sense of pictorial articulations is more intrinsic to pictorial expression than the referential function.

35. For example, it is not difficult to find reasons for the deployment of different kinds of brushstrokes in Jean François Millet's rural scenes, Seurat's pointillism, or Cézanne's late paintings of the Bibemus quarry and the Mont Saint-Victoire.

36. An "act of thought, having been expressed, from then on has the power to outlive itself" (*PdP* 449). This fact, incidentally, highlights the naïveté of Nietzsche's (1980b, 260 [I §2]) suggestion that the coining of concepts is an "expression of power by the dominant."

37. This is what licenses Austin's talk of the "natural economy of language" and the "long test of the survival of the fittest" (Austin 1979, 190, 182; cf. note 23).

38. A notational scheme that is "syntactically dense ... provides for infinitely many characters so ordered that between each two there is a third" (Goodman 1976, 136).

39. For an effective illustration of this see Gombrich's (1960, 309) juxtaposition of a van Gogh copy of Millet's *Cornfield* with the original.

40. I discuss the way paintings have nonconceptual content in more detail in Inkpin 2011.

41. Which Goodman (1976, 135–136) defines as follows: "For every two characters *K* and *K′* and every mark *m* that does not actually belong to both, determination that *m* does not belong to *K* or that *m* does not belong to *K′* is theoretically possible."

42. Merleau-Ponty shows some awareness of this in suggesting that literary language differs from routine language use while standing in a relationship of "homonymy" (*S* 96; cf. *PdM* 19 and *S* 113–114). But despite often referring to the "coherent deformation" it introduces, he does not detail how literary use distinguishes itself from common use.

43. Incidentally—though to argue this is well beyond the scope of this book—it seems to me that, for all its naïveté, this view underpinned Heidegger's own philosophical methods throughout his work.

44. *PdM* 58. Such presence, for Merleau-Ponty, is of course "of the perceptive order" (*PdM* 41, 57, 53)—that is, perspectival, partial, or indirect.

45. *S* 109; cf. *PdP* 450. Merleau-Ponty goes as far as to suggest that "the present exercise of speech" involves taking up "all the preceding experience" sedimented in language (*PdM* 59).

46. For example, in §134 of *Daybreak* (Nietzsche 1980a, 127–128).

47. Charles Guignon, for example, suggests that for Heidegger "the *true springs* of the meaning" that Dasein discloses "are historical," such that authentic being

"brings about a new relationship to the past and history" in which Dasein "*remembers* its historical roots and can find the underlying meaning of what is passed down in the tradition" (Guignon 1983, 141, 139, 138).

48. *SZ* 28. As well he might, given that the term was an eighteenth-century coinage, rather than a genuinely ancient Greek term. For an informative survey of the term's history see Spiegelberg 1960, vol. 1, 8–20.

49. This has an air of paradox: Heidegger, his atemporal model of language notwithstanding, apparently values etymological facts. Yet it is Merleau-Ponty, who disavows the historical perspective to focus on the living present of language, that better does justice to the temporality of language and the relevance of etymological facts.

50. *PdM* 46. As Merleau-Ponty elsewhere points out, to start picking up the system of language the child must already have mastered—as a scheme of differentiation—its phonetic basis (Merleau-Ponty 1968, 34).

51. Complexity of linguistic forms, and with this their power to differentiate features of the world, arises through the combination of simpler forms. Assuming words to be the bearers of conceptual properties, this explains why the logic of words' construction cannot be "put into concepts."

52. Cf. Ricoeur 1969a, 86. One such process being Gadamerian dialogue with texts of the past, as the merging of horizons amid "the inconcludable openness of the occurrence of meaning" (Gadamer 1990, 476).

53. In taking creative expression as paradigmatic he even suggests that to understand language one should precisely *not* attend to the "immediate practice of language" (*PdM* 165; cf. 43).

6 Language and the Structure of Practice

1. Haller (1979, 1984) applies the term *praxeological* to Wittgenstein. Skirbekk (1983, 9), describing a tendency in Scandinavian philosophy, characterizes praxeology fittingly as "theory (logos) of action (praxis) ... a conceptual analysis and reflective discussion of the way human activities are interwoven with their agents and with the things at which they are directed within our everyday world."

2. Baker and Hacker (1985, 25–26) are an influential example: "Clarification of the concept of understanding a rule is the pivotal task in W[ittgenstein]'s reflections. ... W[ittgenstein]'s conception of meaning would disintegrate were it not that using a word correctly is a criterion for understanding it."

3. Cf. *PU* §§90, 104, 109, 111, as well as 116, 94, 106, respectively.

4. Cf. *PU* §§89, 415, and 109, 126, 128, 599.

5. As is well illustrated by Ayer's (1969) response to Austin's (1962) *Sense and Sensibilia.*

6. See, for example, Diamond 2000 and Conant 2000.

7. For a good survey of the difficulties faced by resolute readings see Hutto 2003, 90–98.

8. This seems to be Cavell's (1962, 91–92.) own point in talking of Wittgenstein's "method of knowledge" and his "exhortation ... to self-scrutiny."

9. The exegetic need to account for this feature of *PU*'s discussion is nicely highlighted by Hutto (2003, 97–98). Elsewhere I argue that Rorty makes use of a strategy similar to that attributed to Wittgenstein by Stern (Inkpin 2013).

10. Think of the often epigrammatic or aphoristic character of Wittgenstein's comments, their arrangement in a quasi-dialogical form reminiscent of both Plato and Augustine, or the loose overall arrangement of fragmented considerations into what is presumably not supposed to be a systematic whole. For a perspicacious comparison of the aims and style of Wittgenstein and Augustine see Thompson 2000.

11. See, for example, Munson 1962, Spiegelberg 1981, Gier 1981, Guest 1991. Developing the views of Hintikka and Hintikka (1986, in particular 148–154), Park (1998) offers an extensive and more philosophically focused interpretation of Wittgenstein's philosophy as a form of phenomenology. For a dissenting voice, see Reeder 1989.

12. The best-known traces of this are found in the manuscripts known as *Philosophical Remarks* (1929–1930) and the *Big Typescript* (1933). For anecdotal evidence of Wittgenstein's description of his work as "phenomenology" see Rhees 1981, 131, and Spiegelberg 1981, 214.

13. Such parallels are suggested by Guest (1991), and with some caution by Spiegelberg (1981, 212, 215–216), but most extensively—and most speculatively—by Gier (1981, 91–134).

14. See in particular manuscripts 105 (esp. 114, 116) from 1929 to 1930 and 113 (esp. 245–246) of 1931–1932, which are now widely available in the Vienna Edition (Wittgenstein 1994, vol. 1, 193; vol. 5, 133). For a sound and informative reconstruction of "Wittgenstein's phenomenology" see Kienzler 1997, 105–142.

15. It might be thought that some comments Wittgenstein makes in discussing the possibility of a private language—for example, his denial that concepts can be grounded in first-personal experience (cf. *PU* §293)—make his position incompatible with a phenomenological approach. However, in this context Wittgenstein is denying the importance of "phenomenology" only in relation to "inner" private, introspective, first-personal experience. This denial clearly does not extend to our experience of language in an interpersonal space. (After all, his own claims about

the need for external criteria presuppose that experience of language in a shared space can be described.)

16. Wittgenstein 1960, 17. The term also occurs several times in *PG*, which is based on a revised 1933 typescript (TS 213; see note 39). Although this use might antedate the *Blue Book*, the term is simply used in these passages without being explained.

17. See Wittgenstein 1960, 17, 81; *PU* §§5, 7.

18. §5 recommends studying "the phenomena of language in primitive kinds of its use," explaining that "the child uses such primitive forms of language when learning to speak."

19. For example, Black 1996, 79, 81; Baker and Hacker 1980, 26–27; Rhees 1970, 76, 81; Kenny 1973, 169–170. The same point is sometimes made with reference to the idea of a primitive "form of life" (see Garver 1990, 181).

20. Cf. *PU* §18 and Wittgenstein 1960, 19. An alternative strategy is Marie McGinn's (1997, 50) claim that since there is "no essential structure or function against which the notion of completeness can be defined," "the idea of completeness simply doesn't apply" to language-games. In that case, why does Wittgenstein so insistently talk of "complete" languages? A better answer is that he intends a different notion of completeness to apply, characterized by the absence of gaps. In this sense a small town, a child, and a language might all be complete, even though they will subsequently develop further. (On this kind of gap-free order in transition see Wittgenstein 1960, 44.) Wittgenstein's hope, expressed in the *PU* preface, that his thoughts should form a natural sequence with no gaps also hints at this kind of completeness.

21. *PU* §130; cf. §5 and Wittgenstein 1960, 17.

22. See particularly *PU* §§69–71 and section 6 below.

23. *PU* §11. Cf. §§10–13 and 22–23, 65, respectively.

24. *ÜG* §56. Cf. also §§51, 501.

25. *PU* §37. Hence Hintikka and Hintikka's (1986, 212) emphasis on the role of language-games in establishing "*the basic semantic links between language and reality* in Wittgenstein's mature philosophy." This somewhat misleadingly hints that there is still a problem of contact between language and (nonlinguistic?) reality and such things as "basic" links. However, as Harris nicely argues, an important implication of Wittgenstein's later views is to break with the foundationalist idea that "language touches or connects with the world only at certain points" since "language is embedded in reality and vice versa" (Harris 1986, 43, 47).

26. This can again be seen to parallel Heidegger's view that individuation of entities comes only through the imposition of an *as* structure in Setting-out.

27. For example, this would be the conclusion for a Quinean anthropologist working with stimulus meaning (see Quine 1960, 26–79).

28. Note that in Wittgenstein's presentation of §2 the semantic function of the terms *cube* etc. looks indeterminate. Of course, on the position Wittgenstein is opposing these words function as names, as labels standing for objects. But, so far as Wittgenstein's sketch of the builders' language-game goes, there is nothing to prevent us from thinking of these words as descriptions of form, or as imperatives directed specifically to the action rather than the objects to be brought. This indeterminacy is perhaps intentional, and no doubt congenial to Wittgenstein's aims.

29. Incidentally, this is why reference is neither indeterminate nor inscrutable in the way Quine (1969b, 35) worries about. Knowing how to participate in the language-game entails understanding what "the referent" is.

30. This feature of instruments is nicely applied to Wittgenstein's language-games by Meggle (1985, 78ff.).

31. For example, Wittgenstein talks of the "point" (*Witz*) of language-games in *PU* §§564, 142, and of "techniques" (*Technik*) of word use in §§125, 262, 557 (cf. also §§150, 199).

32. For example, because optimization would then apply across and among different language-games and be causally determined by what agents typically understand or confuse.

33. Cf. *PU* §§11–12, 360, 421, 569. This pragmatist tendency is arguably—as Wittgenstein seems to recognize in §422—more pronounced still in *On Certainty*. For arguments of this kind see Moyal-Sharrock 2003 and Rudd 2005.

34. Cf. *PU* §19; Wittgenstein 1960, 17. The use of the expression "forms of life" at *PU II* 489 (cf. *LSPP* §862) suggests that it was also intended to refer to recurrent components—including language-games—of an overall "form of life."

35. For example, they have interrelated consequences, shared input and output, and respectively embody certain kinds of operation. The "roughness" of this analogy is indicated by Wittgenstein's critique of the calculus model of language (see the next section).

36. *ÜG* §§115, 150, 160, 354, etc. *LSPP* §803 points out the "logical" character of such functional links.

37. *PU* §108; cf. §31 and *PG* §§13, 23.

38. *PG* §§11, 13, 27. Cf. also §§2, 11, 19, 33, 39, 63, 68, 72, 80, 84, 110, 111.

39. The term *language-game* appears eight times in total in *PG*, in §§5, 26, 81, 123, and appendix 4 B. Even then §26 intimates that in describing language-games one might not "still want to call them calculi."

40. *PG* §§32, 44. A notorious difficulty of this interpretation is the vagueness of Wittgenstein's notion of grammatical rules—for example, are these rules syntactic (as the term *grammatical* suggests) or semantic, constitutive, or regulative? (See Wittgenstein 1980, 97–98; Moore 1954, 292; Searle 1969, 33–42.)

41. *PG* §76. A similar image—the light of a reading lamp—is used in the *Blue Book* (Wittgenstein 1960, 27).

42. One complication is what counts as "explaining" a rule. If this includes merely giving examples of use (see *PG* §§43, 28), explaining the rule scarcely differs from just applying it.

43. Fluctuation in the real use of words is another anomaly, on this model, acknowledged in *PG* §36.

44. This is the essence of Kripke's (1982) view. As the following will suggest, I believe this view to be both exegetically and philosophically problematic. Beyond setting out my alternative ("nonskeptical") interpretation of this argument, however, considerations of scope and relevance here preclude detailed engagement with Kripke's "skeptical paradox"—which, after all, §201 dismisses as a "misunderstanding"—and the debates it has spawned.

45. Incidentally, this argument in no way turns on whether rules are "explicit" in agents' awareness or merely "implicit" in the practice. In particular, there is no intimation that the regress problem afflicts only "explicit" rules, such that either "implicit" awareness (Brandom 1994, 20–30) or "knowing how" (Ryle 1990) block the regress.

46. Hence shortly after *PG* Wittgenstein changed his mind, seeing the comparison with exact rules as the source of, rather than the answer to, philosophical misunderstandings of language (Wittgenstein 1960, 25).

47. This is congenial to McDowell's insistence that the regress of "interpretations" should never be allowed to start. However, the present interpretation of the regress-of-rules argument has the advantage of making clear that this is not dogmatic insistence on a supposed bedrock of normativity. My overall approach here is distinguished by taking Wittgenstein's talk of practices, customs, and institutions as a significant commitment, a view to which McDowell was initially sympathetic, but later—appealing to the no-thesis thesis—rejects as un-Wittgensteinian (see McDowell 1984, 342; 1992, 50–51).

48. Cf. *PU II* 573. This point is highlighted by Baker and Hacker (1985, 249–250).

49. Cf. Kripke 1982, 86ff., especially 96–97.

50. Cf. *PU* §§199, 207, as well as chapter 7, section 1.

51. "*UNGEFÄHRE Gesetzmäßigkeit*" (*LSPP* §968; cf. similarly §211).

52. *LSPP* §969. "That we *calculate* with certain concepts, [but] not with others, merely shows how different in kind conceptual tools are (how little reason we ever have here to assume uniformity)" (*BPP I* §1095).

53. *LSPP* §347. A corresponding claim is found at *PU II* 575.

54. *PU* §568. For further examples of physiognomic imagery see §§235, 167, *PU II* 547, and *BPP I* §654.

55. The empirical attunement of rules is also in evidence in *On Certainty*. While emphasizing attunement to "normal conditions" (§27), Wittgenstein there allows that rules and empirical claims differ gradually (§§51–52, 98, 318–319, 321, 454) and acknowledges that rules of language are ultimately empirical (§519).

56. *PG* §133. The actual term *autonomy* is found in *Zettel* (§320), paralleling part of *PG* §133. "That grammar is autonomous, arbitrary, not justified by reference to reality" has been emphasized by Baker and Hacker as "a deep leitmotiv of Wittgenstein's later work" (Baker and Hacker 1980, 76; cf. 1985, 329ff.). It is significant, however, that explication and substantiation of this claim typically rely not on *PU*, but on *PG* and *Zettel*—the latter being a source of particularly uncertain status (cf. von Wright 1982, 136).

57. *PG* §§133, 140; 135–136, 140; 134; 133–134, 138, respectively. The first two claims are presumably part of Wittgenstein's attempt to distance his view from pragmatism.

58. Cf. *PU* §§149, 498 (also the unnumbered note around §30); 355.

59. There is also an obscure invitation to "consider" the claim that rules of grammar are arbitrary in §372. The apparent justification for the above claim is that grammar does not tell us "how language must be built to fulfill its purpose" and simply describes the use of signs (*PU* §496). But this can easily be understood in the sense, discussed above in section 2, that differently functioning words can fulfill the same purpose.

60. As Mulhall puts it, the late Wittgenstein recognizes the "embeddedness of the normative within the natural," and "relocates normative or logical determination within the realms of time, history, and society, without conflating it with empirical determination" (Mulhall 2001, 112, 121).

61. The same point is nicely made at *Zettel* §440 with regard to traffic regulations.

62. Cf. *ÜG* §329. Each "kind of language-game" has its characteristic "kind of certainty" (*PU II* 569).

63. Cf. Wittgenstein 1976, 396, 305. The implication is that to have a doubt presupposes a conception of what it is to have that doubt, of what that doubt amounts to in practice (cf. *ÜG* §§89, 120). This is a highly nontrivial requirement that is routinely neglected by those troubled by the "indeterminacy" of rules.

64. It is a principal theme of *PU* §§65–80, 99–101, but cf. particularly §§69–71.

65. *PU* §§31, 54; cf. §89 and *ÜG* §95.

66. Cf. the unnumbered note around *PU* §139. The possibility of a mismatch and a dialectic between these two aspects of linguistic competence is nicely highlighted by Burge (1989, 182): "In attempting to articulate one's conception of one's concept, one's conceptual explication, one naturally alternates between thinking of examples and refining one's conceptual explication in order to accord with examples that one recognizes as legitimate."

67. Evidence of the latter can also be seen in Wittgenstein's reliance on the idea that sense can be made of bizarre-sounding sentences by specifying the particular conditions in which they might be used (*PU* §117; *ÜG* §§ 25, 413, 423, 622). Recall also Wittgenstein's polemics in the *Blue Book* (Wittgenstein 1960, 18) against philosophy's "craving for generality," its "contemptuous attitude toward the particular case."

68. By contrast in a (mathematical) calculus the aim is precisely to eradicate such particularity and nonregulation, so that each step is rationally defensible in the sense of being necessarily and sufficiently legitimized by adducing a rule.

69. It "lies in the *essence* of language" that if "the language-game, the activity ... fixes the use of a word, then the concept of use is elastic with that of the activity" (*LSPP* §340).

70. Incidentally, there is no implication that individual, or groups of, aestheticians cannot adopt highly precise terms. Wittgenstein's point (*PU* §77) is simply that definitions could not be given for the overall field of activity.

71. At first glance, this might seem to conflict with Wittgenstein's own examples of language-games, all of which seem to be straightforwardly characterizable in terms of rules. However, Wittgenstein's qualification of his own language-games as "clear and simple" (*PU* §130) is not redundant and clearly allows real language-games to be more complex.

72. For example, as regulative principles. Even where there are practical heuristics: there are no *rules*, though much informed opinion, that tell the England cricket captain who should bowl next, whether he should take the new ball, declare, enforce the follow-on, send in a night watchman, and so on.

73. Cf. Aristotle's discussion of Prudence (*φρόνησις*) and the deliberative excellence it involves (Aristotle 1934, 351–357 [1142a25–1142b30]).

74. *PU* §108—"nonentity" here translates *Unding* (literally "nonthing"), which in German conveys that one is referring to something absurd.

75. Theirs is the domain of "unwägbare Evidenz" (*PU II*, 575–576).

7 Coping with Language

1 Wittgenstein 1960, 13 (cf. *PG* §43). The analogy is with Kant's (1983d, 33) distinction between acting out of duty and merely in accord with duty.

2. This supports my earlier claims about the no-thesis thesis (chapter 6, section 1) and the interpretation of Wittgenstein's regress-of-rules argument (chapter 6, section 4).

3. David Bloor understandably describes Wittgenstein's commitments as "disturbingly minimal," before less understandably—without any exegetic foundation—attributing to Wittgenstein a theory of institutions as "performative utterances, produced by the social collective" (Bloor 1997, 28, 32).

4. For example, Wittgenstein 1960, 4, 13, 53; *PG* §§92, 99; cf. *PU* §§53, 301, 397.

5. Cf., e.g., "Don't now ask yourself 'How is it with *me*?'—Ask: 'What do I know about the other?'" (*PU II* 539).

6. It follows, of course, that one applies the same criteria in attributing concepts to oneself, or in understanding one's own utterances, as one does to others. But this is because, as the Private Language Argumentation shows, there is no such thing as an essentially first-personal perspective on language, so that the distinction between first- and third-personal perspectives cannot do any work regarding the understanding of language. Here there is a parallel with Merleau-Ponty's view of style (see chapter 5, section 3).

7. *PU* §199. This exclusion of a pure singularity from the concept of rules is often reiterated—for example, at *BGM* VI §§21, 34; III §67.

8. That rule-following pertains to behaviorally manifested cognitive abilities is highlighted by Shanker (1996) and McGinn (1984, 32–40).

9. On the one hand is the view that rule-following is an ability which, though usually exercised in shared social practices, is individually possessed (e.g., Baker and Hacker 1984, 1990; McGinn 1984; Shanker 1996). On the other hand are those who see a "language community" as playing some essential role—such as contributing proper normativity or rational stability—in the phenomenon of rule-following (e.g., Kripke 1982; Malcolm 1986, 1995; Wright 1980, esp. 220).

10. If anything, comparison with the community would be more demanding, because it would involve identifying which is the community standard.

11. Wittgenstein acknowledges several possibilities in *PU* §§54, 82.

12. *PU* §164. The family-resemblance character of reading is highlighted in §168. Wittgenstein had already made an equivalent claim about "understanding" in *PG* §35.

13. Cf. *PU* §65. "Does that mean 'following a rule' is indefinable? No. I can define it in countless ways" (*BGM* VI §18).

14. Insofar as they have any persistence and regularity over time, language-games are institutions in this sense. Incidentally, the idea that rules are upheld is reflected in the etymology of the German term Wittgenstein uses for "custom": a *Gepflogenheit* is that which is cared for or maintained (*gepflegt*, from *pflegen*).

15. Cf. *PU* §§69, 71, 75, 208.

16. Cf. *PU* §65. *PG* §80 had claimed that a "general propositional form determines the proposition as a term in a calculus."

17. *PU II*, 572 (cf. *BPP I* §630), 529; *PU* §654, where Wittgenstein's term *Urphänomene* is almost certainly an allusion to Goethe (e.g., Goethe 1996, vol. 12, 366–367).

18. *ÜG* §§204, 110; cf. §166: "The difficulty is to see [*einsehen*] the groundlessness of our beliefs." (Similarly *PU* §§211, 212, 217.)

19. Cf. *BPP I* §949, where Wittgenstein likens his conceptual investigations to Goethe's morphology. There are several such indications of Goethe's influence on Wittgenstein (see Schulte 1984; Monk 1990, 509–512, 303–304), which, incidentally, is also a plausible source of Wittgenstein's apparently phenomenological emphasis on description over explanation (Goethe 1996, vol. 13, 123; cf. vol. 12, 432, which *BPP I* §889 cites).

20. Wittgenstein sometimes (e.g., *ÜG* §§253, 262, 612) hints at the—implausible—view advocated, for example, by Kuhn (1970) and Rorty (1979, 315ff.; 1989, 8–9) that rationality is system-internal, so that changes between systems (or forms of life) are governed by nonrational procedures.

21. Recall the stratification and foundation of language-games discussed at end of chapter 6, section 3.

22. The builders language-game of *PU* §2 is an example of a routine in which no rule stating is involved. It is interesting to note that although Wittgenstein thinks of the builders as having knowledge they cannot express (*ÜG* §396), he elsewhere denies that "concepts" are involved in this game (*BGM* VII §71).

23. Characterizing generally those language-games that involve predicative awareness would parallel, and face the same difficulties as, the task of specifying language-games that satisfy the agent's grasp condition for rule-following.

24. For detailed discussion of problems concerning Ryle's apparent identification of knowing-how with abilities, his regress-based arguments in favor of the distinction, and its relevance to his argument against the "intellectualist legend," see Stanley and Williamson 2001 and Snowdon 2003.

25. Though not impossible, it is plausible to think of use as governed by bivalent correctness conditions only in the limiting case of strictly regulated language-games such as mathematics.

26. As Dreyfus (1991, 67–75) does, based on his reading of SZ §§15–16. See chapter 9, section 4 for critique of this view, and chapter 1, section 2 for the basis of this critique.

27. Further, on the Wittgensteinian view suggested here, the fact that not all language-games involve reason-giving is not only an experiential claim. Rather, since justification is finite and founded in prejustificatory practices, not all linguistic practice *can* be construed in terms of reason-giving.

28. In this respect it seems to me that Brandom's (1994, 83) motto "semantics must answer to pragmatics" misrepresents his own priorities.

29. In the spirit of Merleau-Ponty's own emphasis, it might be said that everyday practical language use often comes closer to the model of creative expression than his view of "direct" language use allows.

30. Cf. *PU* §§94, 104, 110–114.

31. Some of Wittgenstein's later reflections point in this direction. For example, choosing between terms, he says, is like choosing "between similar but not identical pictures," as if "according to fine differences in their smell"; a word can have a "familiar face," exhibit "fine aesthetic difference[s]," leading to the "feeling that it has absorbed its meaning" (*PU* §139[a]; *PU II* 560–561). But these comments remain characteristically elliptic and metaphorical, and Wittgenstein makes no attempt to develop a positive account of the role of linguistic form.

32. One might think this an unacceptably antirealist claim, and believe the extension, say, of "natural-kind" terms such as *gold* to be determined in a language-game transcendent manner. But the stipulation that reference is to be determined by physical properties, characteristic of scientific practices, is itself a language-game-relative determination. The objection that my position is an "antirealist" one is considered at greater length in chapter 8.

33. See the discussion of Heidegger's view of Statements in chapter 1, section 4.

34. Examples of the former might include cultic, religious, or superstitious practices, and fashion. Scientific practices would exemplify the latter kind: even if we think his views are, strictly speaking, false, Newton was getting something significantly right about gravitation.

35. For discussion of how presentational sense and pragmatic sense respond to the requirement of phenomenological accountability, see chapter 5, section 6 and chapter 6, section 7, respectively.

36. A partial analogy to this kind of dual functioning is found with many designer or fashionable goods—for example, clothes, furniture, cars, mobile phones—which serve both to fulfill some practical need and to present their owner in a certain way.

37. Despite obvious proximity, my suggestion here differs in at least two important ways from Derrida's idea of "différance" as the temporalizing/spatializing (nonconceptual) "movement of play [*mouvement de jeu*] that 'produces'" the differences "in the *system* of language [*langue*]" (Derrida 1972, 8, 12). Whereas Derrida emphasizes the intractability of this "movement" and treats it (in all but rhetoric) as an ultimate metaphysical principle (*archi-écriture*), I am here highlighting that something *can* be said about the underlying structuration of propositional/conceptual content. Further, whereas Derrida (1967, 89) dismisses the "concept of experience" as "highly embarrassing" due to its supposed inseparability from the idea of full presence he discerns in Husserl, experience seems to me the requisite point of reference for conceiving language. This need is also highlighted by Merleau-Ponty (*S* 149), whose emphasis on perception as a paradigm aims to reconceive experience as partial presence.

8 The World Disclosed

1. *SZ* 207, 202. This charge seems fairly directed at the customary (e.g., Putnam's) talk of "mind-independent objects."

2. *SZ* 205. See also the introduction, section 2, as well as chapter 2, section 1.

3. *SZ* 206. See in particular §43b, which takes the resistance of "innerworldly" entities as a phenomenon that, when properly understood, leads beyond the picture of an immanent mind opposed by transcendent objects.

4. *SZ* 209 (i.e., "das Sein des innerweltlich *vorhandenen* Seienden (res)"), 211.

5. As he puts it at SZ 28: "That-which-is" can "show itself" without distortion in different ways, depending on the "kind of access" to it. ("Seiendes kann sich nun in verschiedener Weise, je nach der Zugangsart zu ihm, von ihm selbst her zeigen.")

6. See the emphasis at *SZ* 71 that *Zuhandenheit*, for example, does not simply refer to the way an entity is grasped. Indeed, Heidegger even suggests that readiness-to-hand is the "in-itselfness" [*An-sich-sein*] of entities (*SZ* 75). Whatever is made of this suggestion, the claim is certainly not that it stands in opposition to the idea of being "for us."

7. Cf. SZ 202. Heidegger frequently emphasizes the need to find the correct mode of "access" to that which is (*Seiendes*), often in relation to his own analysis (e.g., SZ 6, 15–16, 28, 37), and criticizes Cartesian (or Husserlian) epistemology's focus on the wrong kind of access to entities (i.e., the intellect, SZ 95).

8. *SZ* 212, cf. 230, where being [*Sein*] but not what is [*Seiendes*] is linked to Dasein's existence.

9. These terms are used by Dreyfus (1991, 253), Carman (2003, 157ff.), and Blattner (1994, 2004), respectively.

10. As Blattner (1994, 198; see also 2004) puts it.

11. Given Heidegger's definition of being (*Sein*) at the beginning of *SZ* as "that which determines that-which-is as that-which-is" ("*das, was Seiendes als Seiendes bestimmt,*" *SZ* 6), my personal suspicion is that Heidegger is simply saying that without Dasein that-which-is would remain without determination—that is, without any *as* structure. However, the above discussion does not rely on this claim.

12. As Cerbone (1995, 413) nicely points out.

13. As Cerbone (2007; 1995, 418–419) plausibly argues.

14. See also Taylor's (2013) illuminating discussion of the "mediational picture" originally derived from Cartesian assumptions and the need to replace this with "contact" theories. While I am largely sympathetic to Taylor's views, I suggest below that the problem is not mediation as such but how mediation is conceived of.

15. See chapter 4, section 1, as well as chapter 6, section 3, respectively.

16. In saying this I am not denying that nonlinguistic forms of disclosure are possible; it suffices that language can (and often does) play a role in constituting our thoughts and actions.

17. See introduction, section 2. Elsewhere I argue that this is an important indication of the nontranscendental nature of Merleau-Ponty's approach (Inkpin, forthcoming).

18. A commitment common to Merleau-Ponty (chapter 4, section 3), Wittgenstein (chapter 7, section 1), and Frege (chapter 9, section 2), respectively.

19. Our perspective might be expanded in something like the way Nagel thinks of the pursuit of "objectivity" as a "method of understanding" in which "we step back from our initial view of it and form a new conception which has that view and its relation to the world as its object" (Nagel 1986, 4).

20. Williams (1977, 64–68) and Nagel (1986, 5) describe improvements to a perspective as conceiving successively higher points of view that make lower ones commensurate, and are tempted by the thought that this generates a dialectic tending toward an "absolute" conception or point of view as a somehow implicit norm of inquiry. Although perhaps naturally tempting, the appeal to an ideal terminus is, as indicated above, superfluous.

21. This point might be made by saying that the appearance-reality distinction does not apply to artifacts, such as language, in the way it is sometimes thought to apply to mental representations.

22. This is the kind of mediation Hegel, for example, attributes to a "medium" or "tool" of knowledge (*das Erkennen*) at the start of the introduction to the *Phenomenology of Spirit* (Hegel 1989, 68–69).

23. In building a house, for example, the water used in mixing the cement is catalytically involved, the bricks materially so. What I am calling "material mediation" corresponds to the Aristotelian material cause, as "that from which a thing is made and continues to be made" (Aristotle 1996, 39 [194b23]). Of course, in another (non-Aristotelian) sense catalysts might be equally well described as material causes.

24. Indeed, Heidegger's view that language serves to pick out features of the real world seems to require some kind of correspondence—though of course the difficulty is to state what kind. An interesting proposal for the direction to take is Randall's idea of "functional realism," which is based not on a "structural correspondence," but a "functional 'correspondence' between factors in the instrument and in the materials" that is "not discoverable except as the instrument does what it is intended to do" (Randall 1963, 55).

25. Which is duly criticized by, for example, Wright (1992, 44–46).

26. As Rorty (1989, 13–15) has emphasized, Davidson's critique of the idea of conceptual schemes leads by a different route to the same conclusion (see Davidson 2001, 198; 1986, 445–446).

27. Most prominently in the *Tractatus* (e.g., introduction, 5.6–5.632), but also at *PU* §119.

28. See chapter 6, section 5 above. Sacks (2000, 198–218) describes Wittgenstein's position as relying on "transcendental features," understood as empirically and hence contingently generated features that are treated as having transcendental status. Although exegetically plausible for *PU*, it seems to me, as Sacks highlights, that this empirical-transcendental hybrid is ill-suited to underwrite genuinely transcendental claims. Wittgenstein himself seems to have seen debates between idealists and realists simply as verbal disagreements, suggesting in one passage that they differ only in their "battle cries" (Wittgenstein 1989i, 369 [§414]; cf. *PU* §402).

29. That is, I am claiming there are "limits of language" only in what Williams (1981, 151) calls the "tautological" and "empirical" sense.

30. As a matter of course Putnam treats the alternative to an "external" perspective as an "internal" one. See also his suggestion that the main problem of twentieth-century analytic philosophy was to work out how language "hooks onto the world" (Putnam 1990, 43)—a formulation bizarrely intimating that language is not itself part of the world and could conceivably become "unhooked."

9 Phenomenology and Semantics

1. This approach is summarized in chapter 7, section 4.

2. In describing this set of commitments as "common sense" the suggestion is not that anyone defends them in this form, but rather that they would be considered by many too obvious and too indistinct to require explicit formulation or defense. As Dummett (1991, 2–3) says, "analytical philosophy is written by people" who "take for granted the principles of semantic analysis embodied in [the language of mathematical logic]"—"however little many of them may know of the technical results or even concepts of modern logical theory."

3. That is, in Descartes's (1996, 378) words, "a general science explicating everything that can be investigated concerning order and measure without being attributed to any particular material."

4. Sean Kelly (2001, 4) suggests these questions are distinct, with analytic philosophy of language focusing on *what* sentences mean ("meaning"), whereas phenomenology considers what it is for a sentence to mean what it does ("meaningfulness"). The problem with this, otherwise appealing, reconciliation is the tacit assumption that these questions can be answered independently. Indeed, it seems to presuppose the semantic notion of content to be satisfactory, whereas the point of the phenomenological approach's foundational claims is that the notion of propositional/conceptual content cannot be made sense of independently (unphenomenologically), so that the phenomenological approach is involved in answering "what sentences mean." The thought that the two approaches differ in the questions they address therefore cannot form the starting point in considering their relationship, as I am doing here.

5. This term nicely suggests—in Merleau-Pontian manner—that language is something we can have a sense of or for, rather than always involving the exercise of full-blown rational capacities.

6. This fact is nicely brought out by Ricoeur (1969a, 93) in highlighting the role of words as "an exchanger between the system and the act, between the structure and the event."

7. As for example in the case of "possible worlds semantics" or attempts to conceive linguistic meaning in terms of abstract entities.

8. Davidson (2001, 60) suggests that the use of T-sentences provides a criterion that gives "the empirical study of language ... clarity and significance" and outlines an approach that he believes empirical studies of meaning might employ (e.g., ibid., 135–136).

9. Dummett 1976, 74, 82. Dummett's interpretation of such a theory of speakers' abilities as corresponding to Frege's notion of sense is by no means unpresupposing,

and might be challenged exegetically. However, such concerns can be neglected here.

10. Cf. Frege's (1994b, 45) analogy with the "image in [a] telescope," which "is dependent on its location, yet it is objective insofar as it can serve several observers." His platonism is most clearly in evidence in Frege 1986.

11. Searle 1983, 143; cf. 154: Background is "a set of skills, stances, preintentional assumptions and presuppositions, practices, and habits."

12. Searle 1983, 146; cf. 143–144, 145–150; also Searle 1979.

13. Searle 1992, 160–161; for this revision to his earlier (1983) view cf. 186–191.

14. Since the prepredicative use of language can be seen as involving a form of purposive (practical) intentionality, Searle's later modification might be thought congenial to the position proposed here. Such purposive intentionality would nonetheless not be intentional in the (semantic-content-involving) sense specified by Searle.

15. Incidentally, on Searle's account, consciousness of satisfaction conditions does not entail predicative ("explicit") awareness of these conditions and can be manifested in ("implicit") sensitivity to correctness of use. Perhaps due to his unorthodox use of the term *representation*—on which such sensitivity is "representational" in the sense of having conditions of success—Searle's claims have been misinterpreted as concerning the phenomenology of action and failing to do justice to "skilled coping" (Wakefield and Dreyfus 1991; Dreyfus 1999). He forcefully and convincingly clarifies his position in Searle 2000.

16. As Searle explains, a "crucial step in understanding the Background is to see that one can be committed to the truth of a proposition without having any intentional state whatever with that proposition as content." Such commitments are typically practical, such that to deny certain propositions (e.g., that objects are solid) would be contradicted by—that is, would be inconsistent with—one's actions (Searle 1992, 185).

17. In other words, far from being an innovation, this difference in functional topology is the very feature sometimes thought to show up the need for semantic analysis to eliminate the supposed imperfections of language's syntactic and practical formation.

18. See the later Wittgenstein's treatment of vagueness (chapter 6, section 6).

19. See sections 4 and 6 of chapter 5.

20. Cussins 2003, 138. The same strategy is endorsed by Peacocke (1992, 61–98) and Crane (1992). As an example of nonconceptual content Cussins (2003, 150) cites skilled performances, such as knowing how fast one is riding a motorbike.

21. For a sophisticated defense of this idea, see Stanley and Williamson 2001. Similarly directed arguments against the knowing-how/knowing-that distinction are offered in Snowdon 2003. Incidentally, I am not suggesting that nonconceptual content stands in the same relation to conceptual content as knowing-how does to knowing-that; the thought is merely that the characterizations of the nonconceptual and knowing-how share an analogous lack of independence.

22. Harrison (1975) argues in this spirit that Husserl's conception of prepredicative experience is structurally dependent on predicative determinations. Whether or not this applies to Husserl's conception, I contend here that it is not a general problem of claims about the prepredicative.

23. A somewhat dogmatic response would be that this means one is still "owed an account" of them in semantic terms. However, in revealing antecedent commitment to the semantics approach, this would beg the current question of the relationship between semantic content and prepredicative sense.

24. In the terms I have been using here, the conceptual-dependency objection assumes a relationship of functionally continuous foundation, whereas these differences show it to be a functionally discontinuous foundation.

25. See Dreyfus 2005, 2007a, 2007b, 2013, as well as McDowell 2007a, 2007b, 2013.

26. As will become clear, the focus of the debate between Dreyfus and McDowell differs somewhat from my concerns here, and at the time of writing may not have run its full course.

27. This wide-ranging view of expertise is set out by Dreyfus and Dreyfus (1986, 30ff.). Dreyfus (2007a, 373) even suggests, somewhat bizarrely, that the "existential phenomenologist's view" is that "human beings are at their best when involved in action."

28. Given his use of the Kantian opposition of spontaneity and receptivity, a distinctive feature of McDowell's position is the thought that our receptivity to the world involves the "passive operation of conceptual capacities" (McDowell 1996, 67).

29. For example, Dreyfus and Dreyfus 1986, 17. His general view that thinking impedes action is epitomized in the claim that "I only deliberate when coping is blocked" (Dreyfus 2002, 381).

30. This is well highlighted by Searle (2000, 81–82). According to Dreyfus, "monitoring" is "aloof and detached" and impedes the "flow" of expert performance (Dreyfus and Dreyfus 1986, 40).

31. See Dreyfus's (2013, 33–35) discussion of attempts to specify what experts respond to in their absorbed coping. He makes similar claims elsewhere (Dreyfus 2005, 54, 58; 2007b, 357).

32. McDowell 2013, 55. See also McDowell's (2007b, 349) earlier label, the "Myth of the Disembodied Intellect."

33. See chapter 1, section 2. Dreyfus (2012, 71) repeats the claim that our "basic experience" of entities in the everyday world "has no *as-structure*," citing a passage from *Logik* (144) that describes this experience as centered on what entities are for (their *Wozu*). However, he fails to note that in the following paragraph, on the same page, Heidegger emphatically states the exact opposite, namely, that all such experiences *do* involve an "*as* structure" (*Als-Struktur*), albeit a prepredicative one. Though I will not pursue the question here, it may be that motor intentionality of the kind described by Merleau-Ponty is functionally distinct from cases of the kind Heidegger focuses on.

34. See chapter 1, section 4. As Heidegger's label indicates, circumspection (*Umsicht*) "looks around" an entity's surroundings rather than fixing on it in isolation.

35. Another factor—though I will not pursue this exegetic issue further here—is perhaps that Merleau-Ponty identifies the use of concepts with objective thinking and hence predicative awareness. Note, however, that this cannot be Heidegger's view, because it would exclude the development of concepts that address the question of being (which is assumed prior to predicative categories). Heidegger was clearly aware of this methodological difficulty, as reflected in his notion of formally indicating "philosophical concepts" in contrast to predication-based "taxonomical" concepts (see chapter 3, section 2).

36. This might appear to align Heidegger's view with McDowell's, but such an alignment would be purely verbal and would not affect the underlying difference—the assumption of a functional discontinuity, as described above—that McDowell would take issue with.

37. Similar considerations apply to Dreyfus's contrast between "mindless" coping and "minded" use of concepts. Heidegger famously and conspicuously avoids framing his discussion in terms of the traditional notion of "mind." If we disregard this fact and insist on applying the notion of mindedness, it seems wrong to characterize coping as nonmental or mindless—at least insofar as this exhibits the same kind of intelligence as Heidegger's example of hammer use. Conversely, if we take Heidegger's avoidance of the term seriously, then higher-level predicative awareness is equally "mindless." The more general lesson—nicely highlighted by Gardner (2013, 130)—is the need to read historical authors in their own terms before situating their position in relation to our own concerns.

38. McDowell 2007b, 348. See also his claims about experience yielding knowledge that "*can* figure" in high-level rational behaviors, and the extended range of possible descriptions that are potentially applicable to basic world-responsiveness in the human case as opposed to that of nonrational animals (McDowell 2013, 42; 2007b, 344).

39. McDowell 2007a, 366; 2007b, 347–348. See also: "what it means for capacities to be conceptual in the relevant sense" is that "they are capacities whose content is of a form that fits it to figure in discursive activity" (McDowell 2013, 42).

40. McDowell 2007b, 346, 348. McDowell sometimes appears to suggest that conceptual form is in play because new experiences always take their place against the holistic background of a "rationally organized network of capacities" (McDowell 1996, 29; cf. 32). Note, however, that what is at issue between a Heideggerian approach and McDowell's—as McDowell (2007b, 344) recognizes—is not whether there is a holistic background playing a place-fixing role, but whether that background can be described as pervasively rational (as opposed to purposive and practical) in its constitution.

41. Accordingly, Dreyfus (2005, 59–61) also discerns a corresponding "conceptualization" step.

42. See McDowell 1996, 47, 58, paralleling the above definitions of conceptual form/shape.

43. The temptation is presumably to think of a shared conceptual form as a "condition of possibility" for the application of concepts. But this is fallacious, as the above examples show. (Consider also representation on a weather map of air pressure or rainfall levels using different colors, or the audible binary form in which smoke detectors respond to the continuously variable concentration of smoke particles, etc.) It would be more plausible to suggest that for a representation to be informative it is necessary that whatever it refers to be at least as rich in differential structure as the representational means deployed.

44. This would provide an explanation for McDowell's saying little to elucidate—thus leaving open—what conceptual form or shape is supposed to be.

45. For statements of this aim see, for example, McDowell 1996, 67; 2013, 41–42.

46. Although it is of peripheral interest to my concerns here, the above discussion hints at a tension in McDowell's notion of conceptual form. His apparent assumption that this notion imposes a determinate form on the possible contents of experience is difficult (without a conceptualization step) to reconcile with his argument against Evans, which implies that the application of conceptual capacities imposes no structural or formal constraints on whatever these are applied to.

47. As Wittgenstein himself appears to have done later (cf. *ÜG* §§97, 99).

48. As even Leibniz (1982, 8 [§5]), in one of his more down-to-earth moments, puts it, "les hommes en tant qu'ils sont empiriques, c'est à dire dans les trois quarts de leurs actions, n'agissent que comme des bêtes."

49. This should not, incidentally, be seen as undermining or endangering the notion of rationality. Rather this intermediate status is an indication of how

higher-level rational feats are functionally integrated with and properly anchored in the actual world. This is the functional aspect of viewing language as embedded in the world.

50. The analogy with physics might suggest the possibility that a later theory could succeed in unifying two such distinct but complementary theories. However, in the case of language this possibility appears to be ruled out by the way prepredicative factors undermine the prospects for systematicity.

10 Phenomenology and Beyond

1. See, for example, Husserl 1992d, 20–23 (§8).

2. See, for example, Husserl 1992d, 72–75 (§34).

3. In dealing with the "constitution of objectivities of possible consciousness," phenomenology "seems to be rightly characterized also as *transcendental theory* of knowledge" (Husserl 1992d, 84 [§40]).

4. Husserl 1992d, 89 (§41). Heidegger is the obvious target of Husserl's 1931 lecture "Phenomenology and Anthropology" (Husserl 1989, in particular 179).

5. For example, *SZ* 11, 38 and *SdC* 217, *PdP* 74–77. Heidegger frequently claims explicitly to be identifying "conditions of possibility" for such and such (e.g., *SZ* 11, 19, 53, and passim).

6. See Heidegger's frequent emphasis on the methodological importance of choosing the right "starting point [*Ausgang*]," "outset [*Ansatz*]," and "kind of access [*Zugangsart*]" (e.g., *SZ* 16, 36–37; 43, 53; 16, 37, respectively). Such considerations underlie his choice in *SZ* of Dasein's everydayness as the entry point for the investigation (*SZ* 16). See also chapter 8, note 7.

7. See the discussions of "deformalization" and "originality" in sections 2 and 3 of chapter 3.

8. Yet, as Austin (1970, 48) asks: "Why does the answer always turn out to be one or two, or some similar small, well-rounded, philosophically acceptable number? Why, if there are nineteen of any thing, is it not philosophy?"

9. Indeed this may reflect a tension in Heidegger's own views, as he often uses the phenomenological rhetoric of avoiding "free-floating" results and "constructions" (*SZ* 27–28; cf. 9, 19).

10. For the above example see Clark 2008, 3–9; for a much more general and highly persuasive discussion see in particular Gallagher 2005.

11. This claim about embeddedness is made by Rowlands (2010, 67ff., 83), who further claims that much the same can be said about enaction, while

presenting embodiment and extension as the core features of properly non-Cartesian mindedness.

12. The idea of enaction is a development by Varela, Thompson, and Rosch (1991, 172ff.) of Merleau-Ponty's interactionist view in *SdC*. The most notable recent examples of the enactive approach are from Thompson (2007) and the sensorimotor accounts of perception by O'Regan and Noë (2001) and Noë (2004).

13. The "extended-mind" hypothesis was proposed by Clark and Chalmers. The most prominent advocate of this approach is Clark (1997, 2008). Menary (2010) collects some critical responses to this thesis.

14. For example, Merleau-Ponty is an important influence for Gallagher (2005) and Varela, Thompson, and Rosch (1991); Heidegger's influence is central to Wheeler (2005) and plays some role for Clark (1997, esp. 171) and Haugeland (1998); Noë's (2004) views clearly reflect Husserl's influence.

15. In their influential essay on the extended mind, Clark and Chalmers (1998, 12ff.) focus on the example of an Alzheimer's patient reliant on the use of a notebook in performing everyday cognitive feats.

16. See the debate between Adams and Aizawa (2001) and Clark (2008, 85–110).

17. The label "extended mind" serves as a somewhat oxymoronic reminder that no sense—either ontological or functional—is to be made of the interiority previously supposed to define the mind.

18. The so-called parity principle: "If, as we confront some task, a part of the world functions as a process which, were it done in the head, we would have no hesitation in recognizing as part of the cognitive process, then that part of the world is (so we claim) part of the cognitive process" (Clark and Chalmers 1998, 8; cf. Clark 2008, 77).

19. As hinted above, the classic example of a cognitive division of labor is provided by Hutchins (1995, particularly 117–228).

20. See Wheeler 2005, 9. The "language of thought" label is due to Jerry Fodor.

21. For introductory discussions of the challenges presented by connectionism and dynamic systems theory, as well as the nature of the "classical," "standard," or "orthodox" positions these were opposing, see for example Clark 2001, 28–42, 62–84, 120–139; Shapiro 2011, 7–50, 114–157; and Wheeler 2005, 89–120, 6–14. Dreyfus (2002, 2012) is a well-known advocate of antirepresentationism.

22. Language is obviously a prime example of cognitive scaffolding, described by Clark (1997, 193–218) as the "ultimate artifact," while Wheeler (2005, 240) suggests that "a properly embodied-embedded account of language" is the "holy grail" for cognitive scientists interested in cognitive technology.

23. See in particular the discussion of antirepresentationalist arguments by Clark and Toribio (1994), who diagnose the need for a "continuum" of representational kinds (ibid., 427), and Clark (1997, 143–175). Similarly, Wheeler (2005, 128–144, 225–248) argues that Heideggerian phenomenology and cognitive science agree on the need for a spectrum of cases ranging from representation-free smooth coping, via context-dependent representations, through to action-neutral theoretical representations.

24. For example, Wheeler (2005, 195–196) or Clark (1997, 11–33; 2008, 3–29); the latter routinely uses such robotics examples in illustrating and motivating his 4e approach to cognition.

25. This is often summed up in Brooks's (1991, 140) maxim of using "the world as its own best model."

26. "Our goal ... is simple insect level intelligence," as Brooks (1991, 156) puts it, while pointing out that natural evolution took "3 billion years to get from single cells to insects."

27. Wheeler (2005, 198), for example, suggests that "natural selection" is "likely" to have "discovered" action-oriented representation.

28. See in particular O'Regan and Noë 2001 and Noë 2004.

29. See chapter 1, section 4.

30. Wheeler 2005, 233; see in particular 121–160, 225–248. Note that although it has many merits and is of great interest, I am not here endorsing or relying on the details of Wheeler's position.

31. Even now the appeal of phenomenology is sometimes thought to lie in its opposition to naturalism (e.g., Glendinning 2007, 6, 11).

32. As Gallagher (2005, 6), for example, puts it.

33. In fact, the present point does not even require that the cognitive scaffold succeeds in performing its supposed function. Rather, grasping these aspects is part of *attempting* to use something as a scaffold. Despite this, scaffolds should not be thought of (on the model of intervening representations) as generating the possibility of systematic deception for reasons discussed in chapter 8, section 2.

34. Many of the cognitive technologies that 4e cognitive science focuses on are not natural entities but artifacts—that is, entities humans have produced (for specific purposes). Strictly speaking, 4e cognitive science is therefore not a natural science, but a scientific approach that incorporates the study of artifacts. Hence, it is at best debatable whether 4e cognitive science could even attempt to "naturalize" phenomenology or claim that cognition somehow "supervenes on nature."

35. We might, of course, have only a limited grasp—for example, a prepredicative grasp—of what a particular word means, or how it functions in language. But limited understanding is not the same as being generally mistaken. Cf. also note 33.

36. From the cognitive science perspective, one gain from a phenomenological conception of language might be a differentiated description of the articulatory and disclosive function of linguistic signs, liberating it from an overreliance on computational metaphors or (phenomenologically ungrounded) assumptions inherited from the semantics approach.

References

Adams, Fred, and Ken Aizawa. 2001. The bounds of cognition. *Philosophical Psychology* 14 (1): 43–64.

Aristotle. 1934. *Nicomachean Ethics*. Trans. H. Rackham. Cambridge, MA: Harvard University Press.

Aristotle. 1938. On interpretation. In *Categories, On Interpretation, Prior Analytics*, trans. H. P. Cooke. Cambridge, MA: Harvard University Press.

Aristotle. 1996. *Physics*. Trans. R. Waterfield. Oxford: Oxford University Press.

Ashton, Ernst B. 1971. Translating *Philosophie. Delos* 6:16–29.

Austin, John L. 1962. *Sense and Sensibilia*. Oxford: Oxford University Press.

Austin, John L. 1970. Intelligent behaviour: A critical review of *The Concept of Mind*. In *Ryle*, ed. Oscar P. Wood and George Pitcher, 48–51. London: Macmillan.

Austin, John L. 1979. A plea for excuses. In *Philosophical Papers*, ed. James O. Urmson and Geoffrey J. Warnock, 175–204. Oxford: Oxford University Press.

Ayer, Alfred. 1969. Has Austin refuted the sense-datum theory? In *Metaphysics and Common Sense*, 126–148. London: Macmillan.

Baker, Gordon P., and Peter M. S. Hacker. 1980. *Wittgenstein: Meaning and Understanding. Essays on the Philosophical Investigations*, vol. 1. Oxford: Blackwell.

Baker, Gordon P., and Peter M. S. Hacker. 1984. *Scepticism, Rules, and Language*. Oxford: Blackwell.

Baker, Gordon P., and Peter M. S. Hacker. 1985. *Wittgenstein: Rules, Grammar, and Necessity*. Oxford: Blackwell.

Baker, Gordon P., and Peter M. S. Hacker. 1990. Malcolm on language and rules. *Philosophy* 65:167–179.

Baldwin, Thomas. 2004. *Maurice Merleau-Ponty: Basic Writings*. London: Routledge.

Baldwin, Thomas. 2007. Speaking and spoken speech. In *Reading Merleau-Ponty: On the Phenomenology of Perception*, ed. Thomas Baldwin, 87–103. London: Routledge.

Barbaras, Renaud. 1993. De la parole à l'être: Le problème de l'expression comme voie d'accès à l'ontologie. In *Maurice Merleau-Ponty: Le philosophe et son langage*, ed. François Heidsieck, 61–81. Paris: Vrin.

Barbaras, Renaud. 2005. A phenomenology of life. In *The Cambridge Companion to Merleau-Ponty*, ed. Taylor Carman and Mark Hansen, 206–230. Cambridge: Cambridge University Press.

Benveniste, Émile. 1939. Nature du signe linguistique. *Acta linguistica* 1:23–29.

Black, Max. 1996. Wittgenstein's language-games. In *Ludwig Wittgenstein: Critical Assessments*, vol. 2, ed. Stuart Shanker, 74–88. London: Routledge.

Blattner, William. 1994. Is Heidegger a Kantian idealist? *Inquiry* 37:185–201.

Blattner, William. 1999. *Heidegger's Temporal Idealism*. Cambridge: Cambridge University Press.

Blattner, William. 2004. Heidegger's Kantian idealism revisited. *Inquiry* 47 (4): 321–337.

Bloor, David. 1997. *Wittgenstein, Rules, and Institutions*. London: Routledge.

Bouquet, Simon. 2004. Saussure's unfinished semantics. In *The Cambridge Companion to Saussure*, ed. Carol Sanders, 205–218. Cambridge: Cambridge University Press.

Brandom, Robert. 1983. Heidegger's categories. *Monist* 66:387–409.

Brandom, Robert. 1994. *Making It Explicit*. Cambridge, MA: Harvard University Press.

Brentano, Franz. 1995. *Descriptive Psychology*. London: Routledge.

Brooks, Rodney. 1991. Intelligence without representation. *Artificial Intelligence* 47:139–159.

Burge, Tyler. 1989. Wherein is language social? In *Reflections on Chomsky*, ed. Alexander George, 175–191. Oxford: Blackwell.

Carman, Taylor. 2003. *Heidegger's Analytic*. Cambridge: Cambridge University Press.

Carnap, Rudolf. 1935. *Philosophy and Logical Syntax*. London: Kegan Paul.

Cavell, Stanley. 1962. The availability of Wittgenstein's later philosophy. *Philosophical Review* 71:67–93.

Cerbone, David. 1995. World, world-entry, and realism in early Heidegger. *Inquiry* 38 (4): 401–421.

Cerbone, David. 2007. Realism and truth. In *A Companion to Heidegger*, ed. Hubert Dreyfus and Mark Wrathall, 248–264. Oxford: Blackwell.

Charron, Ghyslain. 1972. *Du langage: A. Martinet et M. Merleau-Ponty*. Ottawa: Éditions de l'université d'Ottawa.

Chomsky, Noam. 1986. *Knowledge of Language: Its Nature, Origin, and Use*. New York: Praeger.

Clark, Andy. 1997. *Being There: Putting Brain, Body, and the World Together Again*. Cambridge, MA: MIT Press.

Clark, Andy. 2001. *Mindware: An Introduction to the Philosophy of Cognitive Science*. Oxford: Oxford University Press.

Clark, Andy. 2008. *Supersizing the Mind*. Oxford: Oxford University Press.

Clark, Andy, and David Chalmers. 1998. The extended mind. *Analysis* 58:7–19.

Clark, Andy, and Josefa Toribio. 1994. Doing without representing? *Synthese* 101 (3): 401–431.

Collingwood, Robin G. 1958. *The Principles of Art*. Oxford: Oxford University Press.

Collingwood, Robin G. 1960. *The Idea of Nature*. Oxford: Oxford University Press.

Collingwood, Robin G. 1994. *The Idea of History*. Oxford: Oxford University Press.

Conant, James. 2000. Elucidation and nonsense in Frege and early Wittgenstein. In *The New Wittgenstein*, ed. Alice Crary and Rupert Read, 174–217. London: Routledge.

Condillac, Etienne Bonnot de. 1971. *An Essay on the Origin of Human Knowledge*. Gainesville, FL: Scholars' Facsimiles & Reprints.

Crane, Tim. 1992. The nonconceptual content of experience. In *The Contents of Experience: Essays on Perception*, ed. Tim Crane, 136–157. Cambridge: Cambridge University Press.

Crane, Tim. 2001. *Elements of Mind*. Oxford: Oxford University Press.

Culler, Jonathan. 1986. *Ferdinand de Saussure*. Ithaca, NY: Cornell University Press.

Cussins, Adrian. 2003. Content, conceptual content, and nonconceptual content. In *Essays on Nonconceptual Content*, ed. York H. Gunther, 133–163. Cambridge, MA: MIT Press.

Dahlstrom, Daniel. 1994. Heidegger's method: Philosophical concepts as formal indication. *Review of Metaphysics* 47:775–795.

Dahlstrom, Daniel. 2001. *Heidegger's Concept of Truth*. Cambridge: Cambridge University Press.

Dastur, Françoise. 2001. *Chair et langage: Essais sur Merleau-Ponty*. La Versanne: encre marine.

Davidson, Donald. 1986. A nice derangement of epitaphs. In *Truth and Interpretation: Perspectives on the Philosophy of Donald Davidson*, ed. Ernest LePore, 433–446. Oxford: Blackwell.

Davidson, Donald. 2001. *Truth and Interpretation*. Oxford: Oxford University Press.

Davies, Martin. 1981. Meaning, structure, and understanding. *Synthese* 48:135–161.

Derrida, Jacques. 1967. *De la grammatologie*. Paris: Éditions de minuit.

Derrida, Jacques. 1972. La différance. In *Marges: De la philosophie*, 1–29. Paris: Éditions de Minuit.

Derrida, Jacques. 1978. *Spurs/Éperons*. Chicago: University of Chicago Press.

Derrida, Jacques. 1993. *La voix et le phénoméne*. Paris: Quadrige/PUF.

Descartes, René. 1996. Regulae ad directionem ingenii. In *Oeuvres de Descartes*, ed. Charles Adam and Paul Tannery, vol. 10, 349–488. Paris: Vrin.

Dewey, John. 1988. *The Quest for Certainty: Later Works*, vol. 4. Ed. Jo Ann Boydston. Carbondale/Edwardsville: Southern Illinois University Press.

Diamond, Cora. 2000. Ethics, imagination, and the method of Wittgenstein's *Tractatus*. In *The New Wittgenstein*, ed. Alice Crary and Rupert Read, 149–173. London: Routledge.

Dillon, Martin C. 1997. *Merleau-Ponty's Ontology*. Evanston, IL: Northwestern University Press.

Dilthey, William. 1990. *Der Aufbau der geschichtlichen Welt in den Geisteswissenschaften*. Frankfurt am Main: Suhrkamp.

Dreyfus, Hubert. 1991. *Being in the World: A Commentary on Heidegger's Being and Time, Division I*. Cambridge, MA: MIT Press.

Dreyfus, Hubert. 1999. The primacy of phenomenology over logical analysis. *Philosophical Topics* 27:3–24.

Dreyfus, Hubert. 2002. Intelligence without representation—Merleau-Ponty's critique of mental representation. *Phenomenology and the Cognitive Sciences* 1:367–383.

Dreyfus, Hubert. 2005. Overcoming the myth of the mental: How philosophers can profit from the phenomenology of everyday expertise. *Proceedings and Addresses of the American Philosophical Association* 79 (2): 47–65.

Dreyfus, Hubert. 2007a. Response to McDowell. *Inquiry* 50 (4): 371–377.

Dreyfus, Hubert. 2007b. The return of the myth of the mental. *Inquiry* 50 (4): 352–365.

Dreyfus, Hubert. 2012. Why Heideggerian AI failed and how fixing it would require making it more Heideggerian. In *Heidegger and Cognitive Science*, ed. Julian Kiverstein and Michael Wheeler, 62–104. Basingstoke: Palgrave.

Dreyfus, Hubert. 2013. The myth of the pervasiveness of the mental. In *Mind, Reason, and Being-in-the-World*, ed. Joseph Schear, 15–40. London: Routledge.

Dreyfus, Hubert, and Stuart Dreyfus. 1986. *Mind over Machine*. New York: Free Press.

Dummett, Michael. 1976. What is a theory of meaning? (II). In *Truth and Meaning*, ed. Gareth Evans and John McDowell, 67–137. Oxford: Clarendon.

Dummett, Michael. 1991. *The Logical Basis of Metaphysics*. London: Duckworth.

Evans, Gareth. 1982. *Varieties of Reference*. Oxford: Clarendon.

Falck, Colin. 1986. Saussurean theory and the abolition of reality. *Monist* 69:133–145.

Fontaine–de Visscher, Luce. 1974. *Phénomène ou structure? Essai sur le langage chez Merleau-Ponty*. Brussels: Facultés universitaire Saint-Louis.

Frank, Manfred. 1999a. Style in philosophy: Part I. *Metaphilosophy* 30:145–167.

Frank, Manfred. 1999b. Style in philosophy: Parts II and III. *Metaphilosophy* 30:264–301.

Frege, Gottlob. 1986. Der Gedanke: Eine logische Untersuchung. In *Logische Untersuchungen*, ed. Günther Patzig, 30–53. Göttingen: Vandenhoeck & Ruprecht.

Frege, Gottlob. 1994a. Funktion und Begriff. In *Funktion, Begriff, Bedeutung*, ed. Günther Patzig, 18–39. Göttingen: Vandenhoeck & Ruprecht.

Frege, Gottlob. 1994b. Über Sinn und Bedeutung. In *Funktion, Begriff, Bedeutung*, ed. Günther Patzig, 40–65. Göttingen: Vandenhoeck & Ruprecht.

Frege, Gottlob. 1994c. Über Begriff und Gegenstand. In *Funktion, Begriff, Bedeutung*, ed. Günther Patzig, 66–80. Göttingen: Vandenhoeck & Ruprecht.

Gadamer, Hans-Georg. 1990. *Wahrheit und Methode*. Tübingen: Mohr.

Gallagher, Shaun. 2005. *How the Body Shapes the Mind*. Oxford: Oxford University Press.

Gardner, Howard. 1977. *The Shattered Mind: The Person after Brain Damage*. London: Routledge & Kegan Paul.

Gardner, Sebastian. 2013. Transcendental philosophy and the possibility of the given. In *Mind, Reason, and Being-in-the-World*, ed. Joseph Schear, 110–142. London: Routledge.

Garver, Newton. 1990. Form of life in Wittgenstein's later work. *Dialectica* 44: 175–201.

George, Alexander. 1990. Whose language is it anyway? *Philosophical Quarterly* 40:275–298.

Gier, Nicholas F. 1981. *Wittgenstein and Phenomenology: A Comparative Study of the Later Wittgenstein, Husserl, Heidegger, and Merleau-Ponty*. Albany: SUNY Press.

Gleick, James. 1987. *Chaos: Making a New Science*. London: Cardinal.

Glendinning, Simon. 2007. *In the Name of Phenomenology*. London: Routledge.

Goethe, Johann Wolfgang. 1996. *Werke: Hamburger Ausgabe*. 14 vols. Munich: Deutscher Taschenbuch Verlag.

Gombrich, Ernst. 1960. *Art and Illusion*. Oxford: Phaidon.

Goodman, Nelson. 1976. *Languages of Art*. Indianapolis: Hackett.

Grice, Paul. 1989. *Studies in the Way of Words*. Cambridge, MA: Harvard University Press.

Guest, Gérard. 1991. La phémoménologie de Wittgenstein. *Heidegger Studies* 7:53–74.

Guignon, Charles. 1983. *Heidegger and the Problem of Knowledge*. Indianapolis: Hackett.

Haller, Rudolf. 1979. Die gemeinsame menschliche Handlungsweise. *Zeitschrift für Philosophische Forschung* 33:521–533.

Haller, Rudolf. 1984. Lebensform oder Lebensformen? Eine Bemerkung zu N. Garvers "Die Lebensform in Wittgensteins Philosophischen Untersuchungen." *Grazer philosophische Studien* 21:55–63.

Harris, James. 1986. Language, language games, and ostensive definition. *Synthese* 69:41–49.

Harris, Roy. 2001. *Saussure and His Interpreters*. Edinburgh: Edinburgh University Press.

Harrison, Ross. 1975. The concept of prepredicative experience. In *Phenomenology and Philosophical Understanding*, ed. Edo Pivčević, 94–107. Cambridge: Cambridge University Press.

Hass, Lawrence. 2008. *Merleau-Ponty's Philosophy*. Bloomington: Indiana University Press.

Haugeland, John. 1989. Dasein's disclosedness. *Southern Journal of Philosophy* 28:51–73.

Haugeland, John. 1998. Mind embodied and embedded. In *Having Thought,* 207–237. Cambridge, MA: Harvard University Press.

Hegel, Georg W. F. 1989. *Phänomenologie des Geistes. Werke* 3. Frankfurt am Main: Suhrkamp.

Heidegger, Martin. 1954. *Vorträge und Aufsätze*. Stuttgart: Neske.

Heidegger, Martin. 1959. *Unterwegs zur Sprache*. Stuttgart: Neske.

Heidegger, Martin. 1962. *Being and Time*. Trans. J. Macquarrie and E. Robinson. Oxford: Blackwell.

Heidegger, Martin. 1975. *Grundprobleme der Phänomenologie. Gesamtausgabe* 24. Frankfurt am Main: Klostermann.

Heidegger, Martin. 1976. *Logik: Die Frage nach der Wahrheit. Gesamtausgabe* 21. Frankfurt am Main: Klostermann.

Heidegger, Martin. 1977. *Holzwege. Gesamtausgabe* 5. Frankfurt am Main: Klostermann.

Heidegger, Martin. 1978a. *Frühe Schriften. Gesamtausgabe* 1. Frankfurt am Main: Klostermann.

Heidegger, Martin. 1978b. *Metaphysische Anfangsgründe der Logik. Gesamtausgabe* 26. Frankfurt am Main: Klostermann.

Heidegger, Martin. 1979. *Prolegomena zur Geschichte des Zeitbegriffs. Gesamtausgabe* 20. Frankfurt am Main: Klostermann.

Heidegger, Martin. 1983. *Die Grundbegriffe der Metaphysik. Gesamtausgabe* 29/30. Frankfurt am Main: Klostermann.

Heidegger, Martin. 1985. *Phänomenologische Interpretationen zu Aristoteles. Gesamtausgabe* 61. Frankfurt am Main: Klostermann.

Heidegger, Martin. 1987. *Die Idee der Philosophie und das Weltanschauungsproblem (Kriegsnotsemester 1919). Gesamtausgabe* 56. Frankfurt am Main: Klostermann.

Heidegger, Martin. 1988. *Ontologie (Hermeneutik der Faktizität). Gesamtausgabe* 63. Frankfurt am Main: Klostermann.

Heidegger, Martin. 1989a. Phänomenologische Interpretationen zu Aristoteles: Anzeige der hermeneutischen Situation. In *Dilthey Jahrbuch für Philosophie und Geisteswissenschaften* 6:237–269.

Heidegger, Martin. 1989b. *Überlieferte Sprache und technische Sprache*. St. Gallen: Erker.

Heidegger, Martin. 1992. *Platon: Sophistes. Gesamtausgabe* 19. Frankfurt am Main: Klostermann.

Heidegger, Martin. 1993a. *Grundprobleme der Phänomenologie. Gesamtausgabe* 58. Frankfurt am Main: Klostermann.

Heidegger, Martin. 1993b. *Phänomenologie der Anschauung und des Ausdrucks. Gesamtausgabe* 59. Frankfurt am Main: Klostermann.

Heidegger, Martin. 1993c. *Sein und Zeit.* Tübingen: Niemeyer.

Heidegger, Martin. 1995. *Einleitung in die Phänomenologie der Religion. Gesamtausgabe* 60. Frankfurt am Main: Klostermann.

Heidegger, Martin. 1996a. Anmerkungen zu Karl Jaspers. In *Wegmarken,* 1–44. Frankfurt am Main: Klostermann.

Heidegger, Martin. 1996b. *Being and Time: A Translation of Sein und Zeit.* Trans. J. Stambaugh. Albany: SUNY Press.

Heidegger, Martin. 2000. *Zur Sache des Denkens.* Tübingen: Niemeyer.

Heidegger, Martin. 2002. *Grundbegriffe der aristotelischen Philosophie. Gesamtausgabe* 18. Frankfurt am Main: Klostermann.

Herder, Johann Gottfried. 1966. *Abhandlung über den Ursprung der Sprache.* Stuttgart: Reclam.

Hintikka, Merrill B., and Jaako Hintikka. 1986. *Investigating Wittgenstein.* Oxford: Blackwell.

von Humboldt, Wilhelm. 1995. *Schriften zur Sprache.* Stuttgart: Reclam.

Husserl, Edmund. 1965. *Philosophie als strenge Wissenschaft.* Frankfurt am Main: Klostermann.

Husserl, Edmund. 1968. *Husserliana IX: Phänomenologische Psychologie.* The Hague: Nijhoff.

Husserl, Edmund. 1985. *Erfahrung und Urteil: Untersuchungen zur Genealogie der Logik.* Ed. Ludwig Landgrebe. Hamburg: Meiner.

Husserl, Edmund. 1989. Phänomenologie und Anthropologie. In *Husserliana XXVII: Aufsätze und Vorträge (1922–1937),* 164–181. The Hague: Nijhoff.

Husserl, Edmund. 1992a. *Logische Untersuchungen I. Gesammelte Schriften,* vol. 2. Hamburg: Meiner.

Husserl, Edmund. 1992b. *Logische Untersuchungen II. Gesammelte Schriften,* vols. 3–4. Hamburg: Meiner.

Husserl, Edmund. 1992c. *Ideen zu einer reinen Phänomenologie. Gesammelte Schriften,* vol. 5. Hamburg: Meiner.

Husserl, Edmund. 1992d. *Cartesianische Meditationen. Gesammelte Schriften*, vol. 8. Hamburg: Meiner.

Hutchins, Edwin. 1995. *Cognition in the Wild.* Cambridge, MA: MIT Press.

Hutto, Daniel. 2003. *Wittgenstein and the End of Philosophy: Neither Theory Nor Therapy*. Basingstoke: Palgrave.

Imdahl, Georg. 1994. "Formale Anzeige" bei Heidegger. *Archiv für Begriffsgeschichte* 37:306–332.

Inkpin, Andrew. 2011. The nonconceptual content of paintings. *Estetika* 48 (1): 29–45.

Inkpin, Andrew. 2013. Taking Rorty's irony seriously. *Humanities* 2 (2): 292–312. doi:.10.3390/h2020292.

Inkpin, Andrew. Forthcoming. Was Merleau-Ponty a "transcendental" phenomenologist? *Continental Philosophy Review*.

Jakobson, Roman. 1990. *On Language*. Cambridge, MA: Harvard University Press.

Kant, Immanuel. 1983a. *Kritik der reinen Vernunft, Werke*, ed. Wilhelm Weischedel, vols. 3–4. Darmstadt: Wissenschaftliche Buchgesellschaft.

Kant, Immanuel. 1983b. Prolegomena zu einer jeden Metaphysik, die als Wissenschaft wird auftreten können. In *Schriften zur Metaphysik und Logik, Werke*, ed. Wilhelm Weischedel, vol. 5, 113–264. Darmstadt: Wissenschaftliche Buchgesellschaft.

Kant, Immanuel. 1983c. Was heißt: Sich im Denken orientieren? In *Schriften zur Metaphysik und Logik, Werke*, ed. Wilhelm Weischedel, vol. 5, 267–283. Darmstadt: Wissenschaftliche Buchgesellschaft.

Kant, Immanuel. 1983d. Grundlegung zur Metaphysik der Sitten. In *Schriften zur Ethik und Religionsphilosophie: Erster Teil, Werke*, ed. Wilhelm Weischedel, vol. 6, 1–102. Darmstadt: Wissenschaftliche Buchgesellschaft.

Kelly, Sean D. 2001. *The Relevance of Phenomenology to the Philosophy of Language and Mind*. New York: Garland.

Kenny, Anthony. 1973. *Wittgenstein*. London: Penguin.

Kienzler, Wolfgang. 1997. *Wittgensteins Wende zu seiner Spätphilosophie 1930–1932*. Frankfurt am Main: Suhrkamp.

Kisiel, Theodore. 1993. *The Genesis of Heidegger's Being and Time*. Berkeley: University of California Press.

Kleist, Heinrich von. 1990. Über die allmähliche Verfertigung der Gedanken beim Reden. In *Sämtliche Werke und Briefe*, vol. 3, 534–540. Frankfurt am Main: Deutscher Klassiker Verlag.

Kripke, Saul A. 1982. *Wittgenstein on Rules and Private Language*. Cambridge, MA: Harvard University Press.

Kuhn, Thomas. 1970. *The Structure of Scientific Revolutions*. Chicago: University of Chicago Press.

Kwant, Remy C. 1966. *From Phenomenology to Metaphysics: An Inquiry into the Last Period of Merleau-Ponty's Philosophical Life*. Pittsburgh: Duquesne University Press.

Lafont, Cristina. 1999. *The Linguistic Turn in Hermeneutic Philosophy*. Trans. José Medina. Cambridge, MA: MIT Press.

Lafont, Cristina. 2001. *Heidegger, Language, and World-Disclosure*. Trans. Graham Harman. Cambridge: Cambridge University Press.

Lagueux, Maurice. 1965. Merleau-Ponty et la linguistique de Saussure. *Dialogue* 4:351–364.

Langer, Susanne. 1951. *Philosophy in a New Key*. Cambridge: Cambridge University Press.

Leibniz, Gottfried W. 1982. Principes de la Nature et de la Grace fondés en Raison. In *Vernunftprinzipien der Natur und der Gnade, Monadologie*, 2–25. Hamburg: Meiner.

Locke, John. 1975. *An Essay Concerning Human Understanding*. Oxford: Clarendon.

Madison, Gary Brent. 1981. *The Phenomenology of Merleau-Ponty: A Search for the Limits of Consciousness*. Athens: Ohio University Press.

Malcolm, Norman. 1986. Following a Rule. In *Nothing Is Hidden: Wittgenstein's Criticism of His Early Thought*, 154–181. Oxford: Blackwell.

Malcolm, Norman. 1995. Wittgenstein on language and rules. In *Wittgensteinian Themes: Essays 1978–1989*, ed. Georg H. von Wright. Ithaca, NY: Cornell University Press.

McDowell, John. 1984. Wittgenstein on following a rule. *Synthese* 58:325–363.

McDowell, John. 1992. Meaning and intentionality in Wittgenstein's later philosophy. *Midwest Studies in Philosophy* 17:40–52.

McDowell, John. 1996. *Mind and World*. Cambridge, MA: Harvard University Press.

McDowell, John. 2007a. Response to Dreyfus. *Inquiry* 50 (4): 366–370.

McDowell, John. 2007b. What Myth? *Inquiry* 50 (4): 338–351.

McDowell, John. 2009. Avoiding the myth of the given. In *Having the World in View*, 256–272. Cambridge, MA: Harvard University Press.

McDowell, John. 2013. The myth of the mind as detached. In *Mind, Reason, and Being-in-the-World*, ed. Joseph Schear, 41–58. London: Routledge.

McGinn, Colin. 1984. *Wittgenstein on Meaning: An Interpretation and Evaluation*. Oxford: Blackwell.

McGinn, Marie. 1997. *Wittgenstein and the Philosophical Investigations*. London: Routledge.

Meggle, Georg. 1985. Wittgenstein—ein Instrumentalist. In *Sprachspiel und Methode: Zum Stand der Wittgenstein-Diskussion*, ed. Dieter Birnbacher and Armin Burkhardt, 71–88. Berlin: Walter de Gruyter.

Menary, Richard, ed. 2010. *The Extended Mind*. Cambridge, MA: MIT Press.

Merleau-Ponty, Maurice. 1945. *La phénoménologie de la perception*. Paris: Gallimard.

Merleau-Ponty, Maurice. 1959. *Signes*. Paris: Gallimard.

Merleau-Ponty, Maurice. 1960. *Éloge de la philosophie (et autres essais)*. Paris: Gallimard.

Merleau-Ponty, Maurice. 1962. Un inédit de Maurice Merleau-Ponty. *Revue de Metaphysique et de Morale* 67:401–409.

Merleau-Ponty, Maurice. 1964. *Le visible et l'invisible*. Paris: Gallimard.

Merleau-Ponty, Maurice. 1968. *Résumés de cours: Collège de France 1952–1960*. Paris: Gallimard.

Merleau-Ponty, Maurice. 1969. *La prose du monde*. Paris: Gallimard.

Merleau-Ponty, Maurice. 1996a. *Le primat de la perception et ses conséquences philosophiques*. Paris: Verdier.

Merleau-Ponty, Maurice. 1996b. *Sens et non-sens*. Paris: Gallimard.

Merleau-Ponty, Maurice. 2001. La conscience et l'acquisition du langage. In *Psychologie et pédagogie de l'enfant: Cours de Sorbonne 1949–1952*, 9–87. Lagrasse: Verdier 2001.

Merleau-Ponty, Maurice. 2002. *La structure du comportement*. Paris: Gallimard.

Monk, Ray. 1990. *Ludwig Wittgenstein: The Duty of Genius*. London: Penguin.

Moore, George E. 1954. Wittgenstein's Lectures in 1930–33. *Mind* 63:289–316.

Moran, Dermot. 2000. *Introduction to Phenomenology*. London: Routledge.

Moyal-Sharrock, Danièle. 2003. Logic in action: Wittgenstein's *Logical Pragmatism* and the impotence of scepticism. *Philosophical Investigations* 26:125–148.

Mulhall, Stephen. 2001. *Inheritance and Originality*. Oxford: Clarendon.

Munson, Thomas. 1962. Wittgenstein's phenomenology. *Philosophy and Phenomenological Research* 23:37–50.

Murray, Michael. 1973. Heidegger and Ryle: Two versions of phenomenology. *Review of Metaphysics* 27:88–111.

Nagel, Thomas. 1986. *The View from Nowhere*. Oxford: Oxford University Press.

Nietzsche, Friedrich. 1980a. *Morgenröte: Kritische Studienausgabe*, ed. Giorgio Colli and Mazzino Montinari, vol. 3. Munich: Deutscher Taschenbuch Verlag.

Nietzsche, Friedrich. 1980b. *Zur Genealogie der Moral: Kritische Studienausgabe*, ed. Giorgio Colli and Mazzino Montinari, vol. 5. Munich: Deutscher Taschenbuch Verlag.

Noë, Alva. 2004. *Action in Perception*. Cambridge, MA: MIT Press.

Okrent, Mark. 1988. *Heidegger's Pragmatism*. Ithaca, NY: Cornell University Press.

O'Regan, Kevin, and Alva Noë. 2001. A sensorimotor account of vision and visual consciousness. *Behavioral and Brain Sciences* 24:939–1031.

Panofsky, Erwin. 1993. Iconography and iconology: An introduction to the study of Renaissance art. In *Meaning in the Visual Arts*, 51–81. London: Penguin.

Park, Byong-Chul. 1998. *Phenomenological Aspects of Wittgenstein's Philosophy*. Dordrecht: Kluwer.

Peacocke, Christopher. 1992. *A Study of Concepts*. Cambridge, MA: MIT Press.

Plato. 1926. *Cratylus, Parmenides, Greater Hippias, Lesser Hippias*. Trans. H. N. Fowler. Cambridge, MA: Harvard University Press.

Pöggeler, Otto. 1989. Die Krise des phänomenologischen Philosophiebegriffs (1929). In *Phänomenologie im Widerstreit: Zum 50. Todestag Edmund Husserls*, ed. Christoph Jamme and Otto Pöggeler, 255–276. Frankfurt am Main: Suhrkamp.

Putnam, Hilary. 1975. The meaning of "meaning." In *Mind, Language, and Reality*, 215–271. Cambridge: Cambridge University Press.

Putnam, Hilary. 1981. *Reason, Truth, and History*. Cambridge: Cambridge University Press.

Putnam, Hilary. 1990. *Realism with a Human Face*. Cambridge, MA: Harvard University Press.

Quine, Willard V. O. 1960. *Word and Object*. Cambridge, MA: MIT Press.

Quine, Willard V. O. 1969a. Epistemology naturalized. In *Ontological Relativity and Other Essays*, 69–90. New York: Columbia University Press.

Quine, Willard V. O. 1969b. Ontological relativity. In *Ontological Relativity and Other Essays*, 27–68. New York: Columbia University Press.

Randall, John Herman. 1963. The art of language and the linguistic situation: A naturalistic analysis. *Journal of Philosophy* 60 (2): 29–56.

Reeder, Harry P. 1989. Wittgenstein never was a phenomenologist. *Journal of the British Society for Phenomenology* 20:257–276.

Rhees, Rush. 1970. Wittgenstein's builders. In *Discussions of Wittgenstein*, 71–84. London: Routledge & Kegan Paul.

Rhees, Rush. 1981. *Ludwig Wittgenstein: Personal Recollections*. Oxford: Blackwell.

Ricoeur, Paul. 1969a. La structure, le mot, l'événement. In *Le conflit des interprétations: Essais d'herméneutique*, 80–97. Paris: Éditions de seuil.

Ricoeur, Paul. 1969b. Le problème du double-sens. In *Le conflit des interprétations: Essais d'herméneutique*, 64–79. Paris: Éditions de Seuil.

Rorty, Richard. 1979. *Philosophy and the Mirror of Nature*. Princeton, NJ: Princeton University Press.

Rorty, Richard. 1989. *Contingency, Irony, and Solidarity*. Cambridge: Cambridge University Press.

Rousseau, Jean-Jacques. 1990. *Essai sur l'origine des langues*. Paris: Éditions Gallimard.

Rowlands, Mark. 2010. *The New Science of the Mind: From Extended Mind to Embodied Phenomenology*. Cambridge, MA: MIT Press.

Rudd, Anthony. 2005. Wittgenstein, global scepticism, and the primacy of practice. In *Readings on Wittgenstein's On Certainty*, ed. Danièle Moyal-Sharrock and William H. Brenner, 142–161. Basingstoke: Palgrave Macmillan.

Ryle, Gilbert. 1949. *The Concept of Mind*. Chicago: University of Chicago Press.

Ryle, Gilbert. 1990. Knowing how and knowing that. In *Collected Papers*, vol. 2, 212–225. Bristol: Thommes.

Sacks, Mark. 2000. *Objectivity and Insight*. Oxford: Clarendon.

Sapontzis, Steve F. 1978. A note on Merleau-Ponty's "ambiguity." *Philosophy and Phenomenological Research* 38:538–543.

Sartre, Jean-Paul. 1943. *L'être et le néant*. Paris: Gallimard.

Saussure, Ferdinand de. 1972. *Cours de linguistique générale*. Paris: Payot.

Schear, Joseph, ed. 2013. *Mind, Reason, and Being-in-the-World*. London: Routledge.

Schmidt, James. 1985. *Maurice Merleau-Ponty: Between Phenomenology and Structuralism*. London: Macmillan.

Schnädelbach, Herbert. 1983. *Philosophie in Deutschland 1831–1933*. Frankfurt am Main: Suhrkamp.

Schulte, Joachim. 1984. Chor und Gesetz: Zur "Morphologischen Methode" bei Goethe und Wittgenstein. *Grazer Philosophische Studien* 21:1–32.

Searle, John. 1969. *Speech Acts: An Essay in the Philosophy of Language*. Cambridge: Cambridge University Press.

Searle, John. 1979. Literal meaning. In *Expression and Meaning: Studies in the Theory of Speech Acts*, 117–136. Cambridge: Cambridge University Press.

Searle, John. 1983. *Intentionality: An Essay in the Philosophy of Mind*. Cambridge: Cambridge University Press.

Searle, John. 1992. *The Rediscovery of Mind*. Cambridge, MA: MIT Press.

Searle, John. 2000. The limits of phenomenology. In *Heidegger, Coping, and Cognitive Science: Essays in Honour of Hubert L. Dreyfus*, ed. Mark A. Wrathall and Jeff Malpas, 71–92. Cambridge, MA: MIT Press.

Seel, Martin. 2002. Über Richtigkeit und Wahrheit: Erläuterungen zum Begriff der Welterschließung. In *Sich bestimmen lassen*, 45–67. Frankfurt/Main: Suhrkamp.

Shanker, Stuart. 1996. Sceptical confusions about rule-following. In *Ludwig Wittgenstein: Critical Assessments*, vol. 2, ed. Stuart Shanker, 176–182. London: Routledge.

Shapiro, Lawrence. 2011. *Embodied Cognition*. London: Routledge.

Silverman, Hugh J. 1981. Merleau-Ponty and the interrogation of language. In *Perception, Structure, Language: A Collection of Essays*, ed. John Sallis, 122–141. Atlantic Highlands, NJ: Humanities Press.

Skirbekk, Gunnar. 1983. *Praxeology: An Anthology*. Bergen: Universitetsforlaget.

Snowdon, Paul. 2003. Knowing how and knowing that: A distinction reconsidered. *Proceedings of the Aristotelian Society* 104:1–29.

Spiegelberg, Herbert. 1960. *The Phenomenological Movement: A Historical Introduction*. 2 vols. The Hague: Nijhoff.

Spiegelberg, Herbert. 1981. The puzzle of Wittgenstein's *Phänomenologie* (1929–?). In *The Context of the Phenomenological Movement*, 202–228. The Hague: Nijhoff.

Stanley, Jason, and Timothy Williamson. 2001. Knowing how. *Journal of Philosophy* 98:411–444.

Stern, David. 1996. The availability of Wittgenstein's philosophy. In *The Cambridge Companion to Wittgenstein*, ed. Hans Sluga and David G. Stern, 442–476. Cambridge: Cambridge University Press.

Stern, David. 2004. *Wittgenstein's* Philosophical Investigations: *An Introduction*. Cambridge: Cambridge University Press.

Stroll, Avrum. 1988. *Surfaces*. Minneapolis: University of Minnesota Press.

Taylor, Charles. 1985. *Human Agency and Language: Philosophical Papers I*. Cambridge: Cambridge University Press.

Taylor, Charles. 1989. Embodied agency. In *Merleau-Ponty: Critical Essays*, ed. Henry Pietersma, 1–21. Lanham, MD: University Press of America.

Taylor, Charles. 1995. Heidegger, language, and ecology. In *Philosophical Arguments*, 100–126. Cambridge, MA: Harvard University Press.

Taylor, Charles. 2013. Retrieving realism. In *Mind, Reason, and Being-in-the-World*, ed. Joseph Schear, 61–90. London: Routledge.

Thompson, Caleb. 2000. Wittgenstein's confessions. *Philosophical Investigations* 23:1–25.

Thompson, Evan. 2007. *Mind in Life*. Cambridge, MA: Harvard University Press.

Tormey, Alan. 1971. *The Concept of Expression*. Princeton, NJ: Princeton University Press.

Tugendhat, Ernst. 1970. *Der Wahrheitsbegriff bei Husserl und Heidegger*. Berlin: De Gruyter.

Varela, Francisco, Evan Thompson, and Eleanor Rosch. 1991. *The Embodied Mind: Cognitive Science and Human Experience*. Cambridge, MA: MIT Press.

Von Wright, Georg H. 1982. The origin and composition of the *Investigations*. In *Wittgenstein*, 113–136. Oxford: Blackwell.

Wakefield, Jerome, and Hubert Dreyfus. 1991. Intentionality and the phenomenology of action. In *John Searle and His Critics*, ed. Ernest Lepore and Robert Van Gulick, 259–270. Oxford: Blackwell.

Waldenfels, Bernhard. 1976. Die Offenheit sprachlicher Strukturen bei Merleau-Ponty. In *Maurice Merleau-Ponty und das Problem der Struktur in den Sozialwissenschaften*, ed. Richard Grathoff and Walter Sprondel, 17–28. Stuttgart: Ferdinand Enke Verlag.

Waldenfels, Bernhard. 1981. Perception and structure in Merleau-Ponty. In *Perception, Structure, Language: A Collection of Essays*, ed. John Sallis, 21–38. Atlantic Highlands, NJ: Humanities Press.

Watson, Stephen H. 1983. Merleau-Ponty's involvement with Saussure. In *Continental Philosophy in America*, ed. Hugh J. Silverman, John Sallis, and Thomas M. Seebohm, 208–226. Pittsburgh: Duquesne University Press.

Wheeler, Michael. 2005. *Reconstructing the Cognitive World*. Cambridge, MA: MIT Press.

Whorf, Benjamin. 1956. *Language, Thought, and Reality*. Ed. John B. Carroll. Cambridge, MA: MIT Press.

Williams, Bernard. 1977. *Descartes*. London: Penguin.

Williams, Bernard. 1981. Wittgenstein and idealism. In *Moral Luck*, 144–163. Cambridge: Cambridge University Press.

Wittgenstein, Ludwig. 1960. *The Blue and Brown Books*. New York: Harper & Row.

Wittgenstein, Ludwig. 1976. Ursache und Wirkung: Intuitives Erfassen. *Philosophia* 6: 391–408, 427–446.

Wittgenstein, Ludwig. 1980. *Wittgenstein Lectures: Cambridge 1930–1932*. Ed. Desmond Lee. Oxford: Blackwell.

Wittgenstein, Ludwig. 1989a. *Bemerkungen über die Grundlagen der Mathematik. Werkausgabe* 6. Frankfurt am Main: Suhrkamp.

Wittgenstein, Ludwig. 1989b. *Bemerkungen über die Philosophie der Psychologie I. Werkausgabe* 7, 7–215. Frankfurt am Main: Suhrkamp.

Wittgenstein, Ludwig. 1989c. *Letzte Schriften über die Philosophie der Psychologie. Werkausgabe* 7, 347–488. Frankfurt am Main: Suhrkamp.

Wittgenstein, Ludwig. 1989d. *Philosophische Grammatik. Werkausgabe* 4. Frankfurt am Main: Suhrkamp.

Wittgenstein, Ludwig. 1989e. *Philosophische Untersuchungen. Werkausgabe* 1, 225–618. Frankfurt am Main: Suhrkamp.

Wittgenstein, Ludwig. 1989f. *Tractatus logico-philosophicus. Werkausgabe* 1, 9–85. Frankfurt am Main: Suhrkamp.

Wittgenstein, Ludwig. 1989g. *Über Gewißheit. Werkausgabe* 8, 113–257. Frankfurt am Main: Suhrkamp.

Wittgenstein, Ludwig. 1989h. *Vermischte Bemerkungen. Werkausgabe* 8, 445–575. Frankfurt am Main: Suhrkamp.

Wittgenstein, Ludwig. 1989i. *Zettel. Werkausgabe* 8, 259–443. Frankfurt am Main: Suhrkamp.

Wittgenstein, Ludwig. 1994–. *Wiener Ausgabe. Studien Texte*. Vienna: Springer Verlag.

Wollheim, Richard. 1992. *Art and Its Objects*. Cambridge: Cambridge University Press.

Wrathall, Mark. 2005. Motives, reasons, and causes. In *The Cambridge Companion to Merleau-Ponty*, ed. Taylor Carman and Mark Hansen, 111–128. Cambridge: Cambridge University Press.

Wrathall, Mark. 2011. *Heidegger and Unconcealment*. Cambridge: Cambridge University Press.

Wright, Crispin. 1980. *Wittgenstein on the Foundations of Mathematics*. London: Duckworth.

Wright, Crispin. 1992. *Truth and Objectivity*. Cambridge, MA: Harvard University Press.

Zahavi, Dan. 2005. *Subjectivity and Selfhood*. Cambridge, MA: MIT Press.

Index